A BETTER WAY TO BUILD

A History of the Pankow Companies

"Charlie Pankow had an important impact on late-twentieth-century building development. His companies created a powerful niche market that kept his clients happy and the enterprise profitable. Pankow's clients weren't vainglorious developers seeking to create monuments for themselves. They wanted handsome, lasting buildings, delivered on time and at budget. And that Pankow accomplished time and again. Moreover, Pankow became as much an innovator in the systemic use of concrete as a building material as Gustave Eiffel had been in the late nineteenth century with cast iron."

—*Timothy Tosta, Partner, McKenna Long & Aldridge LLP*

. . .

"Charles Pankow was an innovator who embraced the design-build concept, and he served as a champion of the approach while restoring the master builder to the commercial building site. He believed the contractor needed the ability to integrate cost- and time-saving construction methods. This book describes the 'Pankow Way,' a collaborative approach in which the contractor works effectively with the architects, engineers, and subcontractors to meet owners' expectations. It is an interesting history of a company and the man who created a unique business culture, and I recommend it as a great read for engineers, architects, contractors, and businesspeople."

—*Patrick J. Natale, Executive Director, American Society of Civil Engineers*

. . .

"In the history of construction in the second half of the twentieth century, Charles Pankow stands out as the man who led the design-build revolution. Michael Adamson develops a fascinating portrait of a community leader, philanthropist, creative businessman, perceptive art collector, and major figure in the history of civil engineering. Read this book to learn about the visionary after whom the American Society of Civil Engineers recently named its prestigious competition in architectural engineering—and about the 'renaissance man' behind the vision."

—*Jeffrey S. Russell, Professor of Civil and Environmental Engineering, University of Wisconsin*

"*A Better Way to Build: A History of the Pankow Companies* is a story of one man's vision and the companies he created to carry out that vision. As a young architect I was taught that the general contractor was the opponent of good design. Author Michael Adamson shows how Charlie Pankow turned this idea on its head by reestablishing the historic partnership between design and construction. This is an excellent book for architects, engineers, and contractors. It shows how one visionary was able to turn an entire industry toward a better way of working together. I recommend it most highly."

—*Patrick MacLeamy, CEO, HOK Architects*

. . .

"This is a great read—anyone and everyone in the architectural, engineering, construction, or development business should read this book. Design-bid-build as a delivery method is broken! Charlie Pankow saw this many years ago with his colleagues at Kiewit. Over forty years ago he embarked on the formation of the design-build/design-assist delivery method for construction. He succeeded in a way that no one could have ever predicted. I am pleased to be able to have the opportunity to review this book and to advise everyone in the industry to read it."

—*Charles H. Thornton, Chairman, Charles H. Thornton & Company LLC*

. . .

"Michael Adamson chronicles the erratic genius of Charlie Pankow and the construction empire he built in defiance of management conventions. As a 'Master Builder,' Charlie believed in people, not management doctrine. He attracted platoons of individualists, self-starters whom he motivated and inspired. He taught them to cut costs and schedules, to develop stunningly innovative construction techniques, and to enervate centuries-old design-build methods, thrusting them onto the world of modern construction. If you want a gripping, bare-knuckled story about a man who successfully changed an industry, you've selected the right book."

—*Walker Lee Evey, Former President, Design-Build Institute of America*

A BETTER WAY TO BUILD
A History of the Pankow Companies

MICHAEL R. ADAMSON

Purdue University Press
West Lafayette, Indiana

Cover photo: Shoreline Square, Long Beach (Warren Aerial)

Library of Congress Cataloging-in-Publication Data

Adamson, Michael R.
 A better way to build : a history of the Pankow companies / by Michael R. Adamson.
     p. cm.
 Includes bibliographical references and index.
 ISBN 978-1-55753-634-1 (hardback : alk. paper) -- ISBN 978-1-61249-230-8 (epdf)
 -- ISBN 978-1-61249-231-5 (epub) 1. Pankow, Charles. 2. Pankow (Firm)--History.
 3. Architects and builders--United States--Biography. 4. Building. I. Title.
 TH451.P36A33 2013
 338.7'66240973--dc23
                          2012014132

Contents

List of Illustrations

Foreword

Here's how to succeed in business—the Pankow way, with lessons for any innovative entrepreneur. Business historian Michael Adamson tells the Pankow story and spotlights its impact on the construction industry—an impact that lives on through the founder's generous philanthropy and his outstanding ability to innovate.

Charles Pankow was a private person, serving mostly private clients. His preference for confidentiality may have come from formative years in the employ of industry legend Peter Kiewit, who said that he "didn't get to be [a top contractor] by telling people how I do my business." Record keeping was not a company norm. Nor was organization, nor succession planning. Charlie died in 2004 without naming his successor.

More a leader than a manager, Charlie selected projects, or approved their selection, then more or less charged project managers with executing them. Prime example: He opened an office in Hawaii and let it run almost independently for years; it handled about a third of the company's projects.

At the same time, Charlie strictly set company policy and method of doing business. He knew his niche: He would stay with concrete and with the advantage of construction methods and equipment that were his strength. He would also stay with the design-build method of contracting. So good were his methods—and his loyal crews at performing them—that he actually "salvaged" many projects. He saved projects that had been bid above their owner's budget by redesigning them to fit his expertise, then contracting for a fixed price and completion date, which he met—and often exceeded.

While civil engineer Charlie Pankow's skilled teams advanced design-build on the West Coast and in Hawaii, civil engineer Preston Haskell championed the business model from his firm's base in Florida. Haskell, much more the promoter by personality—and perhaps by his Princeton schooling—led the creation of the Design-Build Institute of America in 1993 and served as its first president. A dedicated Purdue alumnus, the typically reserved Pankow supported his firm's Richard Kunnath as the DBIA's second president.

During a 1988 meeting of what later became the Construction Industry Round Table, held in Washington, DC, Charlie extended an invitation to me as manager of the group of some 70 CEOs. "When this meeting comes to San Francisco, you could have this in my house." As his main residence was in Southern California, he explained: "I bought the San Francisco house to display my art collection, and groups have parties there." It was actually his very own Petit Trianon, a copy of the famous palace at Versailles, filled with his spectacular art treasures. Its mirrored ballroom was ideal for our 1989 dinner.

The Trianon and its treasures were sold at auction in 2004. Purdue received a substantial share of the proceeds from the building, and the sale of the art collection funded the Charles Pankow Foundation, established in 2002. Its mission: "To advance innovation in building design and construction so as to provide the public with buildings of improved quality, efficiency, and value."

The company that Charlie Pankow created carries on, having diversified into general contracting as well as design-build, and steel buildings as well as concrete ones. The founder lives on in more than 160 buildings and the methods by which he built them, in the millions of dollars of research funded annually by his foundation, and in the Pankow Award for Innovation given annually by the American Society of Civil Engineers.

Charlie and his way of building live on, as well, in the following pages as a success story to admire. Read on.

Arthur Fox
Editor Emeritus
Engineering News-Record

Preface

This is a book about the restoration of the master builder to the commercial building site. As a sponsored project, it is written as the story of one firm, but one that delivered large commercial projects in ways that have helped to redefine the role of the general contractor.

Charles J. Pankow—Charlie, to those who knew and worked with him—believed that the construction of customized commercial buildings would benefit from the participation of the contractor in their planning and design with the owner, architect, structural engineer, and other members of the building team. He was confident that combining this approach to project management—later popularized as design-build—with mass production techniques in concrete that had been deployed in the construction of factories and warehouses since the early 1900s would allow him to guarantee the delivery of high-quality buildings on time and under budget. In 1963 he bet his career on it. With most of the men who worked for him in the building division that he had established in the Los Angeles District of Peter Kiewit Sons', he started Charles Pankow, Inc. (CPI). Pankow was not the first contractor to deploy design-build to satisfy the needs of building owners, but he pioneered its use in the commercial sector during the second half of the twentieth century. And while the techniques that he deployed had been used in industrial settings, his project teams—working within a company culture that valued accountability, curiosity, ingenuity, and resourcefulness—adapted, improved, and "tweaked" them in the manner of the craftsmen who propelled Britain to technological leadership during the Industrial Revolution.[1]

Over the next four decades, the Pankow companies completed as many as 1,000 projects. More than 160 of them involved the construction or expansion of large structures: office buildings, department stores, hospitals, hotels, multi-unit residential complexes, parking garages, public buildings, and shopping centers (see appendices A and B). Many of these structures would not have been built were it not for the participation of Pankow as contractor in a design-build setting. At the time of Charlie Pankow's death, in January 2004, the company employed more than 200 people and was handling between $300 and $450 million in outstanding contract volume.[2] Five years earlier, *Engineering News-Record* had recognized Pankow among 125 individuals for their contributions to the construction industry during the 125-year history of the trade publication.[3]

In 2004 the design-build niche that CPI initially had occupied virtually alone now accounted for nearly one-third of the market for nonresidential construction.[4] Diffusion of the methodology was slow. CPI had yet to grow into a large firm by the time Pankow reorganized it twice in the mid-1980s. Moreover, Charles Pankow largely confined its operations to the private sector; public sector procurement requirements were a major obstacle to the spread of the practice. Through leadership positions in engineering societies, publication of journal articles and textbook chapters, and other means, Charlie Pankow and his colleagues promoted design-build. These efforts were prologue to the founding of the Design-Build Institute of America (DBIA) in 1993.

This biography of the Pankow companies heeds the call of the editors of *Business History Review* to investigate "important and contentious subjects where intellectual breakthroughs are possible."[5] The empirical observations of a single firm, of course, would seem to offer little opportunity to draw robust conclusions about an entire industry, much less erect new frameworks for understanding business history. As detailed narrative history, however, the book pays particular attention to entrepreneurship and innovation—two of the areas that the editors suggest hold promise for scholars. It reconstructs the stages in the life of the firm, from its incubation within Peter Kiewit Sons' to its growth into a middle-sized firm with the capacity to take on projects normally associated with much larger contractors. It also describes innovation at the project level, which should interest civil engineers, and establishes a close link between these achievements and the overall success of the firm, which should interest students of business. Underlying both entrepreneurship and innovation was the company's culture, referred to internally as the Pankow Way: a client-service-oriented approach to doing business that suffused decision making throughout the company. From its roots in the Kiewit building division, the Pankow firm had more in common with the so-called Great Groups described by Warren Bennis and Patricia Ward Biederman, than the hierarchical organizations that have received the largest share of attention from business historians.[6] Ultimately, this study seeks to explain only one company's success in terms of entrepreneurship and innovation. It leaves to others the task of using the empirical evidence that it presents to build theoretical frameworks that promote greater understanding of the causal links between entrepreneurship, innovation, and economic growth and help to explain where innovation occurs in society.

The empirical observations of a single firm also would not seem to offer a foundation for comparative analysis, either, particularly given the paucity of scholarly studies of construction firms, especially those engaged in the commercial segment of the industry.[7] As Janet Wells Greene has noted, "the construction industry…has not been studied in depth."[8] Corporate culture, however, has been a multidisciplinary topic of interest, allowing this book to compare the Pankow Way to other "ways," such as Hewlett-Packard's. Like the HP Way, for instance, the Pankow Way encouraged individual initiative and innovation within teams working toward a common purpose and was a means of distinguishing the enterprise from larger, more hierarchically structured competitors. Unlike its more famous counterpart, the Pankow Way was neither codified nor used externally—marketing materials, for example, emphasized the "design-build construction arrangement," whose consideration drives the narrative of this book.[9]

This book is the product of a sponsored corporate history program that has also produced a short film and a set of interview recordings and transcripts.[10] Late in 2007, the Board of Directors of the Charles Pankow Foundation initiated the "Charles J. Pankow Legacy Project." Foremost, board members aimed "to memorialize Mr. Pankow's business and professional life, belief, philosophies, values, and accomplishments, so as to inspire an understanding of the engineer as businessman, innovator, and entrepreneur, for the benefit of aspiring construction engineers and the firms that recruit, employ, and mentor them."[11] Clearly, within these parameters, the products of such an effort, in whatever form they materialized, might easily constitute little more than hagiography. Ultimately, the board did contemplate a book that a scholarly press might publish as one of the products of the project, but it was unclear at the outset whether available evidence would support such an endeavor. Conducting a careful assessment of either Charlie Pankow's or his firm's business decision making, for instance, seemed problematic, given the lack of textual records, as elaborated below. Oral history seemed to be the best means available to recover the firm's past.

Early in 2008, the board selected the author and Christopher J. Castaneda, then chair of the History Department at California State University, Sacramento, to act as co-principal investigators. Our task was to conduct oral history interviews of past and present employees, and people associated with the company, as the Foundation's resources allowed. In all, the budget permitted 35 interviews to be recorded and transcribed. (See Charles J. Pankow Legacy Project Oral Histories, 2008–2011, in the List of Archival Collections.) The interviewees represent a subset of an initial list of more than 70 potential informants. A number of these latter individuals were interviewed in the course of writing this book. The interviews took place between April 2008 and September 2009. Owing to Castaneda's responsibilities as chair of his department, the author ultimately conducted all of the interviews. Castaneda served as adviser to the project and reader of the initial manuscript.

The interviewees represent a cross section of employees, architects, structural engineers, and subcontractors who worked with the Pankow companies as members of the building team. They also include a president emeritus of Purdue University, two professors of civil engineering at Purdue, Charlie Pankow's personal physician, one of his commercial property managers, and two of the four children of Charlie and Doris Pankow. Among the employees interviewed for this project, four worked in accounting or administration; two worked in business development; three were hired as managers from other firms; and fifteen began their careers as carpenters, engineers, or superintendents in the field. Many of the latter became managers in the firm. All six of the executive managers who took responsibility for the company after the death of Charlie Pankow were interviewed, as were past presidents, executive vice presidents, and regional managers. While most of the individuals interviewed for the project began their careers (or their relationships with Charlie Pankow) in the 1970s or later, five of the interviewees began working with Charlie Pankow during the latter's time at Peter Kiewit Sons'.

It should come as no surprise that many of the key individuals associated with Charlie Pankow's professional life had passed away before the Foundation initiated the project, including Robert Carlson and Ralph Tice, two of the four founders of CPI; Harold Henderson, a superintendent with both Kiewit and Pankow; and Robert McCarthy, the partner in the San Francisco firm of partner in Bohnert, Flowers, Roberts and McCarthy who served as counsel to the firm. Others, such as Lloyd Loetterle, another of the four founders of CPI, could not be located or were unable to schedule an interview. Still others, such as George Hammond, a managing partner with the architectural firm Welton Becket, and Russell J. Osterman, one of the three largest shareholders in CPI and a development partner of Charlie Pankow's in many projects, were unable to sit for interviews for reasons of health. Most significantly, the list of people unable to give an interview includes Charlie Pankow himself.

Nevertheless, the individuals interviewed for this project were well placed to relate the career of Charlie Pankow and his firm at the levels of both project innovation and execution, on the one hand, and strategy and executive decision making, on the other. Informants were able to discuss Charlie Pankow's interest in shaping civil engineering and construction management education at Purdue University and to discuss his passion for collecting art. Satisfied that the oral histories, supplemented by material supplied by the company and found in trade journals, selected archives, and other sources, provided the evidentiary basis for a substantive corporate biography, the Foundation's board of directors agreed to fund one.

Without oral history, then, writing this book would not have been possible. As *Civil Engineering*, the journal of the American Society of Civil Engineers, noted more than two decades ago when it called on members to conduct taped interviews of "veteran" engineers, "people don't keep diaries and write personal letters the way they did a century ago." This was certainly the case with Charlie Pankow. He was not inclined to put anything pertaining to policies, directives, and other business into writing. He rarely, if ever, explained his decisions, much less put them on paper. Those who executed the business of the firm took their cues from the top: Not putting things into writing seems to have been characteristic of the organization. Certainly, much of the company's business was conducted by telephone. Indeed, the need for more written communication, in the form of memoranda of telephone conversations and daily job diaries, was on the agenda of the off-site managers' meeting held in May 1975. Further, the company was unable to assemble an archive of textual business records on which the student of business history might draw. Many documents were not retained when the company moved its headquarters after Charles Pankow died. Understandably, as a going concern, the company reserved the use of financial and other sensitive records for reasons of confidentiality. Hence, the business records that historians generally rely upon to construct a corporate biography were unavailable for this study.[12]

Textual records capture Pankow's engineering history in far greater detail than its managerial history. For purposes of estimating future work and recording construction methods and techniques, project engineers compiled detailed technical reports on each building with which they were associated. The company made these reports, along with its newsletter and material produced in association with annual and other meetings, available for this study. Both the newsletter and the annual meetings featured articles or presentations on projects, so that best practices might be shared with colleagues. Moreover, technical articles on Pankow's projects regularly appeared in the trade and local press. Robert Law, the company's chief estimator, provided an invaluable guide to the company's innovations that has been adapted as appendix C. Still, the interviews played a critical role in highlighting key projects and providing additional insights into how teams accomplished project goals.

Of course, as the literature on the subject acknowledges, we cannot expect oral sources to compensate for the lack of textual records generated in the course of conducting the business of a company. As Victor W. Geraci of the Regional Oral History Office at the University of California at Berkeley explains, "Placing the burden on any one discipline or research methodology to fill the gap in the written record is problematic."[13] In this case, the reliance on this research methodology is even more problematic, given the centrality of Charlie Pankow to the story. Many aspects of Pankow's professional life, especially as they relate to his business affairs, are known and understood only through the oral testimony of others. While the interviews provided information sufficient to construct a narrative, gaps in the record remain—about Pankow's career at Peter Kiewit Sons' and the early years of the company in particular. And for certain events, only a single source testified to the "facts" surrounding them. Interviewees also painted different portraits of Charlie Pankow, and not all of these perspectives can be reconciled using the rules of evidence. The genesis and development of many business deals, too, remains either unclear or described at a summary level of detail by the interviewees. Memoranda of design meetings, for instance, would have been a particularly rich source of material in assessing design-build in practice, were they available.[14]

At the same time, the interviewees revealed what it meant to work for, or in association with, Charlie Pankow and his company. To be sure, as Geraci adds, "time both clouds and reorganizes the memory of interviewees." Yet the interviewees supplied information that would not have been available, regardless of the extent and detail of company record keeping. They explained individual and collective motives and goals, whether from the perspective of the job site or the executive office. By reflecting individually on their careers, as defined by their interactions with their colleagues and other members of building teams, past and present employees and other contemporaries deepen both their and our understanding of the contributions of Pankow as contractor to the construction industry and the built environment. Collectively the interviews illuminate the Pankow Way in ways that textual records could not have captured. By "reflect[ing] less about the facts surrounding an event than they do about the meaning of the event to the participants," they validate Geraci's observation that oral history has become "an acceptable means to thicken the written narrative."[15]

· · ·

The support and commitment of the board of directors of the Charles Pankow Foundation was essential to this project. As president of the board and head of the Pankow company, the support of Rik Kunnath in particular was crucial in directing the project in a scholarly direction.

Many thanks go to all of the individuals who consented to interviews for this book, especially Rick and Steve Pankow, two of the four children of Charlie and Doris Pankow, who not only gave interviews, but answered many additional queries and shared materials with me. I am grateful to Robert Law, who made himself available to answer myriad questions about the company's history after he gave his interview. Special thanks to him, too, for saving all of his annual meeting and off-site managers' meetings binders and making them available to me. I also thank Conan "Doug" Craker for making available the minutes of the off-site managers' meeting from his personal collection. Al Fink provided numerous photos from his personal collection, a number of which appear in these pages, and showed me his video that recorded the construction of the Pearl II condominium. He also gave me invaluable tours of buildings that the company built in Hawaii.

In the summer of 2009, Red Metcalf and Mike Liddiard co-hosted a reunion of Pankow employees, past and present, at their homes in Carson City and Reno, Nevada. I thank them for giving me the opportunity to attend, enabling me to speak with many of the individuals who appear in these pages, but were not formally interviewed as part of the Charles J. Pankow Legacy Project.

Deborah Lattimore, the person responsible for transcribing the recorded interviews conducted for this project, embodies the definition of client service. She worked with me to ensure that the transcriptions accurately reflected the recordings and adhered to best practices in oral history. Her turnaround time was remarkable.

I am grateful to all of those who read all or parts of the manuscript, including (and especially) Chris Castaneda, Glenn Bugos, Arthur Fox, Vince Drnevich, the Press's anonymous reader, and the Foundation's readers, including Renate Kofahl, Robert K. Tener, Rik Kunnath, and Mark Perniconi. Since no portions of this book were subjected to scholarly scrutiny, either as conference papers or articles in scholarly journals, their contributions were particularly welcome in helping to give shape to the manuscript. They also improved my technical descriptions of construction work in progress and suggested ways of anchoring the Pankow story in the literature.

No one was more involved in the Charles J. Pankow Legacy Project than its manager, Linda Carey Kunnath, with whom I had countless conversations and several enjoyable lunches. She did an incredible job of developing the list of interviewees and helping to locate them. Her ability to uncover the gems in San Francisco's restaurant scene may be unmatched.

Archivists who helped from afar with photos and other material include Sammie L. Morris and Elizabeth Wilkinson at Purdue and Elizabeth Hogan at the University of Notre Dame. Closer to home, Miranda Hambro guided me through the Oakland & Imada collection in the Environmental Design Archives at the University of California, Berkeley. Teresa Shada, managing editor of *Kieways* magazine, provided material on Peter Kiewit Sons' projects. Though my research into the Lloyd Corporation Archive made but a small contribution to *this* project, I cannot pass up the chance to thank all of the staff at the Henry E. Huntington Library for supporting my research during two fellowships and several ad hoc visits. Thanks, too, to Susan Green of the Huntington Library Press, for giving me the opportunity to research Los Angeles history at midcentury, which forms a crucial backdrop to the career of Charlie Pankow.

Many people provided assistance on particular topics. Carol Reese at the American Society of Civil Engineers and Melinda Reynolds at the American Concrete Institute supplied information on membership questions. Thanks to Melinda, too, for providing me with all the "President's memos" written by Charlie Pankow and Dean Stephan during their terms as ACI (American Concrete Institute) president. Susan Hines at the DBIA answered questions about the growth of design-build. Edward C. Wundram provided invaluable insights on design-build competitions. Todd Gish's work on Hollywood's historical landscape is compelling and deserves a wide audience. John King, urban design critic for the *San Francisco Chronicle*, generously responded to my questions on postwar suburban office parks. Jeff Bergmann, a former tax partner at Peat Marwick who worked on the Pankow account, helped me understand Charlie Pankow's reorganization of the company and explained corporate incentives under Reagan-era tax reform. Vic Geraci of the Bancroft Library always had something substantive to say on the value and uses of oral history. Tatyana Koenig and Steve Rafferty accompanied me on a walking tour of downtown Los Angeles that included a glimpse of the lobby of the renovated Eastern Columbia Building.

All writers should have a great editor. I am privileged to have had two of them: Charles Watkinson and Katherine Purple. I want to thank the entire production staff at the Press, too, for seeing this book to the finish line, especially copy editor Kelley Kimm.

I owe a debt of gratitude to W. Elliot Brownlee, my PhD advisor at the University of California, Santa Barbara, who instilled in me an appreciation of the value in pursuing topics that can be supported by archival sources in one's backyard, so to speak. Our discussions of California business and economic history have been points of departure for several productive lines of inquiry that have indirectly or directly informed the story of the Pankow companies.

List of Abbreviations

A&E   architectural and engineering firm
ACEC   American Consulting Engineers Council
ACI   American Concrete Institute
AF   Architectural Forum
AGC   Associated General Contractors of America
AIA   American Institute of Architects
AR   Architectural Record
ASCE   American Society of Civil Engineers
BD&C   *Building Design & Construction*
BI   Proposed Buildings and Installations, Lloyd Corporation Archive, Henry E. Huntington Library, San Marino, California
CM   Construction Management
CPA   Charles Pankow Associates
CPBI   Charles Pankow Builders, Inc.
CPBL   Charles Pankow Builders, Ltd.
CPI   Charles Pankow, Inc.
DBIA   Design-Build Institute of America
DPW   Department of Public Works, Director's Office Records, Director's Records related to the Division of Architecture, State of California Archives, Sacramento
DPWA   Department of Public Works, Architecture Division Records, Work Orders, State of California Archives, Sacramento
EDA   Environmental Design Archives, University of California, Berkeley
EJCDC   Engineers Joint Contract Documents Committee
ENR   *Engineering News-Record*
FHA   Federal Housing Administration
GC   general contractor

GSA	US General Services Administration
HCHA	Hawaii Council for Housing Action
HUD	US Department of Housing and Urban Development
ICBO	International Conference of Building Officials
LACTC	Los Angeles County Transportation Commission
LAT	*Los Angeles Times*
LC	Lloyd Center records, Lloyd Corporation Archive, Henry E. Huntington Library, San Marino, California
LCL	Lloyd Corporation Letters, Lloyd Corporation Archive, Henry E. Huntington Library, San Marino, California
LCR	Lloyd Corporation Real Estate Documents, Lloyd Corporation Archive, Henry E. Huntington Library, San Marino, California
MTA	Los Angeles County Metropolitan Transportation Authority
MWD	Metropolitan Water District of Southern California
NIST	US Department of Commerce, National Institute of Standards and Technology
PCC	Pankow Construction Company
PDC	Pankow Development Corporation
PHMRF	precast hybrid moment-resistant frame
Port	Portland file, Lloyd Corporation Archive, Henry E. Huntington Library, San Marino, California
PRA	Pasadena Redevelopment Agency
PSPL	Pankow Special Projects, Ltd.
RBL	Ralph B. Lloyd Letters, Lloyd Corporation Archive, Henry E. Huntington Library, San Marino, California
RTD	Southern California Rapid Transit District
ROHO	Regional Oral History Collection, Bancroft Library, University of California, Berkeley
SB&C	*Southwest Builder & Contractor*
SFRA	San Francisco Redevelopment Agency
SOMA	South of Market district of San Francisco
TI	tenant improvement
UNDA	University of Notre Dame Archives

INTRODUCTION

The hallmarks of the Pankow companies have been the deployment of design-build methodology to deliver singular commercial projects, combined with innovations in job site automation, generally associated with, but not limited to, concrete as a building material. But Charlie Pankow was neither the first contractor to utilize design-build in the twentieth century nor the first to deploy techniques to mass-produce concrete structural elements at the building site. This introduction establishes the context for relating the story of the Pankow companies by emphasizing "the continuity of inventive activity" identified by economist Nathan Rosenberg as a primary feature of technological change.[1]

It begins by elaborating on the idea of design-build and contrasting it in theory with the "design-bid-build" approach that prevailed in the commercial segment of the construction industry when Charlie Pankow began his career. Pankow concluded that problems associated with productivity in the construction industry were principally rooted in project management under the latter approach. To be sure, "automating" the building site was a crucial factor in giving Pankow the confidence he needed to accept the execution risk associated with the timely and cost efficient completion of projects. At the same time, he recognized that deploying techniques of mass production would not necessarily overcome the bottlenecks and cost overruns that often occurred when responsibility was fragmented among building team members. Problems that plagued design-bid-build projects created opportunities for an innovation entrepreneur to exploit. Moreover, as chapter 6 elaborates, the failure of Construction Management, an alternative approach, to deliver on its promises, would redound to the benefit of design-builders like Charlie Pankow.[2]

This introduction continues by relating design-build precedents set during the first three decades of the twentieth century that were associated with engineering firms principally engaged in the construction of concrete and steel industrial structures. In these cases, the designer and builder were one in the same, facilitating the implementation of design-build in a way not possible under building teams with different designers and builders, as was the case in all of the commercial projects built by the Pankow firm. At the same time, these engineering firms mobilized techniques of mass production that constitute precedents for the Pankow story. Intertwined with the discussion of design-build precedents, then, are descriptions of the advances in the production of concrete structural elements that show why innovation by Charlie Pankow and his firm should be seen as continuous, if relentless, improvement in processes and techniques on behalf of controlling project outcomes.

The early-twentieth-century practitioners of design-build used reinforced concrete (when not using steel) to erect their structures. After World War II, prestressed concrete displaced reinforced concrete as a building material. This introduction continues by describing postwar developments in concrete, for manufacturing prestressed concrete structural elements was one of the principal ways through which Pankow sought to "automate" the job site. A summary of postwar developments in precast concrete rounds out the discussion of the major technological precedents for the "budgeted construction program" that Pankow marketed to building owners. Discussion of other concrete construction techniques deployed by the company is deferred, for their deployment—at least in commercial buildings—occurred well into Charlie Pankow's career.

Design-Build: Definition and Practice from Antiquity to Mid-Twentieth-Century America

Charles Pankow's career was animated by the idea of the master builder. From the time of Ictinus and Callicrates, builders of the Parthenon, to the days of Filippo Brunelleschi, designer of the dome of the Gothic Cathedral, or Duomo, in fifteenth-century Florence—and indeed, to well into the nineteenth century—architecture, engineering, and construction were closely linked. As one who relied on judgment, precedent, and "rules of thumb" to determine structural "stresses and strains," the architect acted as his own engineer. Designers of buildings became intimately involved in their construction. As Frank Miles Day, twice president of the American Institute of Architects, noted, Brunelleschi designed not only the Duomo, but also the scaffolding and tools used on the job. Moreover, he inspected materials (brick and clay), trained workers to set stone, and supervised the work.[3]

With the development of tall and complex commercial structures and large industrial plants, executed in reinforced concrete or steel, responsibility for design, engineering, and erecting buildings devolved to specialized professionals. For instance, by 1900, a civil engineer, whose field was confined almost entirely to surveying at the dawn of the nineteenth century, might have had a decade or more of experience calculating the "stresses and strains" of buildings as a practicing structural engineer. Before World War I, the "tendency" in America to keep architecture and engineering "widely separated" was well established, as Luzerne S. Cowles, an assistant designing engineer with the Boston Elevated Railway System, told his audience at the Congress of Technology. Likewise, as architect Day observed, the "modern [general contracting] system" that had emerged in the context of legal, managerial, professional, and technological developments associated with the Second Industrial Revolution "sees a divorce between the architect and the master builder." Lamented the supervising architect of both Yale and Johns Hopkins universities, "The architect of today puts down his ideas on paper, and has merely to see that his designs are carried out, yet he is not the executive builder."[4]

Under this approach—widely referred to today as "design-bid-build"—the owner was responsible for securing the land and necessary permits, and conceiving and planning the project. The architect, as the owner's agent, designed the project, often in consultation with a structural engineer. Once the owner approved the plans and specifications developed by the architect, general contractors, who had no involvement with the design process, bid on them. The bidding process was either invited, whereby the architect restricted the number of contractors allowed to bid, or open to any firm that responded to the call for proposals. The owner and successful bidder established a contractual relationship that was wholly separate from the one that defined relations between owner and architect. The architect supervised construction, monitoring adherence on the part of the general contractor to approved construction documents and managing the process by which modifications to the design—so-called change orders—were proposed to, and approved by, the owner. The

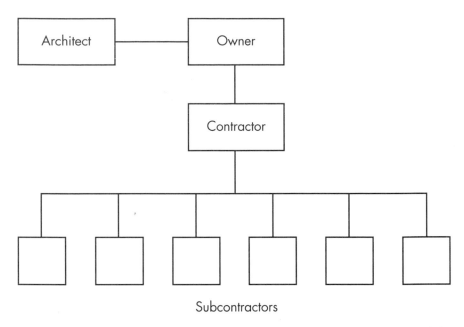

Figure 1. Design-Bid-Build. The owner signs separate contracts with the architect and the contractor. The architect acts as the owner's agent and designs the project. The contractor constructs the building from the architect's completed drawings and specifications.

general contractor was responsible for procuring materials and coordinating operations, but typically subcontracted the building of the project to various individuals and firms (fig. 1).

Design-bid-build was also cemented into public contract law, principally through the efforts of reformers to require city councils and other governmental bodies to procure design services from certified professionals and award construction contracts on the basis of lowest financially responsible bids. As Rebecca Menes notes, municipal contracting during the Progressive Era was "notoriously corrupt."[5] Construction contracts were among the most lucrative, enticing urban bosses and their minions to invest in construction firms. Reformers sought to eliminate bribes, kickbacks, sweetheart and backroom deals, and other forms of graft through competitive bidding. As the editors of *The Architect and Engineer* noted, however, municipalities generally awarded contracts to the lowest bidder rather than the lowest *financially responsible* bidder, thereby negating the promise of reform.[6]

Architects, engineers, and builders alike found common ground in criticizing the general contracting system as a price-driven "race to the bottom" that produced cheap designs, shoddy work, unsightly or unsafe buildings, and delayed or abandoned projects, and that precluded fair competition among responsible operators. The most pressing problem, which these reform-minded members of the building team cast in stark, moral terms, was the bidding system that owners used to procure construction services. According to George E. Burlingame, a San Francisco–based contractor, bidding procedures, which might involve fifteen or more general contractors, encouraged "conscienceless individuals, who are willing to turn heaven and earth to gain a competitive advantage" at the expense of scrupulous contractors. As a result, "business degenerates first into a frank brigandage, and from that into a free fight." Architect Day was less blunt, but no less critical, of a system that pitted "reliable, reputable" contractors against "irresponsible, and perhaps unscrupulous" ones. Bidding, Day charged, did not align the interests of owner and builder. In selecting a contractor on the basis of the low bid, the owner was unlikely to secure the services of a firm "of honor [and] high-integrity, and well equipped by education and experience to carry on building work." The owner was already insulated from the subcontractors who worked at the building site: what confidence could he have in their work, Day wondered, if he hired the firm that would be responsible for supervising them through the mechanism of the low bid? Sullivan W. Jones, vice president and treasurer of the Association of United States Quantity Surveyors, argued that the low bidder was often the contractor who had erred the most in computing quantities from the architect's plans and specifications. Charles Evan Fowler, a Seattle-based consulting engineer, was adamant: "The low bidder, as a rule, has not the necessary experience, plant, or resources to properly carry out a contract." In awarding work on the basis of the lowest bid, both private owners and public officials selected contractors who often were not qualified to complete their projects to their satisfaction.[7]

Reformers within the building community who identified the bidding system as the problem, however, often sought to improve it with measures to eliminate irresponsible and unscrupulous firms rather than replace it with another approach. In language that reflected contemporary urban concerns about hygiene and sanitation, contractor Burlingame called for transparency on the part of all members of the building team on behalf of "a square deal" for all: "The sunlight of discussion is absolutely the best disinfectant . . . ever discovered." Others concurred that owners (or architects, as their agents) should handpick contractors, or at least restrict bidding to a short list of reputable firms. Lamenting the deleterious impact on projects of contractors who subcontracted all the work associated with construction and relied wholly on others to supervise it, Los Angeles architect John Corneby Wilson Austin, designer of Griffith Observatory, the Shrine Auditorium, and Los Angeles City Hall, charged his colleagues with selecting a contractor "who knows his business [and] can do at least some of the work on the building himself." In the public sector, where bidding was required by law, reformers agreed that laws should be modified to allow firms to win contracts based on superior experience, working capital, and other factors.[8]

Architect Day offered two suggestions that were grounded in ideas that postwar design-builders would embrace. First, owners might retain a "quantitative surveyor" to estimate the cost of building the project based on the architect's drawings and specifications. The owner would then have a basis for deciding whether the project was economical. The "quantitative surveyor" might also use the estimate to negotiate the work of subcontractors. Substitute "design-build contractor" for "quantitative surveyor" and here Day was in effect proposing what would become the core of Charlie Pankow's "budgeted construction program," as he called it. Indeed, Day noted that Thomas Brassey, the English railway builder, had deployed such an approach "more or less." He stopped short, however, of suggesting that such estimates might be used as the basis of a negotiated, fixed-price construction contract between owner and contractor. Indeed, here the architect annunciated what would later become one of stronger criticisms of design-build contracting, namely that the approach provided an incentive for the contractor with lump-sum contract in hand to "skimp" on quality of workmanship to increase margins. Secondly, Day suggested that architects might take on the role of master builder: an exceptional suggestion that addressed the bidding question by eliminating the general contractor. How committed Day may have been to either proposal is unclear; above all, he hoped to initiate a conversation on reforming the contracting system. That Day prefaced his advice by stating that reforming the system would not be necessary were all general contractors "men of property, skill, and probity," however, is telling. By prescribing solutions that focused on ensuring the selection of "responsible" contractors within the design-bid-build framework, reformers underappreciated problems rooted in how relationships are configured under design-bid-build.[9]

As design-build advocates within the Pankow firm argue, once construction on a design-bid-build project commences—and it may not, if no bids fall below the cost of realizing the architect's design—problems may manifest themselves as errors or ambiguities in design documents, gaps in plans, or incompatibilities between design and construction methods. Contractually obligated to proceed according to the design documents, the contractor then must issue change orders to address these discrepancies, requiring the owner to supply additional capital. Slips in schedules owing to change orders increase a project's interim financing costs, too. And since project management is fragmented, the argument goes, the members of the building team may adopt adversarial positions out of financial self-interest. (It should be noted, however, that the idea that the architect, engineer, and contractor "should always be kept at daggers' points" to protect the owner's interests historically carried weight.[10]) The owner thus assumes the project's execution risk; designer and builder act to protect their margins by attempting to show that problems associated with the project are not their fault.[11]

Reformers did recognize that cooperation among building team members would work to the benefit of owners. San Antonio, Texas–based construction engineer A. J. McKenzie, for one, told his ASCE colleagues that it was "a well established fact that the interests of the owner are best served when the engineer and contractor cooperate in erecting the best possible structure at the lowest cost."[12] Calls to select contractors on criteria other than the lowest bid, however, foundered on the rocks of custom, professional ethics, and law. After all, as *The Architect and Engineer* noted, awarding contracts to the lowest competing bidder was "an unwritten law among builders." Architects and owners exposed themselves to litigation if they selected a contractor on any basis other than the low bid, whether or not the law was explicit on the matter. Hence, a consensus emerged among the American Institute of Architects, Associated General Contractors of America, and other professional groups, that, as one editorial put it, "it is the height of unfairness, even if not legally wrong, for a bidder to be asked to go through the time and expense of figuring and then fail through no fault of his bid but through the caprice of the architect." In this context, the quantitative bid was the sole objective measure; all others were suspect. Architects, then, should invite to bid only contractors "to whom he would [not] hesitate to award a job in case his figure was lowest." Notwithstanding a robust dialogue within the building industry, the issue of the letting of public contracts, too, was resolved in favor of the right of the lowest bidder to the award. Laws and regulations governing public sector contracting remained on the books essentially unaltered until the late twentieth century. At the same time, design-bid-build prevailed as the project delivery system of choice in the private commercial segment of the construction industry.[13]

Convinced that design-build offered owners a structural solution to the problem of coordinating construction management, Charlie Pankow spent his professional life restoring the master builder to the commercial building site. Under the arrangement that he promoted, contractors assumed responsibility for delivering a project at a price that they determined early in the design process and accepted the execution risk associated with schedule and cost overruns. One contract between owner and contractor replaced the separate agreements that governed the design-bid-build approach: a configuration, Pankow argued, that gave contractors the incentive to work with architects and structural engineers to achieve "more economical and efficient construction" (fig. 2). A contractor who was involved in a project before it broke ground, Pankow was convinced, would have the opportunity to integrate cost- and time-saving construction methods, evaluate alternate design configurations, and advise the architect on cost-effective ways to realize his or her aesthetic vision. "When the builder has the opportunity to join up with a design team, you provide an added dimension to solving [cost] problems," he argued. Moreover, Pankow thought it best to negotiate a lump-sum contract with the owner, and that this could be done on the basis of schematic plans and outline specifications. Pankow believed that lump-sum contracts established a better financial incentive than cost-plus contracts (on either a fee or a percentage basis) for contractors to use their technical and managerial expertise to perform due diligence and solve construction-related problems. He also promised clients that there would be no change orders unless the latter decided to alter the scope of the project. Guaranteeing a project's cost and schedule, Pankow argued, enabled the owner to determine its viability before he—and it was always a he—invested a substantial sum of money in it and helped him to secure financing.[14]

Charlie Pankow neither invented design-build nor was its earliest practitioner in the modern US construction industry. Pankow's predecessors, however, typically acted as both designers and builders on their projects, and thus executed design-build without the architect—perhaps the most prominent member of any commercial building team. The discussion in the next section provides a baseline for assessing the contribution of the Pankow firm to the practice of design-build.

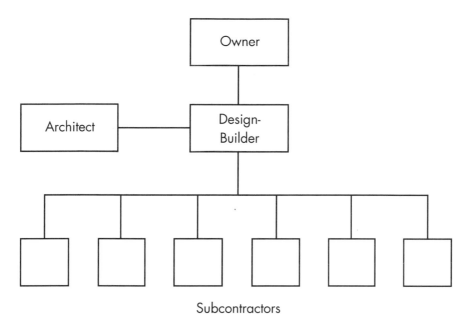

Figure 2. Design-Build. The owner signs one contract with the design-builder, who is most commonly the general contractor, and who acts as the single source of responsibility for design and construction.

Design-Build Linked to Techniques of Mass Production in Early-Twentieth-Century America

Thomas D. Verti, one of three Pankow executives who have served as president of the American Concrete Institute, has argued that "the construction industry is one of the only major industries where the constructed or manufactured product is not designed by the same entity."[15] This blanket assessment, however, does not hold for all segments of the industry across the twentieth century. Indeed, the practice of design-build in America may be linked historically to the construction of standardized reinforced concrete and steel factories and other industrial facilities, such as refineries and chemical plants, by large engineering firms that integrated design and construction within well-capitalized, hierarchical organizations. Legally, bonding requirements that posed a barrier to entry into construction for professional design firms did not prevent their engineering counterparts from offering design services. Taking their cues from Henry Ford and other well-known innovators of techniques of mass production, construction engineers developed building templates that they marketed to industrial owners in the first three decades of the twentieth century.[16] They designed their products to make use of prefabricated elements that could be delivered to the job site. They also mechanized work at the site with techniques that facilitated the substitution of skilled workers previously used to build masonry, stone, or wood structures with laborers whom they might easily train to perform simplified tasks and control through rationalized work flows. In the case of reinforced concrete structures, further mechanization was achieved through the manufacture of elements on site, that is, in yards that were either adjacent to, or near, building sites, as space allowed. Replicating a small number of utilitarian designs enabled construction engineering firms to achieve, in some measure, the economies of scale of the automobile assembly line and the throughput of the oil refinery. While a corporation typically retained an architect to design a unique and complex office tower that projected a desired public image across the modern commercial landscape, there was little incentive for it to do so for an easily replicable, nondescript factory located along a railroad siding. Designing such structures required little or no architectural training.[17]

Amy E. Slaton has focused the attention of scholars on the integration of design and construction services in the erection of standardized reinforced concrete structures along America's East Coast.[18] The deployment of design-build as a project management tool was not restricted to the construction of structures framed by this material, however. For instance, the Austin Company, founded by Samuel Austin in 1878 in Cleveland and arguably the most successful design-builder of the twentieth century, specialized in steel-framed structures whose elements also could be mass-produced. Its Austin Method, which dates from 1901, offered a project delivery system that aligned architecture and engineering in the manner that Cowles contemplated. As distilled in advertisements that appeared in newspapers and national magazines, such as the *Saturday Evening Post*, the Austin Method guaranteed (1) a maximum price—that "the estimated cost of your building project will be the final cost to you," (2) a delivery date, and (3) materials and workmanship of "the highest quality." The guaranteed price policy was a source of competitive advantage, especially after World War I, when integrated engineering and construction firms began to negotiate contracts on a cost-plus basis. Deploying the concept of "undivided responsibility" to promote its approach and professional expertise, the Austin Company sought to convince potential clients to consult its engineers even before they were sure what they wanted to build or where they thought of locating the project.[19]

Like his counterparts who offered clients the utilitarian concrete factories described by Slaton, Samuel Austin deployed techniques of mass production to distinguish his company in the marketplace. Reflected Joseph K. Gannett, vice president for engineering and research, in 1952: "We began to recognize that many of the things we were doing on an individual job basis could be simplified. The idea of standard buildings was the natural result. Ten basic designs were developed. This gave our fabricating shop an opportunity to make and pass along the savings that always come from duplication in any manufacturing process." Standardization of building elements saved owners money by displacing both architects and skilled craftsmen. By developing a limited number of design templates, Austin reconfigured construction as a mass production manufacturing process. The company further reduced costs and shortened schedules through a fast-track delivery system, whereby excavation began before engineers finalized plans and specifications with owners. The company maintained inventories of standard structural steel elements as part of an elaborate distribution network that could deliver prefabricated materials to a building site anywhere in the country as soon as the contract was concluded. To reduce uncertainty in its supply chain, Austin vertically integrated into steel production. It generally avoided municipal projects for the same reason that Charlie Pankow later avoided them: Its engineers could not control the design process for this type of project. Repeat clients attested to the success of the Austin Method. On the basis of design-build delivery and the mechanization of construction, Austin perennially ranked as one of America's largest contractors and its largest design-builder throughout the period of this study.[20]

Austin also applied design-build to the commercial building sector. After World War I, Austin constructed multi-story office buildings, department stores, hotels, theaters, stores, and mixed-use structures. Commercial buildings are invariably architecturally discrete projects that demand more accomplished design expertise than stark, standardized industrial structures. Nevertheless, the Austin Method apparently proved to be a cost-effective means of delivering these projects to owners, as the following example suggests.[21]

Ralph B. Lloyd (1875–1953) was a Southern California oilman who invested much of his wealth in commercial real estate development in Los Angeles and Portland, Oregon.[22] In late 1926, in the wake of his mammoth purchase of almost 800 lots of largely unimproved property around Holladay Park on Portland's East Side, Lloyd began to plan for a mixed-use building in the Walnut Park district that lay a couple of miles to the north. Lloyd controlled all four corners of the intersection of Killingsworth Avenue and Union Avenue (now Martin Luther King Jr. Boulevard) and wanted to make a statement with his "first building of moment" in the city. The project would anchor a major business center—one of Portland's first to lay outside of its central business district. Lloyd envisioned a Class A mixed-use building "that [would] be an ornament to the district." It would feature "one of the finest theaters in the city," with 1,400 seats, and include ground floor retail stores that fronted both streets, and second floor medical offices along Union Avenue. Lloyd expected to spend $200,000 to $300,000 on the project, depending on what his prospective tenant, Multnomah Theaters, a company controlled by New York–based Universal Film Corporation, was willing to pay for a 15-year lease.[23]

Lloyd looked first to local architects John Bennes and Lee Thomas to design the project. The scale of their respective neighborhood theaters matched that envisioned by Lloyd for his project. Bennes had designed the 1,500-seat Hollywood Theatre, which had opened in 1925 in a shopping district along Sandy Boulevard, to the east of Holladay Park. Thomas's Bagdad Theater was under construction in a district south of Lloyd's Holladay Park holdings. Lloyd's tastes did not extend to the exotic themes that both architects incorporated into their designs, however. The Hollywood's elaborate terra cotta tower and "rococo Art Deco façade," as architectural historian Bart King describes it, and the Bagdad's Mediterranean style, were simply too "jazzy." So Lloyd turned to Long Beach, California, architects Parker Wright and Francis Gentry.[24]

The architects, who are best known for their Greek Revival Masonic Temple, which was under construction in Long Beach when Lloyd contacted them, produced preliminary plans and a cost estimate that worked out to a rental of $1.25 per theater seat. Lloyd took this number to Universal Film, which rejected it because the theater would not be located in an established commercial district. To develop an alternate, more financially feasible design, Lloyd turned to the Austin Company. Austin engineers produced a design for a reinforced concrete, steel-trussed structure that would cost $100,000 less to produce than Wright and Gentry's proposal. They achieved the savings in part by reducing the number of theater seats by 400. This reduction in seats also aligned the project with prospective demand. As the new design met Universal Film's leasing criteria, Lloyd could finance it. When the California oilman informed Wright and Gentry that he had retained the Austin Company to act as both designer and builder, he noted that Austin had promised to fix a maximum cost to construct the building on the basis of its estimate; Austin had also guaranteed that the structure would comply with the local building code. Lloyd was describing the essence of design-build, as Charlie Pankow would practice it.[25]

The situation that Lloyd confronted was one that owners faced time and again under the traditional design-bid-build approach, namely that the project, as designed by the architect, did not "pencil," or meet the pro forma budget. Much of the Pankow firm's business would depend on "salvaging" such projects, especially in the first two decades of its existence. For the Austin Company, then, the theater building project was a salvage job.

Suggesting the novelty of design-build in a commercial building context, Lloyd expressed discomfort with using Austin's engineers rather than an architect to design this relatively complex structure, notwithstanding Austin's growing résumé in this sector and Lloyd's prior use of the contractor's services. A few years earlier, Lloyd had contracted with the firm to design and build a standardized warehouse just south of downtown Los Angeles for the Paul G. Hoffman Company, the top distributor of Studebaker automobiles in America. Even as he replaced Wright and Gentry as architects, he retained them to review Austin's design to ensure its safety. He also sought out the Portland office of Robert W. Hunt Company, the Chicago-based firm of consulting, inspecting, and testing engineers that was the largest of its kind in America, to observe the concrete pours, check the strength of the building, and ensure that his project was "properly conceived and constructed."[26]

The extent to which design-build was used to deliver commercial structures during the 1920s is unclear. If Ralph Lloyd's activities as a developer are indicative of the approach that other building owners adopted, however, then deploying design-build in this segment of the industry was exceptional. More often than not, Lloyd took what might be termed a "design-negotiate-build" approach to project procurement and delivery, wherein he hired "responsible" contractors with whom either he or the architect worked on a repeat basis. The selected contractor reviewed the architect's working drawings and specifications and contracted to build the project. On occasion, the contractor made recommendations that resulted in revisions. The contractor received a fee expressed as a percentage of a maximum guaranteed project cost, as agreed to by the developer, architect, and contractor. In these cases, neither Lloyd, as owner, nor his architects advertised for bids or sent out bid packages to pre-qualified contractors.[27]

More research is needed to determine the extent to which contractors deployed design-build on commercial projects before 1930. In any case, the effective moratorium on commercial construction imposed by depression and war (both World War II and Korea) apparently resulted in an institutional loss of knowledge of the methodology among a new generation of owners and contractors. How else to explain why the design-build approach that Charlie Pankow adopted for the American Cement Building project in Los Angeles, as head of a building division within Peter Kiewit Sons', was deemed "novel" and hailed as a "progressive" idea?[28]

However novel and progressive his deployment of design-build may have been in 1961, when the American Cement Building opened, Charlie Pankow nonetheless followed in the footsteps of early twentieth century technical experts in concrete construction who reconfigured work at the building site along modern organizational lines. He shared with his Progressive Era counterparts an enthusiasm for rationalizing work. No less than his predecessors, Pankow sought to achieve, through experimentation and practical experience, precision, predictability, and replicability in building processes. Both design-build and the techniques of concrete construction that Pankow deployed at the building site, to which this book devotes considerable attention, may be seen as best practices rooted in the Machine Age.[29]

The Construction Industry Remains Vast and Fragmented

Amy E. Slaton sees sufficient evidence, both in the application of the techniques of the assembly line to factory building and in the development of elaborate bureaucratic structures by large engineering firms, to argue that Alfred D. Chandler Jr. should have included construction in his pantheon of early-twentieth-century mass production industries. But the scope of her study does not span the industry. The delivery of "thousands of concrete factories . . . almost indistinguishable in appearance" differed fundamentally from the manner in which commercial and residential structures were built, in both the private and public sectors. For the greatest obstacle to throughput in construction lay not in process automation and workflow simplification, but in building team configuration and project management. On a project where designer and builder were one in the same, and the building was a replication of one of a limited number of templates, throughput might have begun to approximate levels achieved in capital-intensive, mass production industries. Throughput inevitably suffered, however, whenever a project was designed by one firm and built by another that had bid on plans and specifications developed without its participation, as was the norm in most of the building industry when Charlie Pankow began his career. The fragmentation of the building team owed in part to the customization that building owners desired for their projects, which, in itself, limited throughput. Buildings are also rooted to place: They cannot be shipped from centralized production facilities in the manner of automobiles and televisions, for instance, no matter how many of their structural elements might be prefabricated. To be sure, as this book elaborates, job site automation was crucial in enabling the Pankow firm as contractor to assume the execution risk of commercial building projects. More important in meeting a project's budget and schedule was the deployment of design-build—a project management methodology that involved all of the members of the building team in the planning and design of the project.[30]

The managerial and technological revolution that Slaton ascribes to concrete factory construction clearly did not transform the construction industry. In 1947, the year before Charles J. Pankow arrived in Los Angeles with a newly minted degree in civil engineering from Purdue University, *Fortune* magazine blasted the US residential construction industry: It deployed "archaic," craft-based production methods; builders were unable to coordinate the work of myriad subcontractors; trades unions dictated wages and "outlaw[ed] technological progress"; and the sector's supply chain was "fantastically expensive." A quarter of a century later, Arthur Sampson, the head of the US General Services Administration, called construction more generally "the worst-managed industry" in the nation.[31]

To be sure, measuring productivity in the construction industry has proved to be a rather elusive task. According to one assessment, for instance, productivity actually increased 2.4 percent annually from 1947 to 1968—a period that covers the formative years of Charlie Pankow's career, when design-build had yet to make a significant impact on the industry. Moreover, as Steven Rosefielde and Daniel Quinn Mills conclude, productivity may have remained close to the US average into the 1970s; they argue that official statistics consistently underestimate output in the industry. At the same time, other analysts insist that productivity began to decline at this time. Problems of measurement aside, the record of myriad projects either left on the drawing board because their cost exceeded the owner's budget or plagued by cost and schedule overruns support the charges of industry critics.[32]

The construction industry also remained "vast and fragmented," in contrast to industries, such as automobile, chemicals, and steel, whose structures became oligopolistic during the twentieth century.[33] From 1947 to 1967, for instance, the number of construction companies increased from 395,300 to 794,858, in step with economic growth in postwar America. Only slightly more than 3,000 firms employed at least 100 people. A quarter century later, only 3,990 of nearly 600,000 firms employed at least 100 individuals, the Pankow firm among them. Construction also has remained fragmented in terms of revenue. In 1967 only 880 of 156,400 commercial and institutional contractors had receipts of $5 million or more; another 5,117 contractors enjoyed revenues of at least $1 million. In 2002, 4,415 of 37,208 contractors had receipts of $10 million or more; another 14,920 contractors had revenues of at least $1 million. This lack of rationalization further suggests that Chandler was right to exclude construction from his compendium of capital-intensive, mass production industries.[34]

This book relates how Charlie Pankow and his colleagues practiced design-build and promoted its diffusion as a solution to problems associated with construction management. It should be stipulated at the outset, however, that the industry-wide transformation that design-build advocates have promised, and continue to promise, remains far from complete. Even if there were consensus that design-build constituted a "best practice" among project delivery methodologies, it should come as no surprise that its diffusion within a "vast and fragmented" industry would be slow. Hence, some three decades after Sampson's withering critique, Barry LePatner, a prominent construction lawyer, could argue that little had changed in the industry. He could cite numerous examples of major projects that were "over-budget and overdue." Even more recently, Barbara White Bryson, associate vice president for facilities engineering and planning at Rice University, and Canan Tetman, principal of Austin, Texas–based CYMK Group, have argued creditably that the construction industry remains "one of the most inefficient on the planet."[35]

Postwar Developments in Concrete That Underlay Charlie Pankow's "Budgeted Construction Program"

Charlie Pankow preferred concrete as a building material over steel—the only practical alternative for mid-twentieth-century designers and builders of tall commercial structures—for reasons that his predecessors engaged in the construction of reinforced concrete factories would have found familiar.[36] It could be molded into many shapes under repetitive processes, and therefore facilitated innovation, adaptation, and "tweaking" at the project level. As Pankow would assert: "There are very few buildings designed in steel that can't be done better in concrete." Concrete was also less expensive than steel, and using it reduced financial risks associated with accomplishing sequential tasks. Long lead times associated with the steel industry's supply chain meant that contractors had to place orders for the material before the owner determined the cost of his project. (Unlike the Austin Company, Pankow had no interest in vertical integration.) Meanwhile, project financing might not materialize. As a result, Pankow would note in 1986, "You see a lot of steel frames standing around rusting because the tail got out ahead of the dog." And deliveries of steel might be delayed, or not fulfilled: Reflecting a contentious struggle between managers and workers over shop floor rights and prerogatives, strikes were common during Charlie Pankow's formative years as a builder. Industry-wide labor actions in 1947, 1952, 1955, and 1956 were prologue to the largest steel industry strike in US history: a 116-day affair that began on 14 July 1959 and ended only when President Eisenhower invoked the provisions of the Taft-Hartley Act. Even when steelworkers were not on strike, the threat of disruption was "in the air," to borrow from classical economist Alfred Marshall's famous explanation of why industries tended to develop in clusters geographically. Suppliers of concrete from local and regional sources were much more reliable.[37]

The Pankow firm stood apart as a contractor that performed concrete work that was normally subcontracted. Its deployment of techniques, such as on-site precasting and slipforming—a method of pouring concrete continuously from the ground to the top of the building, using a system of panels, yokes, and jacks—was not unprecedented. Still, on-site precasting of prestressed concrete structural elements demanded more skill and experience than precasting similar reinforced concrete elements. And few contractors, if any, regularly erected tall commercial buildings using slipforms. By self-performing mechanized concrete work, Pankow as contractor sought to exert cost and quality control over construction. It also sought to use its concrete construction toolkit to streamline construction within a design-build context and improve the finished product delivered to owners.

After World War II, prestressed concrete displaced the reinforced variety, altering the ways in which commercial and industrial buildings were designed and constructed. For more than half a century, contractors had used reinforced concrete, which incorporated rebar, or steel reinforcement bar, inside poured concrete, to frame structures. A cumbersome and expensive material in the late 1880s, reinforced concrete became ubiquitous in commercial and industrial buildings alike by World War I. It allowed internal columns to support structures, floor space to be opened, windows to be enlarged, and buildings to become taller and larger.[38]

Prestressed concrete substituted high-tensile steel tendons—wires, cables, or rods—for rebar. The steel tendons might be pre-tensioned, that is, stretched between anchors before workers poured the concrete, or post-tensioned, whereby workers used hydraulic jacks to stretch the steel after the concrete cured. (As used throughout this book, the term "prestressed" generally refers to elements that have been pre-tensioned; post-tensioned elements are always specified as such. In some instances, "prestressed" is used as a general term, as it is here, to include pre-tensioned and post-tensioned elements.)[39]

Prestressed concrete held significant advantages over both structural steel and reinforced concrete. As T. Y. Lin, who pioneered the use of the material in America from his post in the Department of Civil Engineering at the University of California at Berkeley, noted, prestressed concrete elements were much lighter than those produced with reinforced concrete, and thus reduced the weight of the building. Prestressed concrete also permitted longer spans and more efficient vertical loading, so that elements, such as wall panels, that had been used decoratively, could also be deployed structurally, and designers could lay out floors more simply and efficiently. The material was economical, too. Steel used in prestressed concrete was four times as strong as regular structural steel, yet cost only three times as much. Prestressed concrete was twice as strong as reinforced concrete, but was only 10 to 20 percent more expensive.[40]

The idea of prestressed concrete may be traced to San Francisco–based engineer P. H. Jackson's method of tightening steel tie rods in precast concrete elements, which he patented in 1886. French engineer Eugene Freyssinet is credited with developing post-tensioning technology, beginning as early as 1927, but in 1925, R. F. Dill of Alexandria, Nebraska, filed a patent for a post-tensioning system that deployed high-tensile steel coated in a plastic compound to prevent bonding. In America, structural engineers used post-tensioned girders to span bridges as early as 1949. Over the next seven years, designers specified prestressed concrete elements in 1- and 2-story buildings, including the Midwest Geophysical Laboratory in Tulsa, Oklahoma (1950), and Loveland Elementary School in Colorado (1953). The completion of the 24-story Diamond Head Apartments in Honolulu, in 1956, marked the first recorded use of prestressed concrete in a high-rise structure.[41]

Still, according to Lin, designers were slow to appreciate the implications of the advances in prestressed concrete that had occurred over the previous decade. As the California Division of Architecture—the agency within the Department of Public Works with responsibility for developing, letting, and supervising state building projects—argued, this owed in part to a lack of knowledge about the material, but also to inadequate communication and cooperation between architect and structural engineer. Based on 21 contracts let between 1952 and 1957 that invited designs to make use of prestressed concrete, the Division found that designers were reluctant to use the material, even when presented with an economic argument for doing so. In the instances where they had specified the use of prestressed elements in their structures, Lin observed, designers typically had merely substituted them for conventional materials to reduce costs. As a result, multi-story structures framed in prestressed concrete retained features and floor layouts associated with reinforced concrete. Noting that "the possibilities of prestressing in multi-story buildings have scarcely been explored," Lin called for a "revolution in design" in 1961—about the time that Charlie Pankow initially may have begun to consider starting his own firm. Lin envisioned architects and engineers collaborating to produce a new architecture.[42]

Lin might have added design-build contractors, Charlie Pankow's company in particular. For on many of its early projects, Pankow would work with Lin's consulting firms. Together, builder and structural engineer would promote the use of prestressed concrete on commercial building projects among owners and architects. When Professor Lin would take stock of the "revolution in design" a couple of years after Pankow had begun its corporate existence, he would cite several of these projects to illustrate what was possible with prestressed concrete, even as he would lament the ongoing hesitancy of designers to experiment with the material.[43]

Facilitating the diffusion of prestressed concrete in commercial building was the growth of manufacturing operations during the 1950s, from perhaps half a dozen plants to more than 200.[44] For reasons already mentioned, however, Charlie Pankow would choose to precast prestressed concrete elements on-site.

Precasting concrete elements on the job site or on adjacent or nearby lots, rather than casting them in place or using products prefabricated in distant plants, was not unprecedented. California builder and engineer Ernest Ransome, to take perhaps the most well-known example, integrated on-site precasting of columns, beams, and girders into the Ransome Unit System that he marketed to owners before World War I. After World War II, Los Angeles became a particularly active area for precast concrete construction, with builders routinely casting wall panels on the floor slab of the building, for instance, and having truck cranes lift them into place. Many contractors were reluctant to undertake prestressed concrete precasting, however. As the California Division of Architecture's Allen H. Brownfield observed, it demanded a level of competence and experience that typically they did not possess.[45]

As the following pages elaborate, the Pankow firm specialized in the construction of tall office and residential towers, in contrast to the 1- to 2-story industrial structures of its design-build predecessors. Prior to World War I, the structural frame of such commercial structures was typically cast in place—an operation that involved pouring concrete at elevation. Slipforming was a technique that Charlie Pankow adapted to building in a commercial context. A well-established technique for the erection of grain elevators, chimneys, silos, and other structures that lacked doors, windows, beam pockets, and other irregularities, slipforming was deployed on a high-rise building in America for the first time in 1955. Architect John Doggett and structural engineer S. S. Kenworthy estimated that slipforming saved as much as five months and 10 percent of the cost of constructing the 10-story Madison Towers apartment hotel in Memphis, Tennessee. Charlie Pankow would regularly use the technique to achieve similar results. As he would later write, slipforming offered a way to "substantially lower construction costs, which directly result in lower completed-building costs [making] possible the building of projects that otherwise could never have been built because of limited financing."[46]

Utilizing slipforms and operating yards to precast prestressed concrete elements, which Charlie Pankow's men would begin to perfect as a building division within Peter Kiewit Sons', would immediately distinguish the early Pankow firm from other contractors. These were difficult tasks to perform. Given the newness of prestressed concrete technology, on the one hand, and the rare deployment of slipforms on a commercial project, on the other, it is not surprising that, early in his career, Charlie Pankow would find that relatively few trades workers had experience in them. So he would have his teams, within both Peter Kiewit Sons' and his firm, master both techniques on the job. "Every project is a training process," Charlie Pankow would later note. And yet, he would argue, the investment in self-performing these tasks paid handsome dividends. "When you can mechanize your project [and] put everything on the ground where you can inspect it, you have a much better learning curve and you also have a much better quality assurance program because you can see what's going on right there."[47] Control over construction processes would facilitate the estimating required to determine the contract figures to be negotiated with owners. In Charlie Pankow's construction program, performing concrete work also would be closely linked to design-build; without the participation of the contractor, the architect and structural engineer would not necessarily design a building to fully leverage the techniques of mass production that Charlie Pankow sought to deploy.

The building site that emerged from the modernization of production during the first three decades of the twentieth century, at least as far as the construction of industrial vernacular structures beloved by Walter Gropius, Le Corbusier, and other leading architects of the modern movement were concerned, featured, as Slaton has characterized it, "an extreme division of labor that called for a relatively small coterie of highly trained specialists to supervise the work of many little-trained, and much lower-status, laborers."[48] The commercial building site of the Pankow firm, as self-performing contractor, would be configured somewhat differently. Like its predecessors, the company would deploy techniques that minimized the number of workers—both skilled and unskilled—needed to complete a project. Many of the remaining workers, however, would be engaged in tasks that demanded skill and experience to execute successfully. The demand for carpenters and other trades workers would be high and Pankow typically would recruit them out of the union hall. Supervision would be shared by men who rose through the ranks in the field and newly recruited civil engineering graduates, many of whom would choose to spend their careers in the field, rather than move into the managerial office.

Design-build and engineering innovation were merely the means to the end of satisfying owners, however. For satisfying owners was foremost in Charlie Pankow's mind as a builder; his reliance on repeat business would be a critical element of his business model. The difficulties that owners were experiencing in completing projects on schedule and within their budgets created a potential market opportunity for Pankow. Yet, as architectural historian Reyner Banham has observed, owners historically were reluctant to risk their budgets by allowing designers and builders to experiment with untried materials or construction techniques. As Pankow would learn, it would be much easier to sell design-build to owners whose projects either could not be built as designed or were plagued by cost and schedule overruns.[49] This was the case with oilman Lloyd, who once remarked, more or less matter-of-factly, "We have always found that a building is never built for the estimate or contract price."[50] Yet design-build and techniques of mass-producing concrete elements were not panaceas. Timely project execution still depended on cooperation among all members of the building team. A theme stressed in this book is that the company's culture—referred to within the organization as the Pankow Way—was a key factor in enabling the Pankow firm to fulfill its role as design-build contractor, cooperating with architects, engineers, and subcontractors in meeting owners' expectations.

Plan of the Book

Chapter 1 traces the development of Charlie Pankow as a builder, from his youth, playing and working at his father's building sites, to his training in civil engineering at Purdue University, to his career prior to forming his own company in 1963. Exploring the theme of the engineer practicing in a business context that has interested students of the history of science and technology, chapter 1 focuses on Charlie Pankow's career at Peter Kiewit Sons', a large construction company. It emphasizes not the subordination of the engineer to the commercial needs of management, but the role he played as a corporate entrepreneur in leading a small group of men within a building division that he started with the support of a mentor. Pankow and his colleagues gained expertise in concrete construction techniques on increasingly sophisticated commercial structures. Charlie Pankow also took opportunities as they presented themselves to become involved with architects and engineers in the design, or redesign, of projects. In so doing, Pankow and the men he led developed a culture characterized by a platoon-like mentality and a passion for innovation. The chapter considers the development of this culture in comparative terms. At the same time, builders of concrete office towers were not the "stars" within the Kiewit firm. Though it had a long history in commercial construction, Kiewit was primarily a builder of bridges, dams, roads, and the like. And Peter Kiewit prided himself on bidding jobs. Indeed, the so-called heavy and highway divisions of the company represented an internal enemy that motivated Pankow's men. While he was able to convince Peter Kiewit to allow him to establish a building division, Pankow met resistance from him when he sought to promote his own methods in others areas of the business. According to interviewees, Peter Kiewit was uncomfortable with the type of projects that Charlie Pankow undertook and skeptical of design-build as a project management tool. The treatment of the building division as a step-child within the Kiewit organization would ultimately spur Charlie Pankow to establish his own firm.

The usual setting for the study of the relationship between the engineer and business is the corporation. Less attention has been paid to the engineer as entrepreneur innovating outside of the large organization. Chapter 2 covers the early years of CPI, showing how it replicated the culture and work of the Kiewit building division. Almost to a man, the building division constituted the personnel of the new company. But now Pankow was majority owner of a new firm; chapter 2 shows how he ran it, in effect, as a sole proprietorship. The chapter then uses the firm's early projects in California to illustrate how CPI's project teams applied Charlie Pankow's construction program to various structures for both private and public owners. It shows how repeat clients formed the bedrock of Charlie Pankow's business model. Yet the chapter ends at a moment of crisis, when two key project managers left CPI to form their own company. In doing so, they took with them CPI's most important client.

Chapter 3 covers roughly the same temporal period as chapters 2 and 4, owing to the independent development of the company's operations in Hawaii. In 1965 Charlie Pankow sent George Hutton, one of his engineers in his Kiewit building division, to Honolulu to manage the construction of the James Campbell Building, which was part of the revitalization of the central business district. Owing to a lack of work on the Mainland at the time of the project's completion, Charlie asked Hutton to stay in Hawaii. With little or no direction from the home office, Hutton built a successful business (Charles Pankow Associates [CPA]) that operated for all intents and purposes as a separate entity until Pankow reorganized the company in the mid-1980s. In fact, CPA operated as a subsidiary until that time. Chapter 3 explores how Hutton developed the Hawaii business, and contrasts the ways in which CPA and CPI operated in geographical isolation from one another. Hutton replicated the company culture that had been incubated within the Kiewit building division. Yet the independent path that he charted had a significant impact on the organizational development of the company.

Chapter 4 picks up from chapter 2 to discuss the CPI's work on the Mainland during the turbulent economy of the 1970s and early 1980s. It elaborates on the contractor's relationship with the owner, architect, subcontractors, and labor within the design-build framework. At the same time that George Hutton was building CPA, CPI relied on a small number of clients to fill its order book and recover from the shock of losing its largest client. Of these, Winmar, the real estate arm of Safeco Insurance, was by far the most important. The chapter examines the Winmar relationship, showing how design-build might be deployed as a construction management tool regardless of building type and the material used to frame the structure.

Chapter 4 also examines the important role of development projects in sustaining and growing the company. During the 1970s, Charlie Pankow, George Hutton, and Russell J. Osterman, who also worked in the Kiewit building division, became developers in their own right, with Pankow and Hutton pairing up in Hawaii and Pankow and Osterman investing in projects in California, Oregon, and Washington. The projects generated substantial profits for the partners of the development entities associated with them, but also created opportunities to train newly recruited engineers.

Almost a decade into its existence, CPI began to recruit a new generation of civil engineers and construction managers. It hired experienced professional and newly minted graduates of leading institutions, Charlie Pankow's alma mater in particular. In time, the men recruited during the early 1970s would populate the ranks of upper management or otherwise serve in key positions within the company. Whether hired from industry or university, however, recruits had to be inculcated in the culture that Charlie Pankow and his colleagues transplanted from the Kiewit building division. Chapter 4 opens by discussing the elements of the Pankow Way, a term whose use dates from the early 1970s.

As chapter 5 relates, Charlie Pankow reorganized the company twice in the mid-1980s. Ultimately, the company emerged as a limited partnership, with Charlie Pankow still firmly in control. But though they were motivated above all by tax considerations, the reorganizations helped to ensure that the company would outlive its founder. After two decades in operation, Charlie Pankow, Russ Osterman, and George Hutton owned some 85 percent of CPI's stock. Yet there were many more employees. As a result of the second reorganization, some 40 people had an equity stake in the new firm. At the same time, through the creation of a general partnership, Charlie Pankow retained majority control of the company: necessarily a condition of any restructuring. The reorganization also spelled the end of CPA. Charlie Pankow's assertion of his authority over the company's operations in Hawaii was marked by the breakup of the cadre of employees who had worked for many years for George Hutton. It would mark the beginning of the end of Hutton's career with the company.

Charles Pankow Builders, Ltd. (CPBL), benefited handsomely from late-1980s construction booms in California and Hawaii, but suffered as much as any contractor during the recession that ensued. The severity of the downturn generated changes in the market for contractor services, exposing the limitations of Charlie Pankow's business model and leaving the company's long-term viability in question. Beginning with the reorganization, chapter 5 traces the history of the company to that point of potential crisis.

Chapters 6 and 7 overlap in time. Chapter 6 considers the diffusion of design-build and the role that Charlie Pankow and others in the company played in promoting its use. For many years, Charlie Pankow led by example. His projects demonstrated to owners the value of design-build. But the company did not rely simply on word of mouth. Pankow and others in the firm published articles in trade journals, textbooks, and other publications, and also leveraged positions of leadership in societies like the ACI. Charlie Pankow also took an interest in incorporating design-build in construction engineering curricula. As an engineer in business, he helped to align technical training, at Purdue University in particular, with the practical needs of the marketplace. Efforts to promote design-build gained traction in the 1980s, and culminated in the formation of the DBIA, a national trade association, in the early 1990s. Much of the subsequent growth in the use of design-build occurred in the public sector. Crucially, the relaxation of laws that both separated the provision of design and construction services and restricted the use of design-build enabled public agencies to deploy the methodology for project procurement and delivery.

Chapter 6 also considers the unusual investment of CPBL, as a contractor, in concrete research, analyzing the collaborative effort undertaken by the firm for purposes of improving the ability of concrete structures to withstand seismic events. It examines the commercial application of the technology produced by that research, the precast hybrid moment-resistant frame. This effort foreshadowed the work of the foundation that Charles Pankow founded not long before his death.

Chapter 7 emphasizes the need for owners and executive managers to challenge assumptions about the factors that underpin an organization's success as market conditions change. It considers the efforts of Richard M. (Rik) Kunnath (now CEO) and others to diversify the business in the face of changes in the commercial real estate market produced by the deep recession of the early 1990s. A new entity, Pankow Special Projects, Ltd., was created as an umbrella entity for tenant improvement and other types of work. Showing how Kunnath and others in the company implemented change illustrates the difficulties that closely held and family enterprises often face in surviving beyond the life of the first generation entrepreneur.

Chapter 7 ends by exploring the challenges posed by corporate leadership succession, especially when it involves a founder who retains firm control of an organization in which he or she is heavily invested, not only financially, but also emotionally. Unable to separate himself from the company that provided professional reputation and personal satisfaction, Charlie Pankow never fully envisioned how the firm might carry on in his absence and therefore engaged in no formal succession planning. In this respect, he had much in common with his Anglo-American counterparts in family business. To the extent that he thought about leadership succession, he had in mind investing all executive power in a single individual. In the mid-1990s, he tapped Kunnath to succeed him. In the last year of his life, however, he wavered. Persuaded by a small group of individuals who argued that Kunnath was unfit to lead the company, Charlie Pankow launched an eleventh hour search for another successor. In the wake of Pankow's death, however, senior executives reorganized the firm along more democratic lines and selected Kunnath as its CEO. Those who aspired to leadership of the company had to outlast its founder to achieve it.

An epilogue offers an overview of the company's activities and performance in the wake of the death of its founder. A boom in the commercial market mitigated the trauma of succession. The bust that followed, and which continues as this book is written, has further validated the changes in the business model implemented over the past two decades: Public sector work now accounts for 80 percent or more of the company's order book. And while many commercial contractors have gone out of business or merged, Charles Pankow Builders endures as an independent firm after 50 years in business.

The book closes by relating the establishment of the Charles Pankow Foundation by the founder shortly before his death and tells how it seeks to perpetuate Charlie Pankow's legacy through funding research into better ways to build.

■   ■   ■

Architects and engineers figure in the pages that follow not only because of their prominence on any commercial building team, but because of the building team configuration that Charlie Pankow endorsed. A design-build contractor might retain designers and engineers in-house. The Austin Company and others who have used this approach have pointed to the elimination of communications problems and the fostering of good relations among design and construction professionals as reasons for doing so. It also makes sense to develop a staff of designers and engineers for firms specializing in particular building types, such as factories and refineries. The Pankow firm, however, did not pursue this form of vertical integration. For one, doing so would have been an expensive proposition initially for a firm that remained, as chapter 2 shows, a rather small organization in its first decade. More importantly, as a builder of many types of structures, the company believed that hiring independent architects and engineers on a project basis would provide the best balance between project control and design creativity. Unlike the purveyors of standardized factory types, Pankow built unique structures for its clients. Using independent architects would enhance the quality of the design of such buildings, the company argued. As Dean Stephan puts it, "We sold beautiful buildings. We sold dreams to people. . . . We wanted a set of buildings that looked great" to show an owner. The firm's professionals did not "dictate" designs to architects. Rather, they worked with designers to ensure that they were feasible economically. According to Stephan, retaining in-house designers "promotes a kind of 'incestuous' situation that does not encourage new ideas." Hiring outside designers and engineers also afforded the company the opportunity to select firms on the basis of building type, experience with the materials used to frame the structure, understanding of local ordinances and codes, architectural preferences, and environmental conditions, and other criteria. Such an approach also allowed the company to avoid incurring staff overheads during troughs in construction cycles.[51]

Cases where the company's principals acted as developers aside, the firm typically was brought into a project only after the owner had conceived it, had procured the land and permits, and perhaps had even arranged financing. In many cases, the architect already had proposed a design. What urban planner and designer Allan B. Jacobs has explained regarding his client relationships applies equally to relations between Pankow, as contractor, and the owners who hired the firm:

> We [were invited] to design a specific project, with specific clients, a reasonably specific program, achievable within a set time frame. The clients must have known whom they were bringing in and they were free to reject our designs. They chose to accept them. Presumably they were appropriate.[52]

The company focused on ensuring that projects were executed to the satisfaction of their owners. The quality of its structures has not been called into question, even in cases where architectural and urban design critics have panned the buildings.

Given the dozens of large-scale projects that the company has built—see the appendices for a list—it would be impossible to consider every project in depth. Indeed, entire books have been devoted to single projects.[53] To illustrate how innovation within the Pankow firm unfolded as "a continuous stream of innumerable minor adjustments, modifications, and adaptations by skilled personnel" who developed their abilities and know-how through on-the-job experiences, to borrow Rosenberg's characterization of technological change generally, each chapter focuses attention on particular projects (or related groups of projects). To be sure, as this introduction has argued, there were important precedents for the work of the company's project teams. But because commercial construction had all but ceased during depression and war, deploying design-build and self-performing concrete work constituted an unusual, if not unheard of, project delivery approach in that segment of the industry when it revived after World War II. These mini case studies serve as a basis for understanding the success of the Pankow firm and the diffusion of design-build. They also show how the contractor participated in the transformation of cities by implementing changes to the built environment.[54]

As should become clear in the pages that follow, the accomplishments of the Pankow firm resulted from efforts of small teams of individuals who found better ways of constructing commercial buildings. As more than one of the persons interviewed for this study noted, it takes dozens, if not hundreds, of people to construct a large commercial structure. For all of the many awards and honors that he received for his professional achievements, Charlie Pankow valued collective effort; he was quick to deflect the attention of those who would credit the man to the accomplishments of the firm. As Robert Law, the chief estimator, has reported, "He would never leave [any conversation at] the point where it was what he did. He would always leave it as to what the company had done."[55]

CHAPTER 1

Kiewit Days

In America, the relationship between the engineer and business has been a close one. Said Alexander C. Humphreys, president of the Stevens Institute of Technology, a century ago: "Self-evident should be the truth of the proposition that the engineer ought to be a man of business."[1] Throughout the nineteenth and early twentieth centuries, leaders in the profession urged their colleagues to consider their work in business terms as much as technical ones. As mechanical engineer Henry R. Towne, co-founder of the Yale & Towne Manufacturing Company, put it to engineering students at Purdue University in 1905: "The dollar is the final term in every engineering equation."[2] And with the rise of corporate capitalism, leaders in engineering education urged their students to seize the opportunity to work for, and eventually manage, the large organizations that increasingly dominated the industrial landscape. At a time when female undergraduate students in the field were rare, John Butler Johnson, Dean of the College of Engineering at the University of Wisconsin, asserted that, if an engineer "can make himself a good business man, or as good a manager of men, as he usually makes of himself in the field of engineering he has chosen, there is no place too great, and no salary too high for him to aspire to."[3] And, indeed, during the twentieth century, an increasing share of engineering graduates pursued career paths within large organizations—a trend that accelerated after World War II. As of 1960, industrial concerns and utilities employed 71 percent of all engineers. And the top 1 percent of the firms that employed engineers accounted for three-fourths of the nearly one million engineers working in industry and government. Corporations with 10,000 or more employees alone employed more than one-third of the nation's engineers.[4]

Educator Humphreys left open the possibility that an engineer might act as a professional, much like a medical doctor or lawyer. But even then, business and engineering would be closely intertwined: "The man . . . who is able to bring to the service of his clients . . . a sound technical training and the ability to meet business conditions, proves by his comparative success the material value of this dual capacity."[5] Applying engineering and economic criteria to solve problems faced by commercial building owners, both within the corporation that nurtured his talents and at the firm he founded, animated the career of Charlie Pankow.

Scholars have paid considerable attention to the engineer practicing within the large corporation. In contrast to Humphreys, who saw the possibility of the engineer practicing in harmony with the conditions that "limited and bound" him, they have emphasized the subordination of professional skill to commercial criteria and conflicts between the professional demands of engineering and needs of management.[6] Charlie Pankow spent the formative years of his career within a large organization, Peter Kiewit Sons'. There he had the opportunity to put structural engineering and commercial construction to the test, distilled by Coleman Sellers, president of the American Society of Mechanical Engineers, in 1887: "Will it pay?"[7] Ultimately, he would face commercial conditions and bureaucratic hierarchies that "limited and bound" him, and so he would start his own company. All the while, he and the men who worked for him would maintain a Machine Age vision of engineering as a fountain of progress, even as an "intellectual crisis of technology" gripped the profession in the context of the Cold War.[8]

Early Years

Charles John Pankow Jr. was born in Indianapolis on 6 October 1923. He was the second of three children born to Charles John Pankow and the former Bessie Busey Hoult. Both parents were raised in Illinois: Pankow in Elgin, a medium-sized city located 40 miles northwest of Chicago, and Hoult in Chrisman, a small farming community near the university towns of Champaign and Urbana. Each parent attended the University of Illinois and, upon graduating, took jobs in Champaign. Bessie Hoult received an A.B. in 1912 and taught in the public schools. Pankow *père* received a B.S. in architecture in 1913 and practiced his craft until America entered World War I. During the war, he served as First Lieutenant in the 465th Aero Construction Squadron. Before commencing his duties, he and Hoult married, in Detroit, on 18 October 1917. After the war, Pankow was hired by Ralph Sollitt & Sons Construction, a Chicago-based contractor. At the time of Charlie Pankow's birth, his father was working as an engineer in the firm's South Bend, Indiana, branch. That he was born in Indianapolis suggests that Pankow *père* had been assigned to a project in the capital at the time. The career of the father would exert a powerful influence on the professional life of the son.[9]

During the interwar period, the fortunes of South Bend, where Charlie Pankow and his siblings grew up, were tied to the automobile industry. The city's economy had depended on manufacturing for decades, but until World War I, local firms generally remained relatively small by national standards. Two prominent exceptions, Studebaker Brothers Manufacturing Company and the Oliver Chilled Plow Company, produced horse-drawn vehicles and farm implements, respectively. Other firms produced a range of products, from clothing and toys to paints and varnishes to machine tools and industrial bearings. During 1919–1920, President Albert R. Erskine converted Studebaker wholly to the manufacture of vehicles powered by the combustion engine; the company had begun to produce automobiles shortly after it had gone public in 1911. Owing to troubles with his Detroit-based workforce during World War I—turnover reached 300 percent—Erskine consolidated production of automobiles in South Bend, where labor was more "pliable," and added a truck line. Studebaker was already the sixth most popular marque nationally (among dozens) in 1920. Under Erskine's direction automobile sales nearly tripled by 1923, to 145,167 units. In that year, Vincent Bendix relocated his vehicle parts plant to South Bend. The spectacular success of Studebaker and Bendix during the rest of the decade compensated for the decline of Oliver Chilled Plow, which suffered with the loss of European markets for US agricultural products. A victim of one of America's "sick" industries, the company merged with three other firms in 1929 to form the Oliver Farm Equipment Company.[10]

South Bend's population increased in step with the expansion in manufacturing, which employed more than half of the work force. For its part, Studebaker employed 17,663 workers as of 1922; most of them worked in town.[11] The city's population reached 104,193 in 1930—an increase of almost 47 percent over the course of the decade.

Like most engineers in mid-twentieth-century America, Charlie Pankow was white, male, and middle class, and hailed from an urban area.[12] Charlie Pankow did not, however, share confessional status with many members of his profession, the majority of whom were raised in Protestant families. The Pankow family was Catholic, as was most of South Bend's population. Charles C. Pankow, Charlie Pankow's fraternal grandfather, was born in 1860 in Mecklenburg, Germany. He immigrated to America sometime before 1884, when he married Mary Elizabeth Steizle. An ethnic German, she was born in 1863 in Oshawa, Ontario, on the shores of Lake Ontario. Charlie Pankow's maternal grandparents, William Francis Hoult and Grace Moss, were born in Illinois to families that had settled in Vermont and Virginia as early as the seventeenth century and had migrated to Ohio and Kentucky before they settled in Chrisman.[13]

Ralph Shannon Sollitt, one of Ralph Sollitt's sons, established the South Bend branch of the Sollitt firm in 1920 to take advantage of the opportunities presented by the booming local economy. Sollitt was a third generation contractor. His grandfather, Thomas Sollitt, had built Palmer House and other landmark buildings in the Windy City. For its part, the South Bend branch constructed many of that city's prominent structures of the interwar period, including the Poledor store and office building and the Blackstone, Palace, and Granada theaters. The firm also constructed schools, oil refineries, and factories across Indiana. In 1932 Sollitt completed the landmark US Post Office and Courthouse (now the E. Ross Adair Federal Building) in Fort Wayne. Foreshadowing the approach to cost control that would become embedded in Charlie Pankow's construction program, the designers substituted reinforced concrete for steel in the structural system. The approach saved some $30,000, allowing contractors to submit bids far below the expectations of federal officials, including James A. Wetmore, acting supervising architect of the US Treasury Department. Wetmore and his colleagues took advantage of the windfall to incorporate a number of enhancements into the budget, including stylized art deco ornamentation, marble floors, and an elaborate plaza along the street façade.[14]

Among family members and people associated with the Pankow companies, Charles Pankow Sr. is best remembered for his participation in the construction of Notre Dame Stadium. The 54,000-seat arena was built as a showcase for the football team, which had gained a national following and reputation under coach and athletic director Knute Rockne. Cleveland-based Osborn Engineering designed the project. Founded in 1892, the company had a long record of sports facility design and construction, including baseball's Comiskey Park, Fenway Park, Polo Grounds, and Yankee Stadium, and numerous university stadiums, including the University of Michigan, on which the far smaller Notre Dame Stadium was patterned architecturally (fig. 3). Ralph Sollitt & Sons built the stadium (fig. 4). During 1929–1930, Pankow *père* supervised work at the job site and became acquainted with Rockne, who had shaped the design of the project and took a close interest in overseeing its construction.[15]

Charlie Pankow was introduced to the construction business as a youth. He hung around his father's job sites, "collecting scraps of wood as if they were precious stones," according to one account. During high school summer breaks, Pankow *fils* worked for Sollitt and other construction companies. He came to appreciate the value of physical work from watching workers mix concrete and bend reinforcing steel by hand. By his senior year at South Bend's Central High School, he had decided to study structural engineering. In the fall of 1941, Charlie Pankow entered the civil engineering program at Purdue University.[16]

On the eve of America's entry into World War II, engineering programs put a premium on practical education that trained students for well-paid technical positions in commerce and industry. Attuned to the needs of the marketplace since the Civil War, programs at mid-century still valued hands-on experience, ingenuity, and intuition over scientific theory, even as reformers' efforts to insert the latter into curricula were by then decades old. Heavy teaching loads restricted the amount of time that a professor might devote to research. The research that was undertaken tended to focus on "real world" problems rather than theoretical ones, and typically grew out of instructors' consulting projects. Concentrated in experiment stations, research remained funded largely from industry sources, and devoted to studies of automobile engines, road materials, sewage collection and treatment, and the like. Indeed, owing to the success of Dean A. A. Potter in securing external support, Purdue's experiment station had become the nation's largest by the late 1920s.[17]

Charles A. Ellis, one of Charlie Pankow's instructors in structural engineering, embodied both the preference for the practical over the theoretical in engineering education and the dual role of the professor as educator and consulting engineer. Born in 1876 in Parkman, Maine, Ellis received the A.B. degree from Wesleyan University in 1900. Two years later, he joined the staff of the American Bridge Company, where he distinguished himself in calculating the stresses of subway tunnels under the Hudson River. From 1908 to 1912, he was Assistant Professor of Civil Engineering at the University of

Figure 3. Notre Dame Stadium under construction, 1929–1930. Osborn Engineering Company, architect and structural engineer; Ralph Sollitt & Sons, contractor; altered. (Reproduced by permission of the University of Notre Dame Archives.)

Michigan. He then spent two years as a design engineer with Montreal-based Dominion Bridge Company, before he joined the faculty of the University of Illinois as Assistant Professor. In 1915 he was promoted to Professor of Structural and Bridge Engineering. In 1922 Joseph B. Strauss hired Ellis to redesign the Golden Gate Bridge. A prolific builder of "bascule" drawbridges, Strauss had been approached in 1919 by San Francisco City Engineer Michael O'Shaughnessey—best known for the controversial dam that flooded Hetch Hetchy Valley in Yosemite National Park—to propose a design for a bridge that would span the Golden Gate narrows between the San Francisco Peninsula and Marin County. Strauss turned to Ellis after a panel of experts rejected the former's bascule design and called for a suspension bridge. Ellis applied Leon Moisseiff's newly developed theory to his design, but until recently received no credit for it. Strauss, who was selected in 1929 by the bridge district's board of directors as the project's chief engineer, fired him after Thanksgiving

Figure 4. Workers under the direction of Ralph Sollitt & Sons pour concrete into forms for stand, 25 May 1930. (Reproduced by permission of the University of Notre Dame Archives.)

weekend in 1931. Because he systematically expunged all mention of Ellis in his final report, Strauss received credit for the suspension bridge's design until many years after his death, one year after the bridge opened in 1937. Ellis then worked in Chicago as a consulting engineer before Purdue hired him in 1934. He taught Theory of Structures, Reinforced Concrete and Foundations, Structural Design, and other courses until he retired in 1946.[18]

Ellis elevated hands-on activity over scientific technique as a teaching tool. He eschewed rote memorization of formulae in favor of laboratory work. In the classroom, he preferred geometric notation rather than algebraic terminology as a tool to teach structural behavior—a pedagogical method that, in his view, would help students develop the "mental power to cope with life's problems, the answers to which are not to be found between the covers of any text-book." Ellis took his cues on instilling initiative, resourcefulness, and self-reliance in his students from physicist Charles Riborg Mann, who in 1918 delivered a report on the state of engineering education in America to the Society for the Promotion of Engineering Education in America and the Carnegie Foundation. He believed that engineers could function successfully in the field only if they honed their ability to think quickly and creatively on their feet.[19]

As of 1941, the civil engineering program at Purdue largely adhered to the traditional approach. The school's divisions included Highway, Railway, Sanitary, and Structural engineering, and Surveying, City Planning, and Soil Mechanics. (The state's architectural school was housed at Ball State University.) The school maintained laboratories for materials testing, soil mechanics, and hydraulics. In cooperation with the Indiana Highway Commission, it also maintained bituminous and road laboratories, and a cold room, for highway research. Each year Highway Division faculty joined the Engineering Extension Department in conducting Road School for public officials, contractors, and suppliers. The school's properties also included Ross Camp in Tippecanoe County, home of the nine-week summer surveying camp that civil engineering students attended between their freshman and sophomore years: a hallmark of the program since 1914. In language that foreshadowed the transformation in engineering education that would take place after the war, the *University Bulletin* explained that "instruction in the first two years embraces the fundamental engineering sciences," and included four semesters of mathematics and two semesters of physics. Yet civil engineering majors devoted their junior and senior years "to a study of the principles of practice and to an intensive training in the analysis and design of a variety of engineering structures." The fourth year featured an extended field trip of "important engineering works." The traditional approach to educating engineers was reflected, too, in classes in specific subjects, such as Highway Design, Railroads, and Water Supply & Sewerage, and in several practical laboratories and Shop Work courses.[20]

When Charlie Pankow matriculated, the schools of engineering were rebounding from the effects of the Depression. They had benefited handsomely from the campus-wide building program of President Edward C. Elliott (1922–1945). Each of the major schools of engineering had acquired a new building. The completion of the Civil Building, in 1927, enabled civil engineering students to move out of overcrowded Heavilon Hall. Yet laboratories were overcrowded and nearly obsolete, if they were not already so. And notwithstanding the creation, in 1930, of the private Purdue Research Foundation, external funding of engineering research programs had not returned to 1929 levels. According to H. B. Knoll, Purdue Engineering's historian, professors remained overworked and underpaid.[21]

By 1941, the population of the School of Civil Engineering had been shrinking for a decade. In the spring, it had graduated 45 students, down from a postwar peak of 99 graduates in 1931. For several years now, enrollment in the school had been less than it had been three decades earlier. The cause, according to Knoll, "was hard to determine." For civil engineering was most closely associated with the public infrastructure projects—such as dams, bridges, and highways—that had benefited from various New Deal programs. Ralph B. Wiley, who had succeeded William Hendrick Hatt as head of Civil Engineering in 1937, suggested that conservative (that is, Republican) parents were steering their sons away from the discipline because they did not want them to work in government. On the eve of America's entry into World War II, Wiley feared that the school was falling short of its responsibility of meeting the needs of public employers.[22]

For Charlie Pankow, the problem was turned on its head. Even though his chosen field of study, structural engineering, was "defined as the keystone of Civil Engineering," and was traditionally the most popular senior option in the curriculum, Pankow faced limited career prospects, given his interest in commercial construction.[23] For activity in this segment of the industry had all but ceased with the Depression. His choice to study structural engineering, then, may be seen as an optimistic one, a product of confidence in his abilities, a love of buildings and structures, and a desire to follow in his father's footsteps.

By no means was Charlie Pankow's path to graduation a straightforward one. As a result of his military service, his undergraduate experience was spread over six years. For reasons that remain unclear, he also withdrew from the civil engineering program at the end of his freshman year. Rather than spend nine weeks with his cohort at the summer surveying camp, Pankow returned home, where he took classes at Notre Dame University. He re-entered Purdue in the fall of 1942. In the spring of 1943, he took three required courses that civil engineering students normally completed in the first semester of their junior year. Pankow then attended the summer surveying camp between his sophomore and junior years. When he returned from Ross Camp, Purdue had switched to a wartime schedule of three semesters per academic year. Because of this accelerated calendar, Pankow was technically a senior when he withdrew from Purdue in June 1944 to enter military service. Yet, for reasons that also remain unclear, his coursework had not included several of the courses that the school required for students in their sixth and seventh terms, and so he had been placed on academic probation, where he would remain until he completed the expected coursework upon his return from duty.[24]

When Pankow arrived on campus, Purdue already had been contributing to national defense through a variety of programs. In February 1939, flight training had begun at the university airport. In December 1940, the Defense Training Office had begun to prepare students to work in related industries. For the spring semester of 1940–1941, the Physical Education Department had introduced its "famous and infamous" conditioning course, PE 12. As it happened, Charlie Pankow never endured PE 12, but ultimately was allowed to substitute his military service for it. In the meantime, he completed the course, Military Training 3, in the fall of his sophomore year.[25]

With America's entry into the war, Charlie Pankow tried to enlist in the US Army Air Force, as his father had done in World War I, but he was rejected, owing to hearing that was damaged by a teenage bout with scarlet fever. (The illness also damaged his heart and kidneys, which would have long-term effects on his health.) He continued his studies, as President Elliott asked students to do until their national defense obligations materialized. Two and a half weeks after D-Day, Pankow was able enlist in the US Navy. He thus became one of some 17,500 individuals associated with the university who served in the armed forces during the war. The list of those who served included many of his fraternity brothers at Phi Kappa Psi (fig. 5).[26]

Charlie Pankow spent two years in the Navy. He did not see combat, but for roughly half of his time in the service, he took part in the postwar occupation of Japan. Stationed in Kyoto, the former imperial capital, his experience was apparently unremittingly dull—perhaps an indication of the extent to which the Japanese people "embraced defeat."[27] More than four decades later, Pankow recalled his years in the military as lost and unproductive ones, spent "going through a bunch of routine motions," both at home and abroad. In the context of an interview that focused on the achievements of his eponymous firm, the reflection suggests that his Navy experience may have shaped his thinking generally on the nature of work in the public sector and working with government entities and agencies. Certainly, in his career as a builder, public sector clients would be few. Perhaps the frustration evident in the remark masked disappointment in not seeing action like his father. The time that Pankow spent in Japan was positive from his perspective in at least one respect, however. With the purchase of a fourth-century Ming Dynasty vase, which he traded for a carton of cigarettes, Pankow launched a lifelong passion of art collecting.[28]

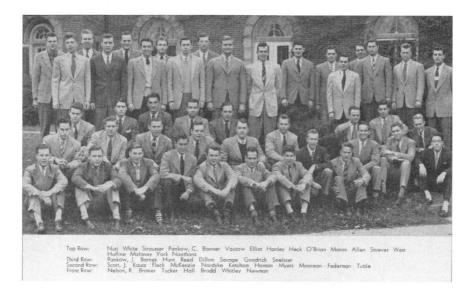

Top Row: Nutt White Strausser Pankow, C. Bonner Vautaw Elliot Hanley Heck O'Brian Mason Allen Stoever West
              Huffine Maloney York Naethans
Third Row: Pankow, J. Barnes Hunt Reed Dillon Savage Goodrich Smeltzer
Second Row: Scott, J. Kautz Fleck McKenzie Nordyke Ketcham Hanson Myers Moorman · Federman Tuttle
Front Row: Nelson, R. Brosier Tucker Hall Brodd Whitley Newman

Figure 5. Charlie Pankow spent all of his time at Purdue University as a member of the Phi Kappa Phi fraternity. In this 1947 yearbook photo, he appears fourth from the left in the top row. (From *The Debris*, 1947, p. 358. Photo courtesy of Purdue University Libraries, Karnes Archives and Special Collections.)

Pankow's time away from the West Lafayette campus was not wholly unproductive as far as his personal life was concerned, either. In September 1945, just before he shipped out, he married Doris Herman, whom he had met six years earlier. A native of Kankakee, Illinois, she was enrolled in the nursing program at a hospital in Evanston, an upscale suburb along Chicago's North Shore, at the time of the wedding.[29]

In September 1946, Pankow resumed his studies at Purdue. As both a married student and a veteran of military service, Charlie Pankow represented the demographic changes in the undergraduate student body that were occurring as a result of the passage of the Servicemen's Readjustment Act—the so-called G.I. Bill of Rights—by the US Congress on 22 June 1944. Purdue had reverted to two semesters per academic year. Despite having completed seven terms before he enlisted in the Navy, Pankow needed a full year to satisfy the School's graduation requirements. During the first semester, his schedule included courses that civil engineering majors normally completed in their junior year: Elementary Structures, Highway Construction (two semesters), and Hydraulics (both the class and the laboratory). Successfully completing these courses permitted him to end his period of academic probation. In his last semester on campus, Pankow took half a dozen technical courses that pointed to a career in building commercial structures: Architectural Engineering, Indeterminate Stresses, Municipal Engineering, Foundation Engineering, Reinforced Concrete and Foundations, and Soil Mechanics. (Incidentally, this course load adhered to the Construction Option in Civil Engineering, one of three options now offered by the School.) In nine semesters over six years (not including summer sessions), Pankow fulfilled most, but not all, of the requirements prescribed by the School of Civil Engineering for students who matriculated in 1941–1942. Nevertheless, the Executive Committee passed him to graduation, five hours short of credit at the senior year level. On 15 June 1947, he received the bachelor's degree in civil engineering.[30]

Charles Pankow began his engineering career with S. B. Barnes Associates, a Los Angeles–based structural engineering firm that Purdue University graduates Stephenson B. Barnes and Mark Deering had founded in 1934. Pankow met Barnes while on naval leave. He recalled that Barnes "knew I was interested in construction rather than straight engineering." Pankow looked up Barnes after he and Doris moved to Altadena, a suburb of Los Angeles in the foothills of the San Gabriel Mountains, in 1948. Doris's father, who worked in the steel business for a Chicago company, had been transferred to the City of Angels, and planned to stay in Southern California when he retired. At the suggestion of Doris Pankow's parents, the couple relocated.[31]

Los Angeles's postwar expansion provided the launch pad for Charlie Pankow's career as a builder. In 1940 the population of Los Angeles County was 2,785,687. Over the next 20 years, it would soar to 6,039,834. The strategic locating of defense production and military bases on the Pacific Coast during World War II, reinforcing a decades-old trend, revived the growth of all of the region's major urban centers, especially Los Angeles, whose business and civic elites were the most successful in wooing federal officials and winning contracts. In 1943 *Life* reported that Los Angeles was "irresistibly attractive to hordes of people," even as the wartime crush produced housing shortages and strained infrastructure and public services. After the war, the city's population continued to swell. By 1949 an average of 3,000 people were moving to Los Angeles per week, adding to the large number of military personnel who had gotten a taste of the local climate and lifestyle while training or on leave and decided to make Southern California their home. Many of the new arrivals joined Angelenos from older sections of town in settling in the San Fernando Valley, whose population increased fourfold from 1944 to 1960. By the early 1950s, popular magazines were predicting that Los Angeles would overtake Chicago as America's second largest city. At the time, the Los Angeles metropolitan area ranked first, second, or third nationally on a range of measures, including employment, income, retail sales, banking, construction, and production of aircraft and automobiles.[32]

Attendant to the growth in population were booms in residential and commercial construction. From 1945 to 1954, the City of Los Angeles issued a record $3 billion in building permits. The city experienced one of its "greatest years" in commercial construction in 1955, as Los Angeles joined a nationwide boom that had been gathering momentum for several years. The office building boom hit full stride in 1961. Thirty-two major structures, representing a record $158.6 million investment, were reshaping the skylines of the downtown, Hollywood, and Wilshire districts. In 1963, the year in which Charlie Pankow and three of his colleagues would leave Peter Kiewit Sons' to establish Charles Pankow, Inc. (CPI), Los Angeles County issued $2.06 billion in permits, up from $1.31 billion in 1955. Pankow's arrival in Los Angeles put him in the right place at the right time for a career in the building industry.[33]

Developing Design-Build and Concrete Construction Expertise within Peter Kiewit Sons'

Charlie Pankow "didn't have any idea that [he] would one day start [his] own firm" when he joined S. B. Barnes Associates. After more than two years, he left Barnes for the Austin Company, where he stayed for about a year. Unfortunately, it is unclear what working for Austin may have entailed. Even in such a brief tenure, however, Pankow surely would have been exposed to the Austin Method, the company's design-build project management and delivery system. Charlie Pankow spent the next twelve years of his career in the Los Angeles District of Peter Kiewit Sons'. For half of that time, he headed a new building division. It was especially in this latter capacity that he and the men whom he led developed expertise in techniques of concrete construction and, within limits set by Kiewit's culture and business model, experimented with design-build as a project management tool.[34]

The balance of this chapter traces how the group of men that Charlie Pankow gathered around him coalesced into an effective building organization with so much trust in their leader that they followed him enthusiastically when he left Kiewit to start his company. Unlike Lockheed's Skunk Works, which has served as the template for myriad corporate initiatives, the building division that Pankow established within Kiewit was not endorsed by upper management as a structural solution to the problem of creating and commercializing innovation.[35] Indeed, in the late 1950s, Kiewit was operating quite successfully in its strategic markets. Nevertheless, as elaborated below, the building division shared organizational dynamics and cultural characteristics with Skunk Works and insulated groups like it. Further, the essential elements of what would become known within the company as the Pankow Way were rooted in the culture of Peter Kiewit Sons' and the management style of its chief executive.

Peter Kiewit Sons' had become one of the largest contractors in America by the time that it hired Charlie Pankow in 1951. The company traces its roots to 1884, when Peter Kiewit established a masonry business in Omaha, Nebraska. His largest contract, indicative of the commercial building roots of the organization, was the 7-story Lincoln Hotel, completed in 1889. By 1912, two of his six children, George and Ralph, had joined the firm, which was then operating as Peter Kiewit and Sons, and was also building small residential and commercial projects. George and Ralph Kiewit assumed control of the company upon the death of Peter Kiewit, in 1914, and changed its name to Peter Kiewit's Sons. Six years later, the youngest child, also named Peter, dropped out of Dartmouth College to join the company. In little more than a decade, he would be running it.[36]

In 1924 the firm landed its first million-dollar contract, the Livestock Exchange Building in Omaha. Contracts for other high-profile commercial projects in the city soon followed, including the tower for the state capitol, in 1927; the Joselyn Art Museum, in 1928; and Union Station, a landmark art deco structure designed by Los Angeles architect Gilbert Stanley Underwood, in 1930. Notwithstanding this success at landing such projects and the personal satisfaction he gained from realizing them, however, Peter Kiewit concluded that the company would have to concentrate on government-funded, heavy construction if it were going to survive the Depression. This was an appropriate change in strategic direction, given the standstill in commercial construction. The value of building permits had plummeted by 44 percent or more in major cities across America from 1929 to 1931—and 1929 was by no means the high watermark of the post–World War I boom in commercial building. Thus, by 1951, the company had all but shed its legacy as a commercial builder, which helps to explain the lack of support that Charlie Pankow would receive for his initiatives as head of a building division.[37]

The proposed shift in strategic direction led to the breakup of the existing Kiewit firm and the creation of a new one. Ralph Kiewit, president of the company, thought that his brother was taking too much of a gamble. He and Peter agreed to dissolve Peter Kiewit's Sons. (Ralph and Peter were the remaining large shareholders in the company; George had sold his interest in 1926.) In 1931 Peter Kiewit formed a new company and named it Peter Kiewit Sons' in honor of his father. Its assets totaled little more than $100,000.[38]

Peter Kiewit liked to say, and is often quoted as saying, "We don't want to be the biggest, just the best."[39] On the strength of public works and defense contracting in the context of Depression, war, and Cold War, however, the company became one of America's largest contractors. During the 1930s, the Public Works Administration was the source of many contracts, validating Kiewit's strategic redirection. By 1940, the company had crews working in as many as eight states at a given time. Wartime contracting proved even more lucrative. The company completed more than $500 million worth of work, including military installations at Fort Lewis, Washington, and Camp Carson, Colorado, and the Glenn L. Martin bomber factory in Omaha. Federal expenditures during the prolonged Cold War provided steady, profitable work. At the same time, public works spending increased with the growth of the domestic economy. Kiewit won many contracts to build infrastructure—the type of work on which the firm had cut its teeth during the Depression. For instance, the company built more than half of the 152-mile Friant-Kern Canal in California's San Joaquin Valley and the mammoth earth embankment Garrison Dam on the Missouri River. Kiewit also built the Santa Ana Freeway in Los Angeles and leveraged that experience to construct more lane-miles of the Interstate Highway System than any other contractor. So-called heavy and highway construction dominated the business of Kiewit and shaped its postwar culture. Kiewit's heavy and highway divisions constituted the real or perceived enemy that the men of Charlie Pankow's building division would use to define and motivate their group.[40]

Kiewit procured its work through bidding. Peter Kiewit took great pride in his own ability to estimate jobs. By the 1950s, the company was widely recognized in the industry for its competitive advantage in bidding work. Peter Kiewit's discouragement of negotiating, rather than bidding, contracts would be one of the factors that would drive Charlie Pankow to start his company.[41]

Yet there was much that Charlie Pankow admired about and learned from Kiewit's organization and culture. Peter Kiewit believed that people made the difference in performance among contractors. He had an eye for talent, recruiting engineers who welcomed challenges and seized the opportunity to perform—and showed them the door, so to speak, if they did not. Like the Manhattan Project's Leslie R. Groves, the Skunk Works's Clarence L. "Kelly" Johnson, and other leaders of successful organizations, however, Kiewit viewed failure as a learning opportunity.[42] He pushed staff to find ways of delivering projects at a cost lower than the estimates on which bids had been made without sacrificing quality. Convinced that reputation held the key to repeat business, he expected his employees "to do what was right" on a project regardless of what it might cost the company financially. Peter Kiewit also encouraged teamwork: Better that a superintendent was able to motivate ten people to work together than do the work of ten people himself. At the same time, he created an environment characterized by fierce competition among managers for promotions, and between engineers for larger projects. This allowed Peter Kiewit to "bet heavily on the quality of the men around him." The heat of competition also invigorated continuous improvement across the company. Peter Kiewit thought that if his people either seemed self-satisfied or were not improving, they likely were already slipping. He frequently said (and, again, is often quoted for having said so), "I'm pleased but not satisfied." Charlie Pankow would take that statement to heart. It lay at the core of what would become the Pankow Way.[43]

Charlie Pankow also took many of his leadership cues from Peter Kiewit, who delegated responsibility without sharing authority. Kiewit often visited job sites to ensure that projects were on schedule, but let managers run their districts much as entrepreneurs would run their own businesses. At the same time, Kiewit did not hesitate to step in with authoritarian ruthlessness if his subordinates failed to meet his expectations. Kiewit respected authority and expected those who worked for him to do so as well. Moreover, he felt no need to explain his decisions. While Charlie Pankow shared these leadership traits with Peter Kiewit, his need to maintain direct control over his company was far greater than Kiewit's, and it expressed itself in the kind of organization that Pankow would create.[44]

Notwithstanding his mantra about not desiring to be the biggest firm, only the best, Peter Kiewit was never against growth, only against growth for its own sake. Consequently, he put in place a decentralized organizational structure that facilitated controlled, continuous growth. Kiewit admired Alfred P. Sloan, who restructured General Motors along product-based divisional lines, enabling the auto maker to become the world's largest manufacturing organization. Kiewit delegated executive authority within a mature organization, but he also tied employees to the company as a whole through a stock ownership program. He ensured that his top executives had the planning tools to achieve coordination of effort across the firm. Organizational hierarchies and employee incentives made possible rapid growth. In 1960 Peter Kiewit Sons' was the largest contractor in America, according to *Engineering News-Record*. Its $290 million in outstanding contract volume in 1963—the year that Charlie Pankow incorporated his company—was more than four times larger in real terms than the Pankow firm's at any point under the leadership of its founder.[45]

In contrast, Charlie Pankow would establish an organization that concentrated executive decision making wholly in his hands. All lines of authority would trace directly to him. He would engage in no strategic or business planning. He would eschew organizational charts altogether. Pankow's company would be little more than the aggregation of platoon-like teams that formed and reformed around projects, the number of which would never exceed Pankow's personal appetite for risk. Whereas Peter Kiewit was chief executive of a large, hierarchical corporation, Charlie Pankow would always remain fundamentally a sole proprietor.

Experimenting with Concrete, Heading a New Kiewit Building Division

What little is known about Charlie Pankow's early years with Kiewit is suggestive of a growing interest in concrete construction and an engineer's desire to assert control over the building site through mechanization. Alan D. Murk, who began working on Kiewit jobs in 1947 as a carpenter out of the union hall, recalls meeting Pankow on a State of California building project in Los Angeles in 1952. (The lack of a building division within the Los Angeles District did not prevent the state's Division of Architecture from inviting the company to bid the project, perhaps on the basis of its reputation as a contractor, its experience with concrete construction, or both.) Consistent with the type of work that engaged Kiewit in postwar California, Murk had worked on freeways and bridges. Murk's father, who spent his entire career with the company, managed the project. Murk was unable to recall the exact project. As Dean Stephan, who would become president of the Pankow firm in the 1990s, notes, the projects that Charlie Pankow worked on at this time "were primarily warehouse-type buildings, a lot of tilt-up, smaller buildings," which fits Murk's description of the project.[46]

Pankow acted as superintendent; it was his first field assignment. To date, Pankow had worked in the office, estimating jobs. He may well have worked on the proposal that responded to the call for bids. That his lack of field experience was no obstacle to Pankow's assignment as superintendent may have owed to the project's lack of complexity. It also may have reflected the elevated status accorded to university training by a contractor like Kiewit. Yet it is telling that the superintendents that Charlie Pankow would prize most highly, in both his building division and in his company, would be men like Murk, who gained their expertise at the building site rather than in the classroom. Further, within the Pankow firm, recruits from Purdue and other leading engineering programs would be expected to learn on the job as field and project engineers before earning a promotion to superintendent. One wonders how Charlie Pankow's experiences supervising jobs drove home the value of on-the-job training. It is more certain that his field experiences during these early years with Kiewit reinforced the appreciation for the worker at the building site that he carried from childhood. And it was a value that he would carry forward: a culture of respect would become a hallmark of the Pankow Way.[47]

The project that brought Murk and Pankow together offered the latter an opportunity to experiment with on-site precasting. The architect incorporated precast concrete features into the design of the building, which was consistent with the Los Angeles area's status as a hotbed of "modern precast concrete construction." (Recall the discussion in the Introduction.) As Murk recalls, "precast in those days . . . was just tilt-up buildings, flat slabs tilted up." According to Murk, the architect assumed that the contractor would use prefabricated beams and other structural elements, but Pankow believed that his team could save money by precasting them on-site. Pankow put Murk in charge of the casting yard.[48]

Another unidentified project also indicates Pankow's interest in on-site precasting of prestressed concrete elements. As he explained in his 1986 Raymond E. Davis Lecture—a lecture series (1972–1987) that honored Davis, a past ACI president who was long associated with the University of California, Berkeley—the design for the roof of a 90,000-square-foot building that he described as a "concrete barn" called for 50-foot beams and 36-foot purlins with a concrete compressive strength of 5,500 pounds per square inch (psi). No manufacturer produced these elements to these specifications, so Pankow decided to fabricate them on-site, post-tensioning the beams. Owing to its limited local availability, the high-tensile steel wire that would be used in the operation had to be shipped from Pittsburgh. Faced with a 90-day completion schedule (and also a looming plumbers' strike that threatened to delay other tasks), crews initiated the precasting of the beams and purlins prior to the delivery of the steel. Workers formed holes in the beams with rubber hose and steel bar to facilitate their picking and placing into storage. When the wire arrived, they fabricated cables from the wire, and pulled them through the beams to allow the elements to be post-tensioned.[49]

By 1956, Pankow recalled, he and the men of the Los Angeles District "were working with post- and pre-tensioned concrete slabs and beams." This work constituted an advance in precast concrete construction over the state of the field described by ACI member J. L. Peterson in early 1954. It is also just about all we know about the first half of Charlie Pankow's career with Kiewit. Sometime during 1956, according to Murk, Charlie Pankow traveled to Omaha and persuaded Peter Kiewit to let him establish his own building division within the Los Angeles District.[50]

Pankow could not have succeeded in this endeavor without the support of Tom Paul. Paul was the first manager of Kiewit's Seattle District, which grew out of a joint venture formed in the fall of 1940 with Seattle-based Sound Construction and Engineering Company to construct the barracks at Fort Lewis. Paul directed the job, whose size the US Army doubled after work had commenced on it, from 760 to more than 1,540 buildings, without granting an extension of the schedule. Organizing the project and finishing it on time made Paul's reputation. By war's end, Sound had become a wholly owned subsidiary of Kiewit. As manager of the Los Angeles District, Paul hired Pankow and mentored him throughout his time with the company. The two got on well and met for lunch at the Derby restaurant, near the Santa Anita racetrack, whenever Paul was in town. Charlie Pankow thought highly of Paul, who not only supported his efforts to establish a building division, but also acted as a buffer between Peter Kiewit and Pankow when the two clashed on bidding and other issues.[51]

A decade after he graduated from Purdue University, Charlie Pankow was an assistant district manager for buildings, with his office in Arcadia, a city of 41,000 people in the San Gabriel Valley, northeast of downtown Los Angeles.

During the next half dozen years, he led a group of capable men who developed expertise in precasting, slipforming, and other techniques of concrete construction. Within the limits set by the Kiewit organization, Pankow became involved in projects as a design-build contractor. By 1963, the group incubated within the company was ready to do business as an independent company.

From the point of view of those who worked for him, Charlie Pankow was a hard-driving and motivational leader who "could be a little intimidating." He was also "very forbearing," "real matter-of-fact," and "a tough guy to read." He peppered people with questions that were virtually impossible to answer. He frowned on team members asking him to solve their problems, expecting them to approach him instead with well-thought-out alternatives as the point of departure for project-related discussions. Nevertheless, Pankow's creativity, understanding of the business, appreciation for hard work and innovation, encouragement of new ideas, intense focus on the project at hand, and drive for continuous improvement in team performance commanded respect for the man as a leader. And if he came off as aloof at first, people generally warmed to him. For he was also sociable and sincere. He put people at ease. Recalls Alan Murk: "He could charm the birds off the trees." As the person who was responsible for generating work for the building division, these latter qualities served Pankow well.[52]

Like Henry Ford and the founders of any number of Silicon Valley computer companies, Pankow surrounded himself with men who could work together creatively and ingeniously without close supervision.[53] While he expected his engineers and superintendents to implement his construction methods and technologies, Pankow showed no interest in micromanagement or, indeed, in management generally. He brought into the building division men who could work imaginatively, not simply to solve technical problems as they arose, but also to discover better ways to erect buildings. It is also telling that the people who worked best with Charlie Pankow were not only creative and hardworking, but also patient and willing to accept his absolute authority. This "Pankow Personality" would manifest itself once Charlie Pankow started his own company.[54] How Charlie Pankow may have screened for these traits in job interviews remains unclear.

The group of men who would follow Charlie Pankow out of Kiewit included David R. Boyd, Robert E. Carlson, Rosser B. Edwards, Tony Giron, John R. "Jack" Grieger, Harold M. Henderson, George F. Hutton, Lloyd Loetterle, Alan D. Murk, Norman L. "Red" Metcalf, Russell J. Osterman, and

Ralph Tice. Boyd and Edwards aside, they included both experienced hires and men who rose through the ranks in the field. Charlie Pankow may have handpicked these men to work with him, but at least one-third of them were hired into the Los Angeles District independently of him. Carlson, Loetterle, and Tice would join Pankow as founders of CPI. Together with Pankow, Hutton and Osterman would become its largest shareholders.

Carlson, Loetterle, and Tice joined Kiewit within a year of Pankow. Carlson arrived in 1950. He was project manager on several of the building division's high-profile projects, including the American Cement Building in Los Angeles and the First and C Building in San Diego, both discussed below. He became the general superintendent for the division, a role that he would fill in CPI. Loetterle joined Kiewit about the same time as Pankow. For two years prior to the formation of CPI, he headed operations in Northern California. Ralph Tice was the type of builder on which Charlie Pankow's company would make its reputation. He began working at Kiewit a year or so after Pankow. Together with Murk and Henderson, who started with Kiewit in 1951 in Rapid City, South Dakota, Tice worked as superintendent. Of Tice, Brad Inman, who headed Pankow's San Francisco office during the 1970s and 1980s, states: "I have met one man in my life who single-handedly could build a building, all by himself, given enough time, and that's Ralph Tice. He could do everything and he knew everything, and what he didn't know he'd figure out. Just an incredibly competent guy, rascally, had a rough edge on him, but he was very seldom wrong."[55]

Russell J. Osterman joined Charlie Pankow in the Arcadia office as an estimator and project manager. He graduated from the University of Michigan in 1952 with a degree in civil engineering. He played defensive right end on the football team; a career that included a Rose Bowl victory over the University of California, in 1951. From 1953 to 1960, he worked with J. A. Utley Construction Company in Detroit as project manager, chief estimator, and assistant general manager. When Utley's order book dwindled at the end of the decade, Osterman looked to relocate to the Pacific Coast to boost his career.[56]

Charlie Pankow hired George Hutton soon after Osterman came on board. Born in Osceola, Missouri, in 1932, Hutton graduated from the University of Missouri in 1954 with a degree in civil engineering. He worked briefly for Standard Oil Company's construction department at its Whiting, Indiana, refinery before he was inducted into the military. Hutton served two years in the US Army's Corps of Engineers and Military Intelligence Group.

He returned to Standard Oil, but the refinery had been destroyed by fire in his absence; the subsequent lack of work persuaded him to leave. He eventually worked for a contractor in Kansas City, where he grew up. In the summer of 1960, Hutton responded to an advertisement to work with Peter Kiewit Sons', the company of which he had thought highly since his university days. He interviewed for the job in Omaha, but Kiewit failed to win the project that had prompted the job search. The personnel manager who interviewed him forwarded his résumé to Charlie Pankow, who, in October 1960, invited Hutton to interview in Los Angeles. Hutton agreed. Pankow hired him as a field engineer and assigned him to a convention center project associated with the El Cortez Hotel in San Diego.[57]

Giron, Grieger, and Metcalf worked for the building division as foremen. Jack Grieger arrived in 1957 and was assigned to a project at El Toro Marine Base, in Orange County. Metcalf joined Kiewit as a journeyman carpenter in 1960, not long after he had left the US Navy. Born in Wellington, Kansas, his father, grandfather, and most of his uncles worked in construction. From middle school on, he helped them build houses. By the time he attended junior college in Oceanside, California, Metcalf had developed the skills of a journeyman carpenter without having served as an apprentice, and demonstrated them as a summer hire building houses in nearby Vista. His first job with Kiewit involved the construction of barracks at Camp Del Mar in San Diego County.[58]

Boyd and Edwards were the only two members of the group without building experience. They represented the professional that Charlie Pankow would recruit as the company expanded in the 1970s: the engineering graduate with a keen interest in learning how to build "from the ground up," as former ACI president and Pankow executive Thomas D. Verti puts it. Both men came to Kiewit from Purdue University, where they received the bachelor's degree in civil engineering in 1960. Edwards started his Kiewit career as an engineer-estimator in the Arcadia office in 1961 after earning a master's degree in construction management. The ranks of the Pankow companies would be filled with more graduates from Charlie Pankow's alma mater than from any other institution of higher learning.[59]

Pankow instilled in his men a belief that they could build anything, and their successful completion of complex commercial building projects validated his confidence. These men saw themselves as embodying Pankow's vision of the master builder. As a group, they developed an esprit de corps that one interviewee likened to that of a platoon.[60] This resulted in part from a feeling that they were not getting the respect that they deserved from the larger Kiewit organization. The group's collective identity was also strengthened by the homogeneity of its members: All were white men in their twenties and thirties; many of them came from small- to medium-sized towns in the Midwest. Sharing faith in his building methods and techniques, few of them would hesitate to follow Charlie Pankow when he decided to form his own company.

In practice, Charles Pankow never "sold" Peter Kiewit on design-build as a construction management methodology, and so he had to work within constraints that included bidding work. Some jobs he and Osterman bid in traditional fashion. Whenever possible, Pankow sought to convince an owner to allow Kiewit as contractor to become involved in the design (or redesign) of the project. If he assented, Pankow worked with the architect and engineer to ensure that the design of the project accommodated his favored concrete construction techniques.

Many of the division's projects were tall concrete structures that provided superintendents, general foremen, and other workers with experience in "automating" work at the building site. The next sections consider four of these projects in greater detail to show how Charlie Pankow's "budgeted construction program," as he came to call it during this period, evolved within the Arcadia-based building division.[61]

Deploying Design-Build (without the Label): The American Cement Building

The executives of the American Cement Corporation decided to build a new headquarters that promoted reinforced concrete as the "building material of the future" in dramatic fashion. Formed in 1957 through the merger of three companies, American Cement was one of America's largest producers of the material. Its executives wanted to project an image of a "young, dynamic, forward-thinking, exciting, bold, imaginative—but eminently practical" organization. Since the turn of the century, skyscrapers had been projecting corporate identity in all of America's major metropolitan areas. As had been the case with Alcoa executives and the Alcoa Building (1953), American Cement's executives would link corporate identity literally to the material used to construct a signature structure. They called for a building that "made fullest use of concrete's aesthetic, functional, and economical characteristics."

The result would exemplify "architecture parlante," a building that explained its function and meaning. [62]

Kiewit won the construction contract through a procedure deemed "novel" at the time. As Charlie Pankow remarked a quarter century later, it was "one of the first major design-build type efforts put out on the streets of Los Angeles." Owners Edward Rothschild and Arthur Gilbert acquired a site opposite MacArthur Park along Wilshire Boulevard, just east of downtown. Together with the owners, American Cement executives selected Daniel, Mann, Johnson, and Mendenhall (DMJM) as architects and structural engineers. They specified a design that provided for column-free office space; met seismic codes with concrete shear walls or diagonal bracing; and was cost-competitive with non-concrete structural systems and materials. The project was bid, but on the basis of only "six or eight sheets of plans," including a "floor plan, the standards of quality and maybe some single-line diagrams on mechanical and electrical systems to illustrate where and what." Recalled Pankow: "It was [pretty] raw, certainly far from beginning of working drawings type of approach." According to Pankow, American Cement executives defined the finished product only to this extent because they "had begun to appreciate the fact that some of the builders in the community had something to offer."[63]

The owner and architect invited prequalified contractors, including Kiewit, to compete for the contract. They included Kiewit because of "its long time interest in reinforced concrete construction." The company's ability to produce $400 million in savings on a gaseous diffusion plant in Portsmouth, Ohio, for uranium enrichment, originally budgeted at $1.2 billion, was compelling. Kiewit had been included in the planning stages of that project because of its size. Owners Gilbert and Rothschild and DMJM's designers expected respondents to suggest ways of lowering the costs of construction in their proposals. There were eight bidders. The prices that they submitted varied by as much as 100 percent. According to Pankow, this indicated "how people do respond when they have to think about something more than taking quantities off of a complete set of plans," as contractors would presumably do under the traditional design-bid-build approach. Kiewit won the contract. Explained one unidentified Kiewit executive: "The critical factor, and the one that means the most to the building owner, is the extra latitude given the contractor in basing his bid on techniques and experience that may be his alone. He is then in a position to take certain risks the architect cannot possibly assume for him." Whether or not Charlie Pankow made the statement—it seems likely that he

Figure 6. The American Cement Building under construction. Daniel, Mann, Johnson & Mendenhall, architect and structural engineer; Peter Kiewit Sons', contractor. The reinforced concrete grill, developed by DMJM partner Phillip J. Daniel, replaces the more typical beam-and-column system on the north and south sides of the building. The parking annex façade is non–load bearing. (Photo courtesy of AECOM.)

did—it expressed his belief in the advantage of an approach that allowed the contractor to participate in planning and design.[64]

DMJM designed the 13-story building as a two-unit structure: a 4-story base for parking, ground floor shops, lobby, and executive offices, and a 9-story tower, with each floor containing 12,000 square feet of speculative office space (fig. 6). A fifth floor transition platform "mated" the base and tower units. A load-bearing and shear-resisting concrete service core, housing the elevators, stairs, and utilities, extended through the tower. The base and tower featured the grillwork that stands out as the building's most distinctive and enduring feature. The tower grill, which clad the wider (north and south) sides of the building, performed dual decorative and structural roles. It was composed of 225 X-shaped reinforced concrete elements that were fabricated under the direction of Alan Murk in a casting yard that lay several miles from the building site and were transported to it by truck (fig.

Figure 7. Alan Murk supervises casting yard operations associated with the construction of the American Cement Building. (From *Kieways* 16 [March/April 1961]: 9. Courtesy of Kiewit Corporation.)

7). In conjunction with concrete floor beams that extended from the central core, the grill substituted for a more conventional beam-and-column system on the two longer sides of the building, allowing the designers to open the office space in the tower. Similar grillwork also wrapped around the base of the structure. Decorative in function, the grillwork nevertheless vented the parking areas.[65]

The DMJM engineers who described the design of the building did not delineate Kiewit's contributions to it, but alluded to discussions that related to the possible use of prestressed floor beams. The design team used reinforced concrete elements because the Los Angeles building code did not permit the use of prestressed concrete in tall office buildings. The latter would have to be tested for fire resistance to satisfy the building inspector, causing the schedule to slip. The slipforming of the central service core and the precasting of the X-shaped members and other concrete pieces, however, indicate the participation of Charlie Pankow in the design process. As the structural engineers noted, Kiewit "used techniques not often applied to buildings."[66]

The American Cement Building broke ground in August 1959, making it the earliest use of slipforming by the new building division. The advertising supplement that company executives placed in the *Los Angeles Times* to celebrate the completion of the building called the technique "risky" and "seldom used" (though other projects in California were also using the technique). Crews under the direction of superintendent Harold Henderson completed the service core of the American Cement Building in four weeks. It was no easy task. The presence of nearby apartment buildings restricted work to daytime hours. The blockouts required to accommodate the windows, floor beams, and other features were complex. Once put in place, however, operators raised the slipform at a rate of 1 foot per hour. DMJM's engineers estimated that the technique saved as much as three months over casting the service core in place, which would have required workers to fabricate a form for each floor.[67]

Precasting concrete elements helped Kiewit submit the lowest bid for the American Cement contract. At the same time, producing high-quality precast concrete structural and architectural elements for a tall, signature commercial building demanded a level of experience and expertise that was apparently not needed for the simple, low-level concrete factories that were mass-produced earlier in the century (see Introduction). For purposes of both cost and quality control, Charlie Pankow did not subcontract the work. Under Murk's direction, Kiewit precast the X-shaped members, garage screen panels, and floor beams. Crews used cast-in-place floor slabs to tie these elements together.[68]

To construct the tower and much of the parking annex that was part of the base, Kiewit used a rail-mounted tower crane that was the largest of its kind—158 feet high to its jib pivot point. The German crane's capacity to handle the 5,000-pound floor beams allowed them to be precast in Murk's yard. Notwithstanding the cost of renting the equipment, the approach saved time and labor over the alternative—casting the beams in place.[69]

The American Cement Building marked the first of many large projects on which Charlie Pankow used Robert L. Heisler and Lee Sandahl as subcontractors for the mechanical and HVAC (heating, ventilation, and air conditioning) systems. Both Heisler and Sandahl attended the University of Southern California on the G.I. Bill. They graduated with degrees in mechanical engineering in 1950 and 1949, respectively. Heisler worked for General Electric for five years in Schenectady, New York, and San José, California, before returning to Los Angeles to start an air conditioning business that would eventually become part of a larger company, Key Mechanical Industries (KMI).

Heisler met Pankow through Gordon Fent, who was one of Kiewit's financial officers and a close friend of Heisler's parents. Heisler agreed with Pankow that a fully articulated design-build approach to project delivery should involve mechanical and HVAC engineering. Charlie Pankow turned to KMI for these services time and again; CPI would account for as much as 30 percent of KMI's business. The relationship would prove to be an important one in the early history of CPI. For Heisler would introduce Pankow to the people who would bond the new company, offer it legal counsel, and become its most important client.[70]

The American Cement project demonstrated the value of incorporating mechanical and HVAC engineering in project planning and design. At the insistence of American Cement executives, the building was cooled with a natural gas–powered absorption refrigeration unit that the subcontractor installed on the roof. The owner negotiated a rate with the gas company that was equal to what the latter was charging its cement plant in Riverside, California. American Cement hailed the system in its *Los Angeles Times* promotional supplement. In fact, Sandahl explains, the absorption unit was "the worst system you could use because they're not too efficient, they're very cumbersome, and they only last about at the most fifteen years." If they were unable to dissuade the owner from installing the system, Heisler and Sandahl warned of its disadvantages. They also were able to plan for its eventual replacement in their design of the air conditioning system, which occurred, Sandahl recalls, "probably ten years later."[71]

How Charlie Pankow's understanding of design-build developed during the first twelve years of his career is unclear. The American Cement Building project is the earliest documented example of Pankow acting as contractor in a building team configuration that approximated design-build. Despite the call for bids, Pankow was able to modify the process: Contractors did not bid on complete working drawings, as was the norm under the design-bid-build approach, but were given the chance to submit their ideas on ways to control costs and schedules in their proposals. On this project, the contractor slipformed the elevator and service core and precast reinforced concrete elements in a nearby yard, but cast the floors in-place. In time, Pankow's men would use other methods to install floors in tall buildings. But not while the group remained a part of Kiewit. The next section illustrates how Charlie Pankow demonstrated to an owner the value of contractor involvement in the redesign of a costly project.

Perfecting Slipforming on a Redesigned Project: First and C Building

A $6.5 million speculative office building in downtown San Diego provided Charlie Pankow with the opportunity to work with the architect and engineer to develop an alternate design that saved the owner several hundred thousand dollars. The redesign achieved these savings by substituting lightweight reinforced concrete for the conventional "hard rock" variety for all above-grade structural members and by utilizing a modified mat foundation that reduced the amount of reinforcing steel required for the project. Developing the alternate design delayed the construction schedule, but Kiewit met the completion date by deploying slipforming and other techniques to erect the building.[72]

Upon its completion in the spring of 1963, the First and C Building became the tallest reinforced concrete structure in a Uniform Building Code (UBC) Seismic Zone 3 area. Los Angeles architects Palmer & Kissel and structural engineer Richard R. Bradshaw, with whom Charlie Pankow would work on more than half a dozen projects, designed the 250-foot structure just before the City of San Diego adopted a 13-story (160-foot) height limit on concrete buildings—a restriction that conformed to the UBC and buildings codes of other US cities in Seismic Zones 3 and 4. (For his part, Bradshaw insisted that reinforced concrete frames could be designed to resist lateral forces and so argued that the code was unreasonable.)[73]

The First and C Building, which was promptly renamed the Electronics Capital Building after the tenant that leased three of its upper floors, featured a 25-story concrete frame, all but two of which of which lay above ground. (It is now known as the Chambers Building.) There were eight levels of parking, two of which were below grade, and a 16-story office tower. A fitness facility occupied the penthouse, with a swimming pool on the roof. The structure featured 24-story-high bearing walls, but not in the usable office space area. The exterior building columns were skewed in the garage to facilitate diagonal parking. The mat foundation was designed to resist localized shear forces during an earthquake. Openings in the mat in the middle of the long spans between columns were designed to reduce both bending moments during seismic events and the amount of rebar needed to reinforce the slab.[74]

Both the First and C Building and the Hillcrest North Medical Center, discussed below, resulted from efforts on the part of downtown business and civic interests to, as Kevin Starr puts it, "reinvent" the central business district as a "high-rise corporate center," even as the metropolitan area sprawled through the canyons and on the ridges of San Diego County. The buildings were constructed concurrently with the Home Federal Savings and Loan and the US National Bank buildings. The downtown boom culminated with the construction of the long-planned civic center, which opened in January 1965.[75]

Kiewit bid the job in January 1961 and immediately turned to redesigning it, both to save the owners money and to improve its own tight margins. This approach was in keeping with Peter Kiewit's expectations that his managers find ways of delivering projects below the bid. The approach was not design-build, strictly speaking, but it did afford the contractor the opportunity to collaborate with the other members of the building team. Construction began in September. Kiewit crews, led by project manager Bob Carlson and superintendent Alan Murk, topped out the building at the end of October 1962.[76]

To facilitate and speed construction, Charlie Pankow chose to slipform not only the service and elevator core but also the stairwells—a more complex operation than the slipforming of the American Cement Building. Notwithstanding his men's relative lack of experience in the technique, Pankow was confident that they could do the job. And they did, deploying one slipform to erect the walls of the five elevator shafts and stairwells at the southeast corner of the building and another to construct the walls of the single stairwell at the northeast corner of the structure. The forms incorporated pockets for slabs and beams at each floor level. Crews raised the slipforms at a rate of 11 feet per day, two floors ahead of the floor slabs.[77]

The twisted exterior towers columns "caused some head scratching" for architect Bradshaw and project manager Carlson. To facilitate construction, they constructed a model of the structure. Twisting the columns 60 degrees between the seventh and eighth floors kept them flush with the walls and maximized the number of stalls in the parking garage. To construct the columns, Kiewit's carpenters divided the casting forms into small sections, which they fashioned to conform to the prescribed surfaces. Carlson's decision to build only three sets of forms slowed the construction of the garage-to-office-tower transition. Nevertheless, construction finished on schedule.[78]

The project featured a visit from Peter Kiewit while crews were laying the foundation. His appearances on job sites could be both inspiring and intimidating. Kiewit took note of the steep banks carved out by the excavation and, fearing that they might cave in, told Carlson to do something about them. Kiewit did not prescribe a particular solution. As half of the rebar was already in place, Carlson decided that finishing the foundation as quickly as possible presented the only practical solution. Crews worked through the weekend and into the next week to pour the foundation. At midweek, Peter Kiewit followed up his visit with a phone call from Omaha and asked Carlson what he had done to address the potential problem. Carlson reported that crews had worked overtime to pour the foundation and that it had set. Kiewit said that he was glad to hear it and thanked Carlson for his efforts.[79]

A Close Approximation of Design-Build: Shorecliff Tower

At the same time that one Kiewit team was topping out the First and C Building, another was preparing to use slipform operations to erect Shorecliff Tower, a 13-story apartment building that would overlook the Pacific Ocean at Santa Monica. *Southwest Builder & Contractor* deemed the use of slipforming in this case to be "important news" because crews were pouring all of the vertical load-bearing walls, and were doing so 3 stories ahead of the floor slabs, which were being cast in place. In other words, the men of Charlie Pankow's building division were becoming more sophisticated in their deployment of the technique. The *Los Angeles Times* featured the project in an article illustrating how slipforming—still a "novel" method of handling concrete—was saving owners, designers, and builders space, labor, and materials on their projects. The article hailed slipforming as one of several new methods that were enabling the construction of bigger and better concrete commercial and industrial buildings across Southern California.[80]

As was the case with the First and C Building, the decision to slipform the structure resulted from collaboration among building team members. The project provided Charlie Pankow with his best opportunity as head of the building division to apply design-build. Ralph Kiewit Sr., brother of Peter Kiewit, and his son, Ralph Jr., were the owners, making the project a captive one in the sense that the contractor did not have to bid the job.[81]

Architects Jones & Emmons and structural engineer Bradshaw joined the contractor in a design-build relationship. A. Quincy Jones and Frederick E. Emmons were already well known for their modernist designs on projects ranging from single-family residences to university master plans. They were in the midst of an 18-year partnership that included designing

Figure 8. Shorecliff Tower, Santa Monica, California, nears completion. Crews used keys and dowels to fasten the balcony floor slabs—each 4 feet wide with a 7-foot cantilever—to the faces of the slipformed walls. A. Quincy Jones and Frederick E. Emmons, architects; Richard R. Bradshaw, structural engineer; Peter Kiewit Sons', contractor. (From *SB&C*, 26 April 1963, 9.)

thousands of single-family homes for builder Joseph Eichler. In 1961 they burnished their bona fides as modernists with a proposal to build 250 houses on a 140-acre tract parcel in the San Fernando Valley, which they submitted to the Case Study House project. John Entenza, editor of Los Angeles-based *Arts & Architecture* magazine, had launched the program in 1945 to enlist the best modernist architects in the region to create prototype homes to help meet the postwar demand for housing. The Jones & Emmons project, Case Study House No. 24, was the only tract-home development represented in Entenza's program. Jones and Emmons shared with Pankow and Bradshaw an interest in exploring innovative ways of constructing concrete structures.[82]

Slipforming was not part of the initial discussion, but architect, contractor, and engineer agreed that the technique would facilitate a more economical design and accelerate construction. Bradshaw recommended its use after he reviewed the preliminary design of Jones & Emmons. As the real estate

Figure 9. Architect A. Quincy Jones joins superintendent Harold Henderson on the Shorecliff Tower building site. (From *SB&C*, 26 April 1963, 76.)

editor of the *Los Angeles Times* noted, slipforming offered relatively greater cost advantages on multi-unit residential than on office projects—the former contained more permanent walls. Workers under the direction of slipform carpenter foreman Red Metcalf built the forms from high-density plywood. Workers used hydraulic jacks to raise the forms at the rate of 15 inches per hour during eight-hour days, completing one 8,000-square-foot floor every four days. Lightweight concrete that set within four hours facilitated the process. As specified by Jones & Emmons, each one-way floor slab contained copper tubing for a radiant heating system, which crews installed when they poured the slab in place. The architects had first experimented with radiant heating several years earlier in the design of their headquarters on Santa Monica Boulevard in West Los Angeles. To facilitate a sequence of tasks that *Southwest Builder & Contractor* called "unusual"—pouring the walls ahead of the floors—contractor and engineer developed special joints in places where the forms did not allow dowels, so that crews could fasten Jones & Emmons's cantilevered floors on the face of the walls (figs. 8, 9). Crews completed the shell in early March 1963. By July, Shorecliff Tower was ready for occupancy. Ralph Kiewit moved his family into the penthouse.[83]

Marrying Design-Build to Building Site Mechanization: Hillcrest North Medical Center

A $1.65 million medical and dental facility in downtown San Diego incorporated all of the techniques of Charlie Pankow's construction program. And though the job was bid, Pankow was able to collaborate with architect and engineer in a manner that approximated design-build. For its combination of "good architecture, unusual design ideas, structural innovations, and outstanding construction techniques," the Portland Cement Association gave the project its Award of Honor in the inaugural year of its Building Awards Program in Southern California.[84]

The owner sought a design that maximized the flexibility of leased space and accommodated radiology, laboratory, pharmacy, medical, and dental supply functions. The team of architect William S. Lewis Jr., a partner in Deems-Martin Associates, designer of the San Diego Convention Center; structural engineer A. J. Blaylock; and contractor Pankow developed a design for the main building that *Southwest Builder & Contractor* deemed "unusual." The medical center featured an 8-story tower supported by an elevator and service core (fig. 10). Precast, prestressed beams, extending from the core, supported the deck slabs that formed the floors. Together with load-bearing exterior walls, the beams allowed interior spaces to be column-free, as was the case with the American Cement Building. An open-air lobby under the tower facilitated access to two 1-story buildings that housed the support service tenants.[85]

Like the American Cement Building, Hillcrest North Medical Center was architecturally distinctive. It featured 828 "channel-shaped" precast panels arranged in a "strikingly Gothic-tracery pattern." Each panel, 14 feet, 8½ inches high and 28 inches wide, framed a recessed area (fig. 11). Every other panel encased a window that was screened on the outside by "anodized tracery." Like the American Cement Building, lightweight concrete panels served as load-bearing walls once crews anchored them with cast-in-place connections. The panels collected and transferred vertical loads to the second floor, where prestressed columns transmitted the loads to the ground. Workers cast the panels on-site, using steel forms manufactured by Food Machinery Corporation.[86]

Project engineer George Hutton devised a slipform operation to build the central core that proved to be unique in the annals of both the building division and the Pankow companies. The core rose out of the foundation, carry-

Figure 10. With the central core of the Hillcrest North Medical Center office building completed, crews use a Linden F30-60 tower crane to lift precast concrete wall panels into place. Deems-Martin Associates, architect; A. J. Blaylock & Associates, structural engineer; Peter Kiewit Sons', contractor. (From *SB&C*, 25 October 1963, 50.)

ing with it a Linden F30-60 tower crane. Its 100-foot horizontal boom could lift 4,400 pounds. Crews used the crane to lift into place both the panels and beams, which crews precast as the slipform operation proceeded. As was the case with all techniques of job site mechanization, the goal was to save time and money. The team certainly saved time on the way up. They slipped the core at a rate of two days per floor, notwithstanding a delay incurred more than halfway up: Part of the operation had to be repeated because the crew raised the slipform faster than the concrete set. Once they had topped out the building, crews cast floor slabs in place from the top floor down. This approach did not save much money, if any at all, over the more conventional,

Figure 11. Superintendent David Boyd, structural engineer A. J. Blaylock, and architect William S. Lewis Jr. (from left to right) occupy three of the precast panel units used in the Hillcrest North Medical Center, completed in 1964. (From *SB&C*, 25 October 1963, 47.)

bottom-up approach, in carpenter foreman Red Metcalf's estimation. But the decision to try it reflected the spirit of innovation that Charlie Pankow tried to encourage in his employees.[87]

■ ■ ■

"Corporate entrepreneur" is a term that, on the face of it, might be applied to Charlie Pankow and the role he played in the Kiewit organization. *Contra* the use of the term to describe a complementary relationship between the "innovating entrepreneur" and the large corporation, however, Pankow was not innovating at the direction of corporate strategists.[88] As discussed below, Peter Kiewit would rebuff Pankow when the latter tried to "sell" him on the potential benefits to the corporation of using design-build to deliver projects. Given the manner in which events would unfold, leading to the creation of Charles Pankow, Inc., it would seem more appropriate to view the larger Kiewit organization as the incubator of a start-up business, however inadvertent the role that Kiewit played in the genesis of CPI may have been. Within the space carved out by Charlie Pankow, with critical support from Tom Paul, the building division developed the expertise and the culture that gave it the chance to succeed as an independent entity.

The American Cement Building, First and C Building, Shorecliff Tower, and Hillcrest North Medical Center projects enabled Charlie Pankow to demonstrate to owners the potential advantages of involving the contractor in planning and design. The First and C Building and Hillcrest North Medical Center projects provided him with the opportunity to offer ideas on redesigning a project after it was bid. While Pankow might preach the benefits of design-build and establish a company on the basis of practicing it, it would be the at-least-once-burned owner who generally would afford him the opportunity to practice design-build in the purest sense of the term. Completing the American Cement Building and Shorecliff Tower projects afforded Pankow the opportunity to demonstrate the benefits of incorporating the contractor into the design process before the architect finalized working drawings. As long as Pankow's group remained under the Kiewit umbrella, however, such opportunities would be limited. The design-bid-build approach favored by Peter Kiewit restricted the extent to which Pankow could apply his construction program.

Conflict with Peter Kiewit: The Los Angeles Music Center

The 3,250-seat Memorial Pavilion, the first building of a $33.5 million Los Angeles Music Center planned for a site atop Bunker Hill, on the western edge of Civic Center Mall, was a signature project that Peter Kiewit wanted to build—notwithstanding the "stepchild" status of commercial building among the company's divisions. In the estimation of project architect John Knight, the future home of the Los Angeles Philharmonic Orchestra would be "a West Coast version of the very best New York houses." From the contractor's perspective, however, its construction was emblematic of the tensions that characterized the relationship between Peter Kiewit and Charlie Pankow. Well before the 9 September 1964 dedication of the building, which would be renamed in honor of Dorothy Buffum Chandler, the wife of the publisher of the *Los Angeles Times*, Pankow would sever his ties with the Kiewit firm.[89]

The project was the culmination of years of effort on the part of Los Angeles's corporate elite to elevate the status of the city as a cultural center. Dorothy Chandler mobilized private funds to build a music center after voters rejected the project, in various guises, on three occasions and, in the wake of these defeats, the Los Angeles County Board of Supervisors failed to mobilize support for a public-private partnership. Chandler organized and chaired a Music Center Building Fund Committee composed of corporate "heavy hitters." It quickly raised $4 million toward the purchase of a site for a facility dedicated solely to the performing arts. Chandler formed the nonprofit Music Center Lease Company and persuaded the supervisors to lease it land adjacent to Civic Center Mall. As approved by the city planning commission in late 1960, the Music Center was planned as a $15 million single-building project financed by $9 million in public funds and $6 million raised from private sources. Construction was set to begin in the summer of 1961. Early in the year, however, the scope of the project expanded. Chandler presented to the supervisors a Music Center that added a 1,700-seat performing arts theater (the Memorial Pavilion) and a 650-seat forum for smaller performances, and an underground garage to accommodate 2,000 vehicles—double the capacity of the original design. The increase in parking capacity responded to a demand from supervisors for additional space for county employees. The project would now cost $23 million to build. Ultimately, the Music Center Lease Company sold $13.7 million bonds and raised another $15 million in private funds for the complex.[90]

Welton Becket and Associates, designer of the Music Center, was an internationally renowned firm with a long list of corporate clients. Known as "the businessman's architect" because of his attention to costs "to make sure clients get exactly what they are paying for," Becket produced functional designs that rarely resulted in salvage jobs. As architect Arthur Love, who began his career with the firm in 1960, notes, "We had a reputation of producing efficient, well-thought-out economical buildings." Under the umbrella of "total design," an architectural mantra of the 1960s that, according to Stuart W. Leslie, "demanded an adherence to aesthetic conformity bordering on the totalitarian," the firm offered clients architectural and engineering services, as well as a range of other services, from land use planning to interior furnishing and decorating.[91] The firm's signature projects included office buildings for Capitol Records, Ford Motor Company, General Petroleum, Gulf Oil, and National City Bank; hotels for Hilton and Sheraton; department stores for Broadway-Hale, Bullock's, Gimbels, Macy's, Meier and Frank, the May Company, Saks Fifth Avenue and many others; and regional shopping centers across the country. The firm's portfolio included public projects, too, most notably the Civic Auditorium in Santa Monica and the Los Angeles Sports

Arena. The firm also had a wealth of experience in master planning, including Century City (see chapter 2) and UCLA.[92]

The expansion of the project sent Becket's designers scrambling to revise their working drawings, which they had nearly completed at the end of 1960. Ultimately, the design of the 5-story, curvilinear Memorial Pavilion featured 92-foot-high fluted columns, made of structural steel sheathed in precast concrete panels, around its periphery. Steel trusses, spanning 125 feet between interior columns, framed the roof with wide-flange purlins and cast-in-place roof slabs of lightweight concrete. Reinforced concrete supported three tiers of seating in the theater. A "continental" seating arrangement, with no seat more than 130 feet from the stage, was designed to permit the theater to empty within 90 seconds. Other design features, including electrically controlled acoustical equipment, aimed to make the building "the finest of its kind anywhere."[93]

In June 1961, architect Becket predicted that excavation would begin in early December. But legal maneuvers, contesting the lease-back financing arrangement approved by county supervisors, delayed the project. Under the plan, Chandler's Music Center Lease Company would build the center and lease it to the county, which in turn would lease it to another nonprofit entity, the Music Center Operating Company. A superior court judge upheld the arrangement in August, but then had to rule on petitions to set aside the ruling. The supervisors finally issued the call for bids in October. Kiewit submitted the low bid of four, at $16,836,937.[94]

Peter Kiewit had approached Charlie Pankow about bidding the project, but the latter had balked, arguing that he did not have enough men for the job. Moreover, because it was a municipal project, Pankow had no chance of influencing the design process. The delays in the project, owing to the change in scope and legal actions, would have further dampened any enthusiasm Pankow may have had for building it. Determined to build the prestigious project, Peter Kiewit assembled a construction team of men who had worked primarily in heavy construction.[95]

Construction did not begin in December, as planned, owing to a suit filed by a taxpayers' group. Construction and lease contracts were signed at the end of February 1962 only after the group dropped its opposition. Excavation began on 12 March, with completion of the Memorial Pavilion targeted for December 1963. The construction contract called for at least 1,400 spaces in the four-level parking garage to be ready by 1 January 1963.[96]

Figure 12. Charles Pankow, assistant district manager for buildings, visits the jobsite of the Los Angeles Music Center. From left to right: Pankow; J. C. Knight, project architect, Welton Becket & Associates; Don Luedke, representative of the Music Center Lease Co.; Dick Rundle, project manager. (From *Kieways* 18 [March/April 1963]: 6. Courtesy of Kiewit Corporation.)

The project began well. Project manager Dick Rundle, a veteran of many Southwest District projects, including the Los Angeles International Airport and the Chico missile site, deployed two daily 10-hour shifts to excavate the job (fig. 12). Then the project fell behind schedule. As Welton Becket himself conceded, "Not many of our projects are this complicated to put together."[97]

Once it became clear that the team would miss the deadline for the completion of the parking garage, Peter Kiewit called Charlie Pankow and insisted that he reassign his men to the project to help complete it. Red Metcalf recalls that his crew was "about halfway up" the slipform of the First and C Building when Pankow pulled project manager Bob Carlson, general foremen Tony Giron and Jack Grieger, and carpenter foreman Harvey Volke off the job. How long Pankow's men remained on the project is not clear. It is certain that they did not see its completion. By 1963, both Grieger and Giron had returned to San Diego, the former to work as general foreman on Hillcrest North Medical Center; the latter to supervise precast operations on a project to build the Rancho Bernardo Reservoir. By August 1963, Carlson had left Kiewit to become vice president of the new CPI.[98]

Construction of the Memorial Pavilion remained behind schedule, even with the addition of Carlson and his colleagues. The portion of the parking garage that was set to open on 1 January 1963 did not open until the beginning of April. In July 1963, by which time Charlie Pankow had left the company and the Hillcrest North Medical Center project was in its finishing stages, the expected completion of the Memorial Pavilion had been pushed back to mid-1964, more than six months after the date originally planned at the start of construction.[99]

With the arrival of Carlson and his colleagues, however, the project made "giant strides," as the *Los Angeles Times* enthused. As of early December 1962, the project was 30 percent completed, owing to "unusual productivity" on the job site. By August 1963, the Pavilion was two-thirds complete. Workers were fastening precast panels to the external steel columns and applying finish plaster to selected surfaces. The parking garage was open. Exactly what Pankow's men did on behalf of "righting the ship" is unclear, as none of the individuals interviewed for this book who worked for Pankow during his Kiewit days were assigned to the job. Hutton does note that, upon his arrival, Carlson stopped work until crews cleaned up the site and put in place procedures to keep it that way. Carlson may also have been responsible for obtaining the Mayco-Weitz tower crane used to accelerate construction. The consensus within the Arcadia-based building division was that Charlie Pankow had rescued the project.[100]

The Music Center project produced the germ of an architect-contractor relationship that would bloom in the 1970s. Arthur Love "was fortunate to design everything at the Music Center," including the three buildings, the parking garage, and the central plaza. Love met Russ Osterman at this time. Love's description of the Memorial Pavilion project as a "quagmire" suggests the difficulties that both architect and contractor faced. In 1970 Love would relocate to Becket's Chicago office, from where, as chapter 4 relates, he would work with Pankow teams on numerous projects in a fruitful design-build relationship.[101]

Breaking with Kiewit: The End of the Building Division

In his six years as the head of the building division, Charlie Pankow developed commercial construction methods and technologies that he sought to promote throughout the Kiewit organization. Peter Kiewit was not receptive to the idea. Indeed, he was uncomfortable with the type of projects that Pankow and his men were undertaking and the methods and techniques by which they were completing them. Ultimately, lack of support from Peter Kiewit prompted Pankow to establish his own firm. All of the men with the Arcadia-based building division felt that the rest of the Kiewit organization treated them as stepchildren. For one, their projects were a tenth the size of those of the heavy and highway divisions, and so, at annual meetings, presentations touted the revenues and profits of the latter groups. It did not matter that Pankow might be able to reply that his projects had much higher margins because of the ways in which his men built them. This perceived treatment of the building division manifested itself on the job site where, as Alan Murk recalls, crews had to make do with secondhand, or even broken-down, equipment. It became increasingly clear to Charlie Pankow that he would have to form his own company if he were to have the opportunity to deliver projects through design-build.[102]

Still, as the Music Center project proceeded, Pankow remained hesitant. For one, he knew that it took capital to bond and finance large commercial projects. Much of the impetus to create a new company came from the men who worked with him.[103]

Proceedings at the Kiewit annual meeting, held in Omaha early in 1963, constituted a moment of truth for Charlie Pankow. With Carlson, Oster-man, and Pankow in attendance, so the story goes, Peter Kiewit handed out awards and otherwise recognized the accomplishments of other groups, but neglected Pankow's building division. Osterman recalls turning to Pankow and suggesting that they should start their own company. The three of them discussed the idea over dinner and drinks. It is unclear whether Charlie Pan-kow had decided to step out on his own before he left Omaha. If not, he did so soon thereafter. For he incorporated CPI in May.[104]

Charlie Pankow concealed his leaving Kiewit. No public or division-wide announcements attended the filing of the articles of incorporation. Neither Pankow nor any of his men were listed as directors of the new company. (The directors were lawyers William M. Poindexter and Alfred B. Doutré, whom Bob Heisler had introduced to Pankow, and Selma Santini.) Apparently no-body left Kiewit at the time, either. Yet Pankow intended to take many of his colleagues with him, or at least invite them to join him, while he sought to make as orderly an exit as possible. Toward that end, he gave Kiewit as much as six months' notice. Osterman also agreed to stay with Kiewit after Pankow departed until the Southwest District's management was satisfied with the transition. Pankow did not announce the incorporation of CPI until he had secured the company's first project, MacArthur Broadway Center in Oakland, California. As the next chapter elaborates, this project was a joint venture with Sollitt, Charles Pankow *pere's* company. Negotiating this arrangement required considerable time and effort on Charlie Pankow's part. Pankow also had to negotiate the construction contract, which was not signed until the last day of August 1963. Employees had several months to speculate about what was happening to their group and what it meant for their careers.[105]

Rumors about Charlie Pankow's leaving Kiewit swirled around the Hillcrest North Medical Center job site during the summer of 1963. Superintendent Alan Murk already knew about Charlie's aspirations: After returning from the Kiewit annual meeting, Pankow had solicited his thoughts on leaving Kiewit to start a new firm. Murk had had nothing to say in reply. "It hit me kind of cold," he recalls. Now Pankow briefed him on the new firm. He wanted Murk to be his superintendent in Oakland. Pankow instructed Murk to train his replacement on the Medical Center job before he resigned. Murk agreed to join the new company, notwithstanding the conviction of his father that Pankow would surely fail. Murk, in turn, informed George Hutton, who found the news "disturbing," because "[I] really felt like I'd found a home, and it wasn't a home with Kiewit, but it was a home with the building division." Murk encouraged Hutton by reporting that Pankow hoped many of his men would join him. In July, Hutton was promoted to superintendent and assigned to the Rancho Bernardo Reservoir project across town. At the same time, Tony Giron and Red Metcalf left the Hillcrest North Medical Center job to join Hutton as field superintendent and carpenter foreman, respectively.[106]

Charlie Pankow made public the formation of the new company in August. Articles that appeared in the *Los Angeles Times* and the *Pasadena Independent Star-News* identified Carlson, Ralph Tice, and Lloyd Loetterle as the other principals in the new firm (figs. 13–16). But the fate of both the building division and its employees remained unclear. Rather than leave all at once, those who joined Pankow did so seemingly one at a time. By the end of August, Alan Murk and Harold Henderson had quit the firm and had relocated to Northern California. In early September, Pankow told Hutton about his vision for the new company and invited him to join it as project engineer on MacArthur Broadway Center. Hutton agreed and reported to the job site within a week. Jack Grieger, who had been working as general foreman on Hillcrest North Medical Center, joined Hutton in Oakland as the project's general foreman.[107]

Figures 13–16. Founders of Charles Pankow, Inc., incorporated 17 May 1963, clockwise from the top left corner: Charles J. Pankow, Lloyd Loetterle, Robert Carlson, and Ralph Tice. (Courtesy of Charles Pankow Builders, Ltd., Pasadena, California.)

For Red Metcalf, this procession out the doors of the Arcadia building division was bewildering. He had barely reported to work at the Rancho Bernardo Reservoir job site when Hutton told him that he, too, was quitting. It was the first that Metcalf had heard about the formation of CPI. Hutton counseled patience: Charlie Pankow would call Metcalf in the near future, as he was negotiating a second job, a multi-unit apartment building in San Francisco, that would utilize slipform operations. Marion Young, who had been one of three carpenter foremen on the Los Angeles Music Center, replaced Hutton. A couple of weeks later, Tony Giron quit the Rancho Bernardo Reservoir project to work on MacArthur Broadway Center as a second general foreman. Young promoted Metcalf to general foreman, replacing Giron in the precast yard. It was his first salaried position, and he eagerly accepted it. In the back of his mind, however, he awaited a phone call from Charlie Pankow.[108] Russ Osterman replaced Charlie Pankow as head of the building group—but not for long. Soon after Hutton departed, Osterman called a meeting to rally the remaining employees. He urged them to stick with Kiewit, and promised that the division would continue operating. But late in November, Osterman informed Metcalf that he, too, was quitting to join Pankow. Metcalf was incredulous. After all, Osterman's stirring endorsement of the Kiewit organization was still ringing in his ears. Not long thereafter, three of CPI's four officers—Pankow, Carlson, and Loetterle—visited Metcalf at his home in Carlsbad. They informed him that the company had lined up the apartment building project in San Francisco. Charlie Pankow offered to pay Metcalf no more than he was making, but promised to promote him to superintendent if he performed well. Metcalf made up his mind to accept the offer. He gave notice to Marion Young, who informed the Arcadia office. But someone at the district level—Metcalf cannot recall who it was—convinced him to stay with Kiewit by telling him that the company would be around long after Pankow went bankrupt. Metcalf informed Pankow, who left the door open for him, whenever he might decide to join the firm.[109]

The Arcadia building division did not fold with the departure of Pankow and many of his men. Soon after Metcalf completed the Rancho Bernardo Reservoir project, Rosser Edwards tapped him to be his general foreman on the Valley Music Theater project in Woodland Hills. In December 1963, Kiewit won the contract to build a 2,900-seat concrete domed theater-in-the-round for a group headed by Art Linkletter, Randolph Hale, and Nick Mayo. Completed in the summer of 1964, it was the group's last project. Edwards and David Boyd joined CPI soon thereafter. Metcalf moved into heavy construction and would work on major projects in the Central Valley until 1969, when George Hutton would invite him to join Pankow in Hawaii.[110]

CHAPTER 2

Executing Design-Build, 1963–1971

Unlike the founders of Hewlett-Packard, Charlie Pankow did not establish his company in a garage. But one had to walk through the garage of his Altadena, California, home to get to the basement, where he set up shop not more than 10 miles to the west of his former Kiewit office. Here off-site managers, accounting staff, and wife Doris Pankow, the sole administrative assistant, worked until early 1965, when the company leased office space in a building on Walnut Street in downtown Pasadena, a couple of miles to the south. Two years later, Pankow relocated his firm's corporate offices near his residence, to a building on North Lake Avenue that once had housed the Altadena post office; there they would remain until his death.[1]

In terms of size, staff, and work, Charles Pankow, Inc. (CPI), remained in its early years essentially a replica of the Kiewit building division, reflecting in part the conservative course that Pankow charted for the new firm. This chapter considers how Pankow and the group of men who joined him fared in a competitive marketplace without the resources of a parent corporation behind them. Charlie Pankow was now free to market design-build as he wished. Moreover, he could be confident that his men could complete projects to owners' satisfaction, as they had done the previous six years. Yet he faced a "Catch-22." Building a reputation largely on the basis of word of mouth rather than on the tools of modern advertising, as Pankow would choose to do, would require the successful completion of projects delivered through design-build. Yet he faced the challenge of convincing owners to hire his firm over any number of large and established contractors with commercial building divisions.

To court the local building community, Charles Pankow, Bob Carlson, Lloyd Loetterle, and Ralph Tice announced at the outset that they "plan[ned] to continue the same type and size [of] work with which they were previously identified" at Kiewit. "Industrializing" the construction of tall concrete commercial structures, they explained, would enable CPI to estimate construction costs and establish schedules. The new company aimed to do business in San Francisco, Los Angeles, and remote locations on a project-by-project basis: "A mobile system with which [we were] well experienced" at Kiewit.[2]

Charlie Pankow articulated a business model that he would not modify, even when changing market conditions threatened to undermine it. Negotiated, lump-sum contracts on private sector commercial projects constructed of concrete would be the norm. That is, Pankow would avoid bid work (though, as this chapter elaborates, several early projects were bid). He would avoid cost-plus contracts, too, arguing that they created incentives that reduced the level of trust that an owner could invest in the contractor. Above all, he would rely on repeat clients to stay in business.[3]

Charlie Pankow sought gross margins of 15 percent in his contracts, although he normally did not achieve this goal. This bar was relatively high for the industry. It reflected both the execution risk that CPI assumed as a design-build contractor and the compensation that it expected to receive from performing work that it might otherwise subcontract. Notes Russ Osterman, CPI's lead estimator in these early years, a subcontractor typically would demand a margin of 15–20 percent for performing tasks like precasting and slipforming. Ultimately, Charlie Pankow would limit the growth of the company by pressing for higher margins than potential clients were willing to concede and by constraining sales staff in other ways. Yet the conservative financial approach allowed him to control CPI and preserve its independence.[4]

For Charlie Pankow, CPI was his firm and he would do nothing that might jeopardize his firm grip on it. He started the firm with minimal capital. Yet he never countenanced a public stock offering. Pankow had no interest in managing a company according to the expectations of stock market analysts and outside investors. Anyone who bought stock in CPI had to agree to sell it upon leaving the firm. The company would grow only through retained earnings. This approach helped him to insulate the company from the business cycle, which publicly held contractors, such as Perini and Turner, found difficult or impossible to manage.[5]

Securing a bonding commitment from Federal Insurance Company (now part of Chubb) was an important moment in the history of CPI. Charlie Pankow sought to secure a bonding capacity commensurate with building structures like the American Cement Building. Lacking the net worth to obtain performance and payments bonds from a surety, Pankow initially used a broker to secure them, without success. So the story goes, he then tried a more direct approach. Late in 1964 or early in 1965, Pankow met with several sureties at a conference in Las Vegas. One of them was Alex Kerner, who headed the bonding department for Federal Insurance in Los Angeles. Federal was considered the "Cadillac" of sureties, according to Doug Craker, who served as project accountant, assistant controller, and controller with the company from 1963 to 1980. Through Bob Heisler, Pankow had been introduced to Kerner. Heisler had used Federal Insurance to bond his work for many years. At the conference, Pankow explained the concept of his "budgeted construction program," including its design-build and mechanized components. Kerner agreed to bond Pankow's projects as long as the contractor agreed to certain parameters, including self-performing at least half of the concrete work on projects. Notes Craker: From that point on, "we never had a problem with getting a bond and getting intent to bond by [Federal Insurance] if we needed it for an owner." The company enjoyed "practically an unlimited credit regarding our bonding limits."[6]

Another factor that influenced Charlie Pankow's thinking on the type of firm he wanted to run was the treatment that his father had received as a non-family manager within a family-run business. In 1935 Ralph Shannon Sollitt incorporated the South Bend branch of the business as Sollitt Construction Company. By the 1950s, Pankow *père* had become vice president, but he held no ownership stake. Ralph Shannon Sollitt apparently had handpicked him as his successor. When Sollitt died in 1967, however, his daughter assumed the helm. It spelled the end of the elder Pankow's career in construction.[7]

Charlie Pankow concluded that nepotism was a disease that would sap the esprit de corps that he and his men had brought with them from Kiewit. As he later explained, "I've seen too many good companies in this industry ruined by nepotism. And when that happens it takes a lot of good people down with it." It did the blood relatives no favors, either, in his view. Pankow would never tell his three sons that they could not work for him; rather, that he preferred that they come in through the front, rather than the back, door, so to speak, after beginning their careers at other firms. (In any case, only one son, Steve, would pursue a career in construction.) Pankow would never contemplate a leadership succession plan that included his children.[8]

Based on his father's treatment, Pankow also concluded that the people responsible for directing a company should share in its ownership. He had benefited handsomely from the stock purchase plan at Kiewit—providing him with the resources to become the majority shareholder of CPI at its founding—and wanted to establish some version of it in the new firm. At the first annual meeting, held in a San Francisco Bay Area hotel in September 1964, Pankow announced that anyone could buy stock (as long as it did not jeopardize his majority ownership). In 1973 the company would initiate a plan to accommodate top employees who had been hired in the interim. Modeled on the Kiewit plan, which Red Metcalf describes as "a hell of a program," it allowed employees to buy stock at the previous fiscal year's closing price and take out five-year notes at imputed interest to finance the purchase. (The Kiewit plan allowed salaried employees to borrow at low rates of interest as much money as they put up themselves for stock purchases.) Employee ownership reinforced the esprit de corps that Pankow's men brought with them from the Kiewit building division.[9]

While he remained at the helm of the Kiewit building division, Charlie Pankow observed that commercial building owners were becoming increasingly dissatisfied in the marketplace. As supply caught up with pent-up demand for office and retail space—produced by controls on nonessential construction during World War II and on selected building materials during the Korean War—cost pressures increased. Competition to lease commercial space heated up in the late 1950s. In the aftermath of World War II, owners had been able to lease space at double, or even triple, the rates that they had charged in 1940. Now they were finding it difficult to lease space at rates that justified their speculative investments. By the time of CPI's founding, Los Angeles was "badly overbuilt." Yet, even with the vacancy rate on Class A office space running around 12 percent, owners were poised to release a backlog of some 3 million square feet of space on to the local market over the next three years. These conditions gave rise to so-called salvage projects, whereby an owner might spend half a million or more dollars on plans and specifications for a project that could not be justified by the rents that he could charge tenants. Salvage jobs would become the bread and butter of Charlie Pankow's

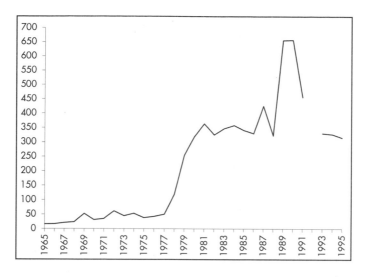

Figure 17. Value of contracts outstanding, 1965–1995 ($ millions). (From ENR Top 400 [annual list].)

company. Indeed, as late as 1974, he would describe to recruits the work of his company in those terms.[10]

A Culture of Innovation Expressed in Early Work

Notwithstanding its low volume of outstanding contracts relative to later years (fig. 17), the work of CPI in its first decade encompassed nearly the entire spectrum of structures that the firm has constructed throughout its history. Office buildings dominated the portfolio, but there were multi-unit apartment and condominium projects, shopping centers, and military installations, too. As he did during his Kiewit days, Charlie Pankow worked with architects who produced designs in the idiom of modern architecture. He also established a relationship with T. Y. Lin and his associates that facilitated the execution of many of these designs in prestressed concrete. CPI's men used the expertise that they had developed at Kiewit to execute projects on time and within owners' budgets, establishing the basis for repeat business that sustained the company and set the stage for growth in the 1970s.

Public sector work comprised a larger share of CPI's order book in its early years than it would later on. Two multi-unit residential projects were examples of the urban renewal efforts of the San Francisco Redevelopment Agency (SFRA). Another project involved the building of the Las Flores Area at Camp Pendleton. The company also built Joe West Hall on the campus of San José State College (now, University). Public work may not have been part of Charlie Pankow's vision as a builder, but the new company was in no position to turn down these opportunities. Yet these projects, like those completed by the Arcadia-based building division of Kiewit, allowed the contractor to participate in design or redesign, deploy techniques to automate concrete construction, or both.

Outside of Hawaii, most of the company's work during this early period was confined to Northern California. While the company was completing these projects, Charlie Pankow developed several key relationships that resulted in repeat business in the San Francisco Bay Area. Work for these clients demanded all of the resources of the new company in its early years: The company was profitable from day one, yet its capacity to take on more than a couple of projects at a time remained limited until well into the 1970s.

Significantly, three of the company's first four projects were joint ventures. Out of necessity: CPI did not yet have the bonding capacity to build these projects alone. These joint-venture relationships were critical to the formation of CPI and its viability in its early years.

The Sollitt Joint Venture: MacArthur Broadway Center

Edmond E. Herrscher's dream of building a shopping center at the intersection of Broadway and MacArthur Boulevard near downtown Oakland provided the catalyst for the incorporation of CPI. It was an unsolicited job. The architect, Irving D. Shapiro, approached Charlie Pankow after Herrscher received no bids that fell within his budget. CPI worked with T. Y. Lin's structural engineering firm to redesign the project, CEO Richard M. Kunnath explains, using their "knowledge about what would be the best way to design [it] from a contractor's point of view, the cheapest, most effective, code-compliant method, and we came in . . . very significantly below the previous costs, well below the owner's budget." The redesigned project cost $4.7 million to build—some $1.8 million below the lowest bid that Herrscher had received. Reflects Kunnath: The project "really was a prototypical example" that demonstrated the advantages of being able to use "what you know about design rather than simply being the recipient of somebody else's ill-conceived ideas." It confirmed that "the only way that you [could get] work [under design-bid-build] was to work cheaper and cheaper for lower and lower fees." Charlie Pankow showed Herrscher that he could solve the owner's dilemma by "throwing out" the architect's design and "fundamentally restructuring" it. Completing the MacArthur Broadway Center helped the new contractor to establish its bona fides.[11]

Before he became a developer, Edmond Herrscher gained regional, if not national, notoriety as a divorce lawyer. A native of San Leandro, a white, working-class city that lay immediately to the south of Oakland, Herrscher obtained his degree from the University of California's Hastings Law School. In his first high-profile case, he "freed" a Maui sugar heiress from a mental institution after her family committed her to it. Then he won a $2.5 million settlement for Fannie May Howard in her divorce from Charles Howard— the man who later bought the fabled racehorse Seabiscuit. Herrscher then married Fannie May. In 1948 Herrscher stopped practicing law to develop commercial real estate. In the mid-1950s, he began dating Wendy Skouras, the widow of Thomas Skouras, a member of the Twentieth Century Fox film family. (Wife Fanny May had died in 1942.) As the story goes, Fox president and chairman Spyros P. Skouras showed Herrscher the studio's back lot that was located between the city of Beverly Hills and the Los Angeles Country Club. In the context of the economically uncertain future that all of the so-called Big Five studios faced in the wake of the 1948 Supreme Court decision and consent decree that had forced them to divest their theater chains, Herrscher offered Skouras his ideas on developing the property.[12] Since the studio acquired the land in 1923, the West Los Angeles and Mid-Wilshire districts on both sides of it had been developed intensively and so the lot was appreciating in value. Impressed by his analysis and recommendations, the studio chief put Herrscher in charge of real estate at Fox. Herrscher, who married Wendy Skouras soon thereafter, hired Welton Becket and Associates to design a master plan for a self-contained, mixed-use urban center (the first structure of which would house Welton Becket's headquarters). Two years in development, Skouras announced plans for the "new city" at the company's annual meeting in 1957. The master plan, in turn, attracted the interest of developer William Zeckendorf (whose many projects included Roosevelt Field shopping center, which opened in August 1956 on Long Island, New York, and which the Pankow firm would expand during the 1990s). Envisioning "an oasis in the midst of a great city," Zeckendorf jumped at the chance to realize an urban project of such scale relatively unencumbered by land use and other restrictions. In 1960 he enlisted Alcoa as a partner. The multinational corporation wanted to showcase aluminum as a building material, as it had done with the completion of its headquarters building in 1953 (see chapter 1), and had ample capital to realize Welton Becket's ambitious master plan. Over the next three decades, Century City materialized as a highly dense cluster of office and residential towers, low-rise condominium complexes, hotels, and a regional shopping center. For his role in developing the property and supplying its name, Herrscher became known as the "Father of Century City."[13]

Surprisingly, then, Herrscher considered MacArthur Broadway Center to be his greatest accomplishment. It was "the culmination of the 'dream shopping center'" that he had envisioned ever since he and Andrew Williams opened a 24-hour, "self-service" supermarket in 1937 on the corner of Broadway and MacArthur Boulevard in Oakland. One of four Andrew Williams stores in the Bay Area, Herrscher bought out Williams's interest in the budding chain in 1939. The following year, *Life* featured the market in a photographic essay on retail shopping trends. In 1949 Herrscher merged the chain with Mayfair Markets. At the time of the construction of MacArthur Broadway Center, Herrscher was chairman of finance for Arden-Mayfair, the parent company, and was part owner of an $8 million shopping center in National City, California.[14]

Long regarded as a prime commercial location, the six-and-a-half-acre site became even more attractive with the construction of the segment of Interstate 580 that lay between Castro Valley and the interchange leading to the Bay Bridge that is now referred to as the "MacArthur Maze." For years, however, Herrscher's shopping center remained a vision of his mind's eye. The problem was parking. MacArthur Broadway Center was one of the few large shopping centers built during the 1950s and 1960s in a dense urban area. In 1950 Herrscher completed his acquisition of the land bounded by MacArthur Boulevard, Piedmont Avenue, Broadway, and the new freeway. Unlike a developer, such as Ralph Lloyd, who could offer architect John Graham some 60 acres to use in the design of the Lloyd Center (which opened on Portland's East Side in August 1960), Herrscher did not control enough property to accommodate the parking needed to make a shopping center economical. As Herrscher lamented, "Thousands of dollars' worth of plans and blueprints have been waste-basketed in the past ten years because the required space for parking just wouldn't allow a realistic amount of store space."[15]

Architect Shapiro suggested the solution: rooftop parking that would accommodate almost 700 vehicles over a single-story center that covered the entire site. The approach, which derived its technical solution from multi-level parking garages, rarely had been applied to retail structures because the price of the land generally did not justify the cost of construction. The few exceptions during the interwar years included the completion, in 1937, by Marshall Field & Company, of a parking lot for its customers at its Evanston, Illinois, branch location. The company constructed the deck, with space for 50 vehicles, atop a single-story auto service station structure that adjoined the department store. Two years later, Sears, Roebuck & Company "brought the concept to a new scale," writes architectural historian Richard Longstreth, with the opening of its largest store to date on the Pacific Coast, at the corner of Pico and West Boulevards in Los Angeles. Rooftop parking for 275 vehicles over the massive store and service wing constituted one of three levels of parking on the hillside site. Conventional parking was impossible, since the ground dropped away from West Boulevard at a 20-degree slope. After World War II, site configurations that might have justified rooftop parking were almost nonexistent, as most large-scale retail projects were developed in outlying locations that provided ample room for surface parking. Nevertheless, it is likely that the Sears store inspired the Southern California–based Shapiro, whose distinctive Columbia Savings & Loan Association Building was located less than a mile away, at the corner of Wilshire Boulevard and La Brea Avenue. Shapiro's solution "brought the concept to [another] new scale." As applied to a major shopping center in early postwar America, it may have been unique.[16]

Shapiro worked with structural engineers Felix Kulka, Herbert Korner, and Y. C. Yang of San Francisco–based T. Y. Lin, Kukla, Yang & Associates on the design of the parking deck, which added only 50 cents per square foot to the cost of the roof. Their structural solution also added 300 parking stalls at the street and basement levels. Moving sidewalks with magnetic strips, dubbed Speedramps, were designed to convey up to 3,000 customers per hour, along with their carts—each containing a retractable magnet—from the top of the parking ramps to pedestrian mall areas, and back again. The designers placed a 55,000-square-foot office building over the parking deck (fig. 18). Shapiro could allocate but 180,000 square feet to leasable retail space, which was small relative to the largest regional shopping centers: The Lloyd Center, Roosevelt Field, and Garden City Plaza in Paramus, New Jersey (1957–1958), contained 1.2 million, 1.15 million, and 1.15 million square feet of leasable retail space, respectively.[17]

Reflecting the expansion in the range of services that shopping centers were offering at the time, the architect included an "international food ba-

Figure 18. Office building atop the post-tensioned concrete parking deck at MacArthur Broadway Center (1963–1965), demolished. Irving D. Shapiro, architect; T. Y. Lin, Kukla, Yang & Associates, structural engineer. (Courtesy of Charles Pankow Builders, Ltd., Pasadena, California.)

zaar," encompassing 16 restaurants and dining facilities with space for 600 office workers and shoppers. A garden court area featured a brass sculptural fountain, "Ten Pagan Gods of Festivity," by Los Angeles artist Taki. Shapiro also placed, as he described it, "special lighting fixtures, bits of art work, landscaped pedestrian malls, a hanging garden, and telephone booths inspired by old English jousting tents." He aimed to attract "club women" to kiosks located in air-conditioned malls for charity fund raising and specialty merchandizing events. Herrscher hoped that these "unique features" would "be copied by others for years to come."[18]

But first the design had to be built, and, after two rounds, none of the bids met Herrscher's pro forma budget. When Shapiro may have approached Pankow about redesigning the shopping center is unclear, reflecting the veil of uncertainty that shrouded the founding of CPI. The groundbreaking ceremony for the project took place on 17 May 1963—the same day that Charlie Pankow filed articles of incorporation. Pankow may have lined up the project at this point. Yet he did not appear in a photograph of the ceremony that appeared in the *Oakland Tribune*. At this point, construction was slated to begin in June; it did not. Pankow did not commit to the project until early August, when he lined up Sollitt as a joint venture partner. In return for a 50 percent interest in the project, Sollitt provided the bonding capacity and a credit line. Once the joint venture, named Charles Pankow & Associate, was in place, Pankow recruited the men he wanted for the job and left Kiewit with a team that was prepared to redesign and build Edmond Herrscher's "dream shopping center." A $3.5 million construction contract for the first four stages of the project was not signed until 31 August—three and a half months after the groundbreaking ceremony. The redesign of the project began in September.[19]

Construction of MacArthur Broadway Center was "fast-tracked," meaning that construction occurred in phases, with each phase proceeding once its design was completed and all required site and building permits were approved. Construction also proceeded in this manner to allow the Mayfair Market to continue doing business for as long as possible. The Pankow team, which included Bob Carlson, Tony Giron, Harold Henderson, George Hutton, Jack Grieger, and Alan Murk, regularly held meetings to discuss design changes and propose construction techniques. They conveyed their ideas and decisions to Y. C. Yang, the structural engineer who was responsible for the project. From their San Francisco office, Yang and his colleagues produced the design documents that the contractor used to build the center. All the while, John N. Stuart, Sollitt's representative, controlled the project's finances. He ensured that suppliers were paid on time so that materials, including 22,000 cubic yards of concrete from Permanente Cement's Santa Clara County facility, were delivered to the job site in a timely manner.[20]

Illustrating the continuity between the Kiewit building division and the new firm, Pankow's men incorporated techniques with which they had become familiar into the redesign and construction of the center. The office building had been designed to use "monumental" 70-foot precast, prestressed, single-tee beams for the floors, but, as George Hutton explains, there was no practical means of deploying a crane to place such heavy structural elements seven stories up. The parking structure was designed to incorporate post-tensioned deck slabs and prestressed concrete beams that spanned 60 feet between columns. T. Y. Lin's structural engineers assumed that these elements would be fabricated by a local manufacturer. Pankow's men saved transportation costs by casting these elements in place. At the same time, Carlson decided to precast the shorter columns, and set up a yard to produce them. To post-tension the slabs, the project team used a Prescon system of cables made of grouped, ¼-inch wire, which crews greased and wrapped to prevent the formation of a bond between the cables and the concrete. Continuous post-tensioning allowed the parking deck slabs to comprise a monolithic structure, eliminating the need to apply a topping to waterproof it. The *Oakland Tribune* called the on-site fabrication of the concrete structure a "most unusual" approach, but noted that it was "one of the things that made M-B Center economically possible."[21]

The redesign and construction of MacArthur Broadway Center took two years—ten months longer than Shapiro had estimated in May 1963. The grand opening, on 9 September 1965, was choreographed to link the shopping experience to the fantastic wonders of the Space Age. Young men conveyed customers and their packages from the tops of the Speedramps to their cars in so-called Astrobuses. To personify the Center's Space Age modernity, the opening featured Astro the Space Clown and Space Girl—a starlet flown in by Herrscher from Hollywood, along with other Fox celebrities.[22]

For the members of the project team, MacArthur Broadway Center was demanding and difficult. In completing it to the satisfaction of the owner, however, team members demonstrated that they could execute Charlie Pankow's construction program apart from the Kiewit organization. T. Y. Lin cited the office building as an example of the design revolution that prestressed concrete was making possible. The Prestressed Concrete Institute gave the structure one of its ten Merit Awards for 1965. The high-profile success of the project helped Pankow secure additional work in the San Francisco Bay Area. And Pankow's relationship with structural engineer Y. C. Yang would lead indirectly to the company's first project in Hawaii.[23]

During the MacArthur Broadway Center project, a three-bedroom unit in Crystal Tower, a 12-story apartment tower in San Francisco's North Beach district, served as the new company's Northern California office. Crystal Tower was a Kiewit project that Lloyd Loetterle had managed. According to Steve Pankow, the third of Charlie and Doris Pankow's four children (and the third of three sons), the client was so pleased with the job that he let the flat to his father rent-free for a year. Space was at a premium. Two of the bedrooms were filled with desks; the third was lined from wall-to-wall with sleeping bags. Subsequently, Pankow rented office space in an otherwise vacant cannery building at the corner of Hyde and Beach Streets in the Wharf area. It lay at the bottom of a cable car line and across the street from the Buena Vista, a popular tourist establishment that is at least regionally famous for its

Irish coffees. Charlie Pankow had matchbooks printed that read, "The Buena Vista: Across from Charles Pankow."[24]

The End of the Sollitt Joint Venture

The Pankow-Sollitt joint venture continued for one more project—the construction of the Las Flores Area of the US Marine Corps training base at Camp Pendleton. On the face of it, it lay outside of Charlie Pankow's business model. As a federal project, it was subject to both bidding and regulations that normally would frustrate the application of design-build. Under a value engineering program that began in 1954 within the US Navy and was formalized under Secretary Robert McNamara's Cost Reduction Program for the US Department of Defense, however, the request for proposal (RFP) offered the contractor the opportunity to redesign the project after it was procured. Value engineering incentives incorporated into construction contracts under the Armed Services Procurement Regulation for Value Engineering permitted a contractor to propose cost-saving modifications, if they did not adversely affect the project. The contractor and the federal government split the savings on a 50-50 basis, realized through the reduction of the contract price by half of the savings. The contractor submitted its ideas as Value Engineering Change Proposals. Approved VECPs were processed as normal change orders. So Charlie Pankow agreed to bid on the contract.[25]

The Camp Pendleton job "was just ideal for precast and tilt-up walls," notes Doug Craker, who was project accountant for both of the Pankow-Sollitt joint venture projects. It consisted of barracks and a chapel. The RFP specified them to be cast-in-place structures. CPI submitted VECPs to construct the buildings as tilt-up structures that used precast elements. Once again, Sollitt provided the interim financing. John Stuart supervised project accounting and job costing. Bob Carlson and Tony Giron relocated from Oakland to run the job.[26]

Ralph Shannon Sollitt died about the time that CPI finished its work at Camp Pendleton, and his passing spelled the end of the relationship between Charles Pankow Sr. and the Sollitt firm. Charles Pankow & Associate dissolved. The joint-venture entity had played a critical role in launching the new CPI. Stuart, who had delayed his retirement to work for the joint venture as a favor to Charles Pankow Sr., now retired in nearby Oceanside. He had spent part of his career in gold mining in Idaho and had wanted to retire in the West. Both Carlson and Craker moved into the Altadena office, as operations manager and assistant controller, respectively.[27]

Redeveloping San Francisco: A Public-Private Relationship

CPI's first high-rise residential projects housed residents displaced by the SFRA in the Western Addition and South of Market (SOMA) districts of the city. The first project, a joint venture effort with slipform construction pioneer Heede International, was built for Eichler Homes, the developer and contractor. As subcontractors, Heede and Pankow erected the apartment building concurrently with MacArthur Broadway Center. Seven years later, CPI constructed Clementina Towers, a turnkey "elderly" (the contemporary term for seniors) housing project for the San Francisco Housing Authority. For several years, its 276 units constituted the only replacement housing for the nearly 4,000 people displaced by the Redevelopment Agency to make way for Yerba Buena Center, a major cultural district whose redevelopment remains ongoing.[28]

California's Community Redevelopment Act, passed in 1945 and amended in 1952, allowed municipalities and counties to create redevelopment areas and establish independent agencies for purposes of reversing the decline of their urban districts—especially downtown—that accompanied the migration of white, middle-class populations to the suburbs. The operative word was "blight," an elastic term used by developers and planners since the 1910s to describe districts where land values were falling (or rising relatively more slowly than elsewhere in town). As originally used, the term referred to buildings that were dilapidated, obsolete, and/or vacant. As codified into law, the term became the fountainhead of urban renewal. Any community redevelopment agency with an interest in addressing visual and socioeconomic decline and demographic change might write a plan that delineated an area and declare it to be blighted, even if most of the buildings within it were structurally sound or even officially designated as historic. This description fits the case of Hollywood, for instance, which the Los Angeles Community Redevelopment Agency (LACRA) declared blighted in 1986. Just two years earlier, the US National Park Service had granted historic district status to the 1,107 acres encompassed by the LACRA plan. With the approval of the plan by the Los Angeles City Council, dozens of buildings that might have been adapted for reuse were razed in the name of revitalization.[29]

Urban renewal became closely associated with slum clearance. Lower-income minority individuals and families had replaced the white populations that had fled town for the suburbs. Developers and civic leaders viewed blighted districts as incipient slums. As such, they saw both types of areas as threats to the social and economic well-being of their cities. Conflating the two, they determined that areas encumbered by physically deteriorating buildings and neighborhoods composed of low-income minorities should be razed to make way for private development. They routinely deployed blight as a legal term to trigger the process. By granting redevelopment agencies eminent domain and subsidies for land acquisition and clearance, federal housing acts passed in 1949 and 1954 assisted local officials in doing just that.[30]

At the same time, federal law contained provisions that developers and their political allies vigorously opposed. Amendments to the National Housing Act, passed in 1937, established a federal program to sponsor public housing projects administered by local agencies. The Redevelopment Act of 1945 required cities that received federal funds to meet federal housing policy objectives. The 1949 housing law authorized the construction of 800,000 units of public housing and required that one public housing unit replace each "slum unit" that cities eliminated. Opponents among local growth machines decried public housing on ideological and racial grounds and mobilized against it. By 1952, more than 40 cities had overturned or terminated public housing programs. Yet redevelopment agencies were still required to adhere to the requirements of federal housing law if they used federal funds to implement their plans.[31]

California's legislators provided the "developer-municipal complex" with the financial means to circumvent the stipulations of federal housing law. Amendments to the Community Redevelopment Act, passed in 1952, enabled redevelopment agencies to fund slum clearance through tax-increment financing. This device allowed agencies to divert tax revenues generated from increases in property values after the start of a redevelopment project for its life, which might last decades. This stream of revenue enabled redevelopment agencies to issue general revenue bonds to fund programs. State lawmakers recognized that cleared urban land was a poor source of tax revenues until it was redeveloped. In assuming that redevelopment, rather than "market forces," would be responsible for property taxes increases, however, legislators allowed private interests to capture all of the rents associated with new development. In San Francisco, redevelopment agency officials and the developers that they served all but ignored the interests of residents displaced by blight eradication and slum removal.[32]

The redevelopment of the Western Addition, which occurred in two phases, was one of the largest urban renewal projects on the Pacific Coast— and one of the most controversial. The first phase, of which the Pankow-Heede joint venture project was part, involved widening Geary Boulevard as an eight-lane thoroughfare that bisected the city from downtown to Golden Gate Park and building up the area around it with new projects. Prior to their forced relocation during wartime, Japanese Americans populated the area. During World War II, African American workers and their families moved in. The Fillmore district within the redevelopment area became the city's most prominent African American neighborhood. In 1948 the SFRA declared the area to be blighted. (Demolition did not begin until 1956, however.) In 1964 SFRA director M. Justin Herman expanded the redevelopment area to encompass 60 square blocks. In all, the SFRA demolished some 2,500 Victorian homes, closed 883 businesses, and forced 4,729 households to relocate. Few were able to return, owing to a lack of affordable replacement housing. As Chester Hartman notes, urban renewal and redevelopment became popularly known as "Negro removal" because of efforts such as this one.[33]

Redeveloping the Western Addition:
A Joint-Venture for Eichler Homes

By 1960, architect-cum-developer Joseph Eichler was erecting some 900 modern, single-family homes annually in the San Francisco Bay Area. Rising land prices "forced [him] into other types of housing" that made more intensive use of urban land. Eichler was convinced that explosive population growth was making high-density housing inevitable, even as local planners and officials resisted the idea. Over the next half decade, he entered the multi-unit residential market in San Francisco and tried to create one in Santa Clara County. He welcomed the challenge of finding economical means of creating high-quality housing, regardless of the class of the people for whom he was building. Indeed, Eichler bet his company on being able to do so. Unfortunately, in building complex projects in a weak rental market, he overextended his company financially, leading to the failure of Eichler Homes before the decade was out.[34]

Eichler built multi-unit residential projects for almost every income group. In Palo Alto and Santa Clara, he built cooperative housing projects under Section 213 of the National Housing Act. His first project in San Francisco was Geneva Terrace, which he designed in 1961. The project's configuration evolved from four high-rise towers to twin 12-story towers with 585 units between them and 221 townhouses. Geneva Terrace was constructed under Section 221(d)(3) of the National Housing Act, which provided below-market interest rate financing for cooperative housing or apartment rent subsidies for lower- and moderate-income families, seniors, and disabled individuals who did not qualify for public housing, but could not afford market-rate rentals. Eichler called Geneva Terrace "one of the finest structures that has been built anywhere" for families at these income levels. The towers included a landscaped plaza, high-speed elevators, and central television and FM cable outlets. Similar in configuration was Eichler's Laguna Heights project in the Western Addition. It included a 15-story tower designed by Jones & Emmons and six 3-story apartment buildings clustered around a common garden area. In his low-rise designs, Eichler provided the amenities of the single-family suburban home in an urban, multi-unit residential setting. The 3-story apartments went some way to restore the scale of the neighborhood cleared by the SFRA. Central Towers, the project built by the Pankow-Heede joint venture, targeted low-income residents with 362 studio and one-bedroom units in twin 15-story towers. At the higher end of the market, Eichler developed 105 lots in Diamond Heights. Envisioned by the SFRA as a self-contained neighborhood, the project combined "suburban ideals with San Francisco's urban amenities," as architectural historian Richard Brandi describes. Its residential portion included 458 units, 20 percent of which were federally subsidized. The high-integrity housing stock was diverse and included low-rise apartments, townhouses, and single-family homes. Churches, stores, businesses, and open recreational spaces realized the city-within-a-city vision of its designers. The SFRA sold the land in Diamond Heights to the highest bidder; homes sold for more than double the median price in the Bay Area. At the high end of the market, Eichler built Eichler Summit atop San Francisco's Russian Hill, a 32-story tower with 112 luxury flats and two penthouses.[35]

With the exception of the Laguna Heights high-rise, Joseph Eichler worked with architect Claude Oakland on all his projects. Beginning in 1950, Eichler retained the San Francisco architectural firm of Anshen & Allen to produce modernist, single-family homes. Oakland, a native of Monroe, Louisiana, who studied architecture at Tulane University, was the firm's principal designer. When he decided to diversify into the multi-unit residential market, Eichler encouraged Oakland to set up his own firm.[36]

* Eichler Homes was the contractor for the Central Towers project. The company acquired the block bounded by Hyde, Leavenworth, Eddy, and Turk Streets from the SFRA and then worked with Oakland and T. Y. Lin, Kukla, Yang & Associates on the design. The ground floors of each tower contained the lobby, mechanical areas, and space for stores. Each of the typical floors, 2 through 15, contained nine studio and four one-bedroom apartments, so that each tower contained 181 units (fig. 19). Oakland placed a two-level parking structure between the towers, the roof of which served as a plaza and recreation area. The work was designed for slipform construction, with all but one of the vertical, bearing walls continuing from the second floor to the top. (A single wall dropped off at the third floor.) The first-floor walls would be poured in place. Post-tensioned floor slabs would provide lateral support for the walls.[37]

On 18 December 1963, Eichler Homes let the $845,000 subcontract for the concrete portion of the project to the Pankow-Heede joint venture. For each tower, the subcontractors would pour in place the walls and columns of the first floor and construct the typical wall-slab structure from the second floor through the roof. Eichler Homes subcontracted to other firms the parking structure and ramps, the foundation work for the towers, the slabs-on-grade, the pouring of the topping over the radiant heating system, and all other nontypical concrete work. Pankow-Heede was responsible for coordinating the work of the subcontractors.[38]

Working one normal eight-hour shift, five days per week, crews under the direction of Harold Henderson—reassigned from MacArthur Broadway Center—turned over 10,000 square feet of new floor area to the electrical, mechanical, and HVAC subcontractors every other day. Heede supplied its patented hydraulic jacking system for the slipform operation and the Linden climbing cranes used on the job. (Heede was the American distributor of the latter.) The Heede system used oil lines to connect hydraulic jacks to a central reservoir. The jacks climbed approximately 1 inch each time crews activated the electric pump that powered the system. Crews raised the 4-foot-high slipforms at rates as high as 15 inches per hour, enabling them to complete one story per shift. Workers were unable to slip the walls continuously, project manager Lloyd Loetterle explained, because of the inserts that were needed for post-tensioning the

Figure 19. Central Towers, between Turk and Eddy Streets, San Francisco, architect's sketch. Completed in 1964. Claude Oakland, architect; T. Y. Lin, Kukla, Yang & Associates, structural engineer. (Reproduced by permission of the Environmental Design Archives, University of California, Berkeley.)

cast-in-place floor slabs. The building code for Seismic Zone 4 further complicated the slipform operation because of the additional reinforcing steel and connections that were required in certain areas, especially at the floor line, to satisfy building inspectors. Charlie Pankow saw an opportunity to improve the Heede jacking system, and would do so on subsequent projects where CPI acted as contractor. Still, crews completed the shells of the two towers in 90 days.[39]

If they provided Charlie Pankow with the opportunity to demonstrate the advantages of slipform construction, Eichler's high-rise projects spelled the ruin of the company. Eichler was a merchant builder who took ownership of the projects that he developed. Hence, from a cash flow perspective, he had to wait until these projects were completed before he could rent any of the units—in contrast to the single-family homes that he built, which could be sold as he completed them. Further, the FHA (Federal Housing Administration) limited rents and returns on investment on projects that it financed. In the case of Geneva Terrace, for example, the FHA limited monthly rents on two- and three-bedroom units to $136 and $157, respectively, and capped the return on investment at 6 percent. Notwithstanding the population pressures that persuaded Eichler to diversify into multi-unit residential construction, San Francisco's rental market faltered just as the Laguna Heights and Central Towers projects were completed. To improve his cash flow, Eichler sought and received permission to convert the low-rise apartments of the Laguna Heights project to cooperative (Section 213) units that he could sell at market rates, but it was not enough to cover the construction of the high-rise tower, a Section 221(d)(3) rental project. Eichler was able to charge top-of-market rents for Eichler Summit on top of posh Russian Hill, but the rents that he collected fell short of compensating for the $2 million overrun in costs incurred during the construction of the project, whose design architectural historian Paul Adamson describes as "ambitious." Geneva Terrace, too, was saddled with cost overruns. The largest project of its kind, Eichler Homes incurred a deficit of almost $1 million to construct its two towers. Before it completed the project in 1966, the company was in trouble. Eichler offered Geneva Terrace to the City of San Francisco as a low-rent public housing project. Officials turned him down. For two years, Eichler Homes operated on the brink of bankruptcy before Eichler dissolved the firm in 1968.[40]

Charles Pankow might have used the demise of Eichler Homes to illustrate the risk to owners of failing to control construction costs. Eichler's system of producing single-family homes, which evolved over the course of a decade in which the company produced close to 10,000 units, allowed Eichler to maintain quality even as he tinkered with ways to cut costs. The market segment into which he branched involved building fewer projects, each of which was large and complex. Eichler had little margin for error and much less opportunity to "learn by doing." Well before Eichler Homes faced bankruptcy, Eichler conceded that his company "had a lot of problems" making the transition to the multi-unit residential segment of the market. Securing approvals from federal housing officials, planning commissions, and city councils proved to be "quite a job." For instance, Eichler and Oakland worked and reworked the design of Geneva Terrace for more than two years before officials cleared the project for construction. The projects that Eichler took on were far more complex in terms of scheduling, materials, and methods than the suburban tract homes with which he was familiar. Each project, he noted, required "a tremendous amount of time and study." Charlie Pankow then might have hammered home the point that the contractor assumed the risks that doomed Eichler under the design-build project delivery system that he was selling.[41]

Slipforming Residential Towers at Yerba Buena Center: Clementina Towers

Yerba Buena Center was San Francisco's most ambitious redevelopment program. It was also no less controversial than the redevelopment of the Western Addition. Because of its late timing relative to the city's other redevelopment efforts, it was subject to great delays. Neighborhood groups used the courts, housing and environmental laws, and other tools to challenge its design and the SFRA's plans for the relocation of residents. Clementina Towers was one of several projects that redevelopment officials undertook in response to their complaints.[42]

Yerba Buena Center was part of a postwar effort on the part of downtown corporate and real estate elites to expand San Francisco's central business district to boost the city's status as a regional, national, and international gateway for finance, distribution, and other services. Downtown redevelopment began in 1959 with the Golden Gateway Redevelopment Project, which replaced the city's produce market with the Embarcadero Center, the Alcoa Building (not the aforementioned Alcoa Building, which was constructed in Pittsburgh), and upmarket housing. Thereafter the only feasible path of expansion was south of Market Street (hence the SOMA name for the district), a low-density area historically associated with manufacturing, shipbuilding, and warehousing.[43]

Planning and design of Yerba Buena Center began in 1961, when the SFRA won approval from the Board of Supervisors and HUD (US Department of Housing and Urban Development) officials for a federal urban renewal survey and planning grant. Five years later, the SFRA and HUD signed a contract that made federal funds available for the project. Acquisition and condemnation of residence hotels that housed some 4,000 mostly single, "elderly," male, and poor individuals and the businesses that catered to their needs commenced six months later. In June 1969, the SFRA unveiled the design of Yerba Buena Center to great fanfare. The so-called Central Blocks—21 acres bounded by Third, Fourth, Market, and Folsom Streets—would contain a 350,000-square-foot convention center, a 14,000-seat sports arena, an 800-room hotel, a 2,200-seat theater, a cultural and trade center, an airline terminal, and two municipal parking garages. Sixty-six surrounding acres would be filled with office buildings, shops, and restaurants, but no housing. As envisioned by M. Justin Herman, who headed the SFRA from 1959 until his sudden death from heart attack on 30 August 1971, a single group would redevelop the Central Blocks. Once completed, it would lease the public facilities back to the City. Bundling together all of the redevelopment projects was risky, however, because multiple uses and sites were involved. To attract investors—Herman refused to finance the project through general revenue bonds that voters would have to approve—the SFRA promised to remove all residents from the redevelopment area. Grassroots resistance to that policy ultimately led to the construction of Clementina Towers for "elderly," displaced residents.[44]

The SFRA's relocation plan generated a backlash that culminated in legal action on the part of the organization established to defend the interests of residents and businesses targeted by the redevelopment plan, Tenants and Owners in Opposition to Redevelopment (TOOR). A lawsuit that TOOR filed in federal district court in November 1969 set off a process that took almost four years to resolve, not least because of SFRA intransigence while Herman was still alive. Finally, a settlement agreement was reached whereby the SFRA agreed to construct some 400 permanently subsidized low-rent units on four sites within the redevelopment area under the auspices of the San Francisco Housing Authority (SFHA).[45]

The SFHA had shifted its attention to providing housing for seniors in the wake of the Housing and Urban Development Act of 1965. The law had increased loan funding for "elderly" and handicapped housing and authorized turnkey development, whereby local authorities contracted to acquire housing projects from private developers upon their completion. The latter provision was an attempt on the part of Congress to expedite construction of public housing; administrative procedures were delaying projects by as much as two years. Policy initiatives on behalf of seniors dated from 1956, when Congress added special provisions under sections 203 and 207 of the Housing Act to give preference to them and also people displaced by urban renewal in public housing programs. (The residents of Clementina Towers would be both "elderly" *and* displaced.) The Section 202 program, established in the 1959 Housing Act, enabled private developers to participate in moderate-income housing for seniors. Rather than insure loans made by private lenders, as other federal housing programs provided, the Section 202 program made low interest loans directly to developers.[46]

As Gwendolyn Wright argues, providing public housing for seniors was more politically and socially appealing to members of Congress and the public than extending the same resources to poor families. In contrast to the urban poor, who were increasingly frustrated by US housing policy, seniors "were likely to be grateful, docile, unseen." They were also more likely to adhere to middle-class behavioral norms and be white. Citing higher construction costs associated with railings, ramps, and other features that met the needs of residents, Congress offered developers allotments for "elderly" housing that consistently exceeded subsidies for family housing on a unit basis. Hence, housing projects for seniors typically looked more like private than public housing developments. Where Congress led, the SFHA followed. Clementina Towers was one of several apartment towers built in the city for seniors between 1966 and 1974.[47]

As both developer and contractor, Charlie Pankow took the opportunity to plan, design, and construct Clementina Towers as design-build methodology prescribed. He formed Yerba Buena Developers to acquire the site from the SFRA and then contracted to turn over the project to the Housing Authority once CPI completed construction. The building team included architect Peter Rookeley and structural engineer Richard Bradshaw, whose relationship with Charlie Pankow dated from the latter's Kiewit days. Rookeley's firm, John S. Bolles & Associates, was best known for its design of Candlestick Park, which opened in time for the 1960 baseball season. In the early 1960s, the firm ushered in a local trend toward designing public housing to look like contemporary private apartments with its Ping Yuen project. Completed in 1964 in Chinatown, the award-winning family housing complex was noted for a large mural on the façade that faced Broadway. Bolles also designed one of the earliest local projects for seniors, John F. Kennedy Towers, a curved highrise in the Fillmore district, completed in 1966. Concurrently with Rookeley's design of Clementina Towers, Bolles was designing another "elderly" housing project in the Lower Pacific Heights neighborhood. Both projects incorporated design features that addressed residents' concerns for safety. Clementina Towers was tucked between narrow Clementina and Tehama Streets in the middle of the large block bounded by Fifth, Folsom, Fourth, and Mission Streets. Set back from these alleyways, the towers were separated from the sidewalks by an iron fence. A covered walkway linked its two towers. At the same time, Rookeley devoted none of the site to gardens or other outdoor areas typically found in local family housing projects at the time.[48]

Clementina Towers consisted of twin 15-story, reinforced concrete towers, each containing 138 studio and one-bedroom units on typical floors 2–13. The first floor of each tower contained a community center. Penthouses housed

Figure 20. Clementina Towers, San Francisco, completed for the San Francisco Housing Authority in 1971. The contractor deployed two slipforms per tower and staggered the heights of the cranes to make possible the completion of one floor per building every four days. John S. Bolles & Associates, architect; Richard R. Bradshaw, structural engineer. (From *Western Construction* 46 [January 1971]: 33.)

the elevator equipment. Architect, engineer, and contractor collaborated to design the project to accommodate slipform construction of each tower from the footings to the top of the penthouse parapet. Construction began in March 1970 and was completed in June 1971.[49]

Under the direction of project superintendent Harold Henderson, the slipform operation took three months (fig. 20). Pankow modified Rookeley's initial design to maximize the productivity of the operation. For purposes of construction, the contractor divided the towers into two identical parts, with vertical gaps for doorways separating the sections. By deploying two slips per tower, Henderson minimized the workers that he needed for the task. Crews worked in smaller units at a higher rate within normal shifts. They completed one floor every four days (in each tower). Workers cleaned forms on day one of the cycle. Ironworkers set rebar and blockouts in the forms on the second day. On day three, electricians installed conduits and other fittings. Crews poured concrete on the fourth day. To avoid conflicts over the use of the two tower cranes employed on the site, Henderson scheduled cycles so that the pours for each of the four slips occurred on different days. In contrast to the Heede hydraulic jacking system utilized in the construction of Central Towers, workers raised the slipforms with compressed air pressure. Jacks were spaced 5 feet apart on both sides of slipform. They were connected to the compressed air circuit by a system of steel pipes. The carpenter foreman increased air pressure to spring-activate the jacks.[50]

For CPI, Clementina Towers represented a pinnacle of achievement in slipform construction for Mainland projects. On 9 February 1971, a 6.6 magnitude earthquake struck the San Fernando Valley, in Southern California, causing the partial or total collapse of modern, code-conforming buildings, including the Olive View Community Hospital, which had opened just one month earlier. The quake prompted major revisions that were ultimately incorporated into the 1988 UBC, including a 33 percent increase in minimum design load levels, mandatory positive and direct interconnections of building components, mandatory anchorage for structural components, and additional structural detailing. Between 1971 and 1988, municipalities in California modified their building codes in anticipation of these revisions. The codes made it difficult to realize time and cost advantages through slipform operations in seismic zones. CPI would use slipforms to erect the cores of buildings on the Pacific Coast through 1983, but its records for slipform construction would be set in Honolulu, as the next chapter shows.[51]

Applying Design-Build (Still without the Label): 1625 The Alameda Building and Joe West Hall

Charles Pankow's "budgeted construction program" was fully developed by the mid-1960s, as a review of the design and construction of two concrete towers in downtown San José, California shows.

A 9-story speculative office building that CPI constructed during 1965 for owners Demmon-Hunter resulted from pre-construction collaboration among architect, engineer, and contractor. CPI used the project to showcase its ability to control construction costs through the mechanization of work at the building site. As Charlie Pankow wrote in the *Journal of the American Concrete Institute*, it had "not been common" to this point for owners to give designers and contractors flexibility in design criteria and construction techniques. T. Y. Lin, Kulka, Yang & Associates cited it as a case example of how to integrate precast, prestressed concrete elements into the design of high-rise buildings. The structural engineers observed that designers of both steel and concrete buildings were still treating precast exterior wall panels as nonstructural elements when they could use them in the frame. Both engineer and contractor hoped to convince owners and colleagues that they offered better ways of designing and constructing tall concrete commercial structures.[52]

Charlie Pankow likely met developer Derk Hunter while his men were directing the construction of MacArthur Broadway Center. A real estate broker with the firm of Cornish & Carey, Hunter founded Demmon-Hunter in 1960 with partner Roy Demmon. Their early clients included high-technology firms that established homes in Stanford Industrial Park. In keeping with the campus-like and suburban character of the Park, which was affiliated with Stanford University and surrounded by leafy Palo Alto, these were low-rise (no more than 2-story) developments with deep, landscaped setbacks.[53] In contrast, two projects that Hunter and Demmon undertook with Charlie Pankow, 1625 The Alameda Building and the First American Building, were part of the postwar high-rise development of downtown San José.[54]

Hunter was receptive to Charlie Pankow's argument that the owner would benefit from including the contractor in the design of the project. In November 1964, Pankow signed a $1.25 million lump-sum contract with Hunter after he reviewed the preliminary drawings of architect Allan M. Walter. Pankow, Walter, and the structural engineers then agreed on ways to automate and simplify construction. Electrical, plumbing, and mechanical subcontractors developed their designs during planning and submitted them to the architect and engineer for approval. The members of the building team believed that their collaborative approach saved some 15 percent on the estimated construction cost and cut the time required to erect the building by one-third.[55]

CPI used slipforms to construct both the service core and the stair towers. It precast all of the concrete elements on-site. The floor system utilized pre-tensioned, precast lightweight concrete flat slabs supported on pre-tensioned, precast beams, also made of lightweight concrete, with a poured-in-place, 2-inch concrete topping. The slipformed core provided lateral force resistance. Columns cast in place between precast decorative panels made of lightweight concrete carried the loads from the building. The non–load-bearing panels served as side forms for the columns; workers clamped forms for the remaining two sides across the space between them. The cast-in-place columns locked the panels together. Integrating the panels into the construction in this manner provided additional reserve strength. To obviate the need

Figure 21. Construction of the shell of the 1625 The Alameda Building, San José, completed in 1965. The elevator and service cores act as shear walls. Allan M. Walter & Associates, architect; T. Y. Lin, Kukla, Yang & Associates, structural engineer. (Photo Ernest Umemoto. Courtesy of Charles Pankow Builders, Ltd., Pasadena, California.)

to paint the exterior of the structure, the concrete for the exterior wall units was mixed to produce the shade of white that Walter specified (fig. 21).[56]

The building team specified that all of the precast elements be made of lightweight concrete so that a traveling crane—with a four-ton capacity at an extension of 90 feet—could place any piece wherever it was needed. CPI located a 200-foot-long prestressing bed in what would become the parking lot and configured it to manufacture two rows of both beams and slabs at a time. Workers precast the pieces on consecutive days, and used a third day to cure the concrete to its requisite strength. Crews in the casting yard worked normal eight-hour shifts, five days a week. (This schedule also applied to slip-form operations.) To prepare the casting yard, workers laid track for the crane alongside the prestressing bed. They placed three steel forms to cast the exterior wall panels at the far end of the track.[57]

CPI also deployed slipform operations and an on-site casting yard in the construction of Joe West Hall on the campus of San José State College (University). The contractor again collaborated with Allan M. Walter & Associates and T. Y. Lin, Kulka, Yang & Associates. The project was a cluster of two towers—a 12-story residence hall that could accommodate more than 650 students and a 2-story dining facility. State of California officials, however, did not allow Walter and Lin to design the project specifically for slipform construction. Moreover, as a public project subject to bid, Charlie Pankow could neither negotiate a lump-sum contract with the owner nor collaborate with the designers during pre-construction, as he had on 1625 The Alameda. Of 12 respondents to the call for proposals, CPI submitted the low bid of $3,363,000. As project engineer Thomas J. Branson noted in his technical report, the three lowest bidders proposed that the project utilize slipforming, suggesting the savings that might be realized by using the technique to erect tall concrete buildings. The building team also saved money by including only two elevators and not installing air conditioning. By using slipforms to erect all of the walls and columns on the top 10 floors and most of the columns on the first 2 floors, CPI topped out the residence hall by the end of January 1968, eight months after the start of construction (fig. 22). Owing its protruding window frames, the residence hall is often referred to as the "Waffle Tower" (figs. 23, 24).[58]

As was the case half a century earlier, the engineer's impulse to control work at the building site was expressed in the substitution of capital and techniques for labor. In his words, Charlie Pankow configured capital and labor "not only [to reduce] the many inefficiencies of large labor forces [but to assure] a more timely schedule." According to Pankow, the use of smaller crews also simplified the on-site education process and therefore improved productivity and quality. CPI's project teams continued to organize work much as they had done as part of Kiewit. They deployed powerful cranes, built intricate slipforms, and used adjacent or nearby land to precast concrete. They found ways to reduce the number of forms needed to cast elements in place. Reflecting the incorporation of prestressed concrete into the design of tall commercial structures, CPI also devised a prestressing bed for use in its field operations. Several interviewees attested to the affinity that Charlie Pankow had for the carpenters and other building trades workers on his job sites. No doubt that this was true. Pankow frequently visited job sites to meet with his clients and engineers, but also the workers hired out of the union hall (fig. 25). At the same time, he sought to minimize the number of the workers to con-

Figure 22. Carpenters fabricated over ¼ mile of slipforms (measured by face length) for the construction of a residence hall at San José City College. Completed in 1968, the dormitory is named for Joseph Henry West, who retired as Dean of Educational Services in 1964 after a 35-year career with the school. Allan M. Walter & Associates, architect; T. Y. Lin, Kukla, Yang & Associates, structural engineer. (Photo Sherman Brazil. Courtesy of Charles Pankow Builders, Ltd., Pasadena, California.)

trol schedules and budgets and therefore increase the chances that an owner would become a repeat client.[59]

The configuration of work on Pankow job sites differed from building sites described by Amy Slaton that featured a small number of engineer-managers directing the work of numerous unskilled laborers (see Introduction). Ironically, the techniques that Charlie Pankow used to erect concrete structures and his decision to achieve control over the construction process by performing tasks that contractors typically subcontracted meant that he relied on men with the skill and experience to execute concrete work of high quality. Pankow employed men with the ability to fabricate forms and perform the precast and slipform operations associated with their use. His project teams worked with carpenters out of the union hall to execute Charlie Pankow's construction program. To be sure, his employees included men who supervised workers at the building site. But Pankow's top superintendents were men who had spent their careers in the field. Newly minted engineering graduates recruited into the firm would have to rise through the ranks as well, if they hoped to become superintendents.

Figure 23. Three-ton concrete window frames stacked in the on-site casting yard for use in the construction of the residence hall at San José City College. (Photo Sherman Brazil. Courtesy of Charles Pankow Builders, Ltd., Pasadena, California.)

A Repeat Client and a Blow to the Firm: The Borel Development Company and the Birth of Webcor Builders

The Borel Development Company was CPI's most important client in its early years, retaining the contractor to construct a business park on the San Francisco Peninsula and three discrete office buildings in Southern California.[60] The relationship terminated abruptly, however: a traumatic blow to a firm of CPI's size less than a decade into its existence.

William Wilson III, Miller Ream, and Gilbert Bovet organized the Borel Development Company to develop the easternmost portion of the Borel Estate that was divided among the grandchildren of Antoine Borel. Borel was a Swiss immigrant who arrived in America in 1861 at age 21 to work in his brother's private bank in San Francisco. He eventually took over the management of the bank and became a leading financier in the city. His many directorships included William Ralston's Bank of California, the Spring Valley Water Company, the Union Trust Company, and the California Street Cable Railroad Company. During the 1880s, Borel acquired about 100 acres of land in San Mateo County on the San Francisco Peninsula that had been part of the vast Rancho de las Pulgas. In 1961 his estate split the property in two. One group of heirs took the lands that lay west of a line drawn near Nadina Street in the city of San Mateo.

Figure 24. Charlie Pankow and Harold Henderson, his project superintendent, stand in front of the "Waffle Tower." (Photo © Vano Photography.)

The other group received the 19 acres that lay to the east, strategically located along El Camino Real, the primary surface road between San Francisco and San José, and the Nineteenth Avenue Freeway, now State Highway 92.[61]

The business park that Wilson and his partners planned was indicative of the postwar suburban development whose economic impacts redevelopment agencies were striving, and, for the most part, failing to counteract (as reflected in commercial property values, which had been rising in San Mateo and other cities on the Peninsula, while they stagnated in the city's central business district.[62]) The developers sought to attract corporate tenants from downtown San Francisco with office buildings arrayed around a circular drive and set in a landscape that preserved the property's numerous redwood, oak, and Monterey pine trees. The office buildings would offer potential tenants high quality office space at competitive rates, with access to nearby San Francisco International Airport. In developing the southeast corner of the property as a community shopping center, the partners targeted another segment of a rapidly growing market: housewives who might otherwise shop in San Francisco. The building of public infrastructure benefited the developers: The opening of a widened San Mateo-Hayward Bridge in October 1967 increased the value of the property by expanding access to it from the East Bay.[63]

Figure 25. Charlie Pankow visits building site at San José State College. Superintendent Harold Henderson and project engineer Thomas Branson stand to his left. (Photo © Vano Photography.)

William Wilson III met Charlie Pankow through Roy Demmon and Derk Hunter. It is likely that the San José developers heard of the Borel Estate Development's plans to build its first office building and recommended Pankow on the basis of the contractor's work on their 1625 The Alameda Building. Wilson was familiar with concrete work from time that he had spent with a Southern California contractor. He was intrigued by the techniques that Pankow was using to construct concrete buildings and impressed with his ability to fix the price of a project based on preliminary plans.[64]

Wilson negotiated all of the contracts with Russ Osterman after Charlie Pankow developed the initial contact, as would be the case with most of Pankow's Mainland clients. San Francisco architect Jorge de Quesada, whom the developers selected before they identified Pankow as a possible contractor, designed all of the projects. A native of Cuba, de Quesada practiced architecture in Havana for a decade before he left the country in the wake of Castro's revolution. In April 1960, de Quesada arrived in San Francisco. Two years later, he obtained a license to practice architecture in California. He opened a practice in San Francisco with commissions to design St. Thomas More Church in San Francisco and buildings for Wilson and his partners. The office buildings that de Quesada designed for the Borel Estate ranged in height from three to eight stories. De Quesada unified the buildings architecturally, however, with sculptured precast concrete panels and tinted solar windows (figs. 26–28). The first building fronted El Camino Real. The second was set behind it. The third and fourth buildings were constructed at the back of the circle, with space for a small plaza between them. There would be ample space for additional buildings, even with the completion of Borel Square, the shopping center. There was plenty of room, too, for CPI's men to precast panels and other concrete elements, in an area later devoted to parking.[65]

Construction of the first building, hailed for its "striking design and advanced construction," showed how the building team might cooperate to improve the quality of the product delivered to the owner under design-build. Together with the structural engineers at T. Y. Lin, Kulka, Yang & Associates, Pankow modified de Quesada's initial design to speed work flow and increase the stability of the structure. The final design joined precast panels to form an enclosure for cast-in-place columns, eliminating the need for forms and shoring. Integrating the architectural panels into load-resisting walls in this way strengthened the building against both wind and seismic loads (fig. 26). The building team applied this approach to the design of the other Borel office buildings.[66]

Wilson and his partners were able to lease space in their buildings to "blue chip" tenants. Owens-Illinois, a leading manufacturer of glass, plastic, and forest products with 4,500 employees on the Pacific Coast, moved its Glass Container Division's Pacific Region offices from downtown San Francisco into the first building (fig. 26). It signed a long-term lease for about half of the available floor area. Phillips Petroleum moved 100 employees from its area sales office in the Crown Zellerbach Building in downtown San Francisco into the smaller second structure (fig. 27). John Hancock Mutual Life Insurance moved its Pacific Coast executive offices into that same building. Fluor Utah Engineers and Constructors moved some 200 engineers and administrators into the third building (fig. 28).[67]

Rosser Edwards ran the Borel Estate projects for CPI. David Boyd was his project engineer—the person responsible "for reducing the project plans and specifications into finite information that can be executed by the craftsmen, insuring that [resources and materials were] on the job when required [maintaining] quantity and cost records," quality control, and supervision of field engineers. As related in the last chapter, Edwards had joined Pankow not long after he completed the Valley Music Theater project in the

Figure 26. Entrance to the Borel Estate Building, the first building constructed by CPI on the Borel Estate, San Mateo, California. Jorge de Quesada, architect; T. Y. Lin, Kukla, Yang & Associates, structural engineer. Completed in 1966. (Photo Teresa Callen, taken October 2011.)

Figure 27. 66 Bovet Road, the second and smallest office building constructed by CPI on the Borel Estate. Completed in 1969. Jorge de Quesada, architect; T. Y. Lin, Kukla, Yang & Associates, structural engineer. (Photo Teresa Callen, taken October 2011.)

Figure 28. Rear view of the Fluor Building, the third office building constructed by CPI on the Borel Estate. Completed in 1970. The fourth office building, on the right, was sited to create a shared plaza between the two structures. An open-plan parking structure extends behind both buildings and wraps around the south side of the building to the right, forming a buffer between it and State Highway 92. Jorge de Quesada, architect; T. Y. Lin, Kukla, Yang & Associates, structural engineer. (Photo Teresa Callen, taken October 2011.)

summer of 1964. Boyd also joined the company in 1964—in time to replace Lloyd Loetterle as project engineer on the Central Towers job site. In 1970, with additional projects on the drawing board, Wilson approached Edwards and Boyd about starting a construction company. He felt that they "knew the routine," as far as working with Borel Development Company was concerned, and would jump at the opportunity that he presented them. As Wilson put it, "It's not often that one can start up with work in hand." Edwards and Boyd agreed. On 19 January 1971, Wilson, Edwards, and Boyd incorporated Webcor Builders.[68]

Charlie Pankow was devastated. On a personal level, he felt betrayed. Edwards and Boyd informed him of their decision as a *fait accompli*. Pankow expected more loyalty from his men. Indeed, mutual respect and reciprocated loyalty was the glue that had bonded Pankow's men since their Kiewit days. Had Edwards and Boyd approached him beforehand, Lee Sandahl suggests, Pankow may have better understood their situation and perhaps may have encouraged the move. After all, had he not taken his building division with him almost to a man to start his firm? As matters stood, Pankow was losing two of his stars. That Edwards and Boyd were fellow Purdue graduates made the sting all the more painful.[69]

On a business level, Pankow lost a repeat client that had provided his company with more work than any other during the 1960s. Indeed, at the time of the departure of Edwards and Boyd, Wilson and his partners had lined up CPI as contractor for their first project in downtown San Francisco—an office tower at the corner of Howard and Spear Streets.[70] For the young Pankow firm, the loss of Wilson and his partners as a client was potentially life threatening.

In the wake of the departure of Edwards and Boyd, Charlie Pankow changed the way that CPI managed its projects. Superintendents—the "on-site chief executive responsible for the proper execution of the work in accordance with the contract documents"[71]—had been responsible for all costs related to a project, and all accounting had been performed at the project level. Now the company separated costs in the field. Superintendents retained responsibility over costs related to materials, but accountants in the head office took responsibility for costs related to subcontractors. Pankow expected the project manager (who typically worked from Altadena) to know as much about the financial aspects of a job as did the superintendent. Charlie Pankow implemented these changes primarily to make it more difficult for another potential Edwards or Boyd to walk away from the company with a client in hand. Superintendents would no longer have total control over a project and therefore would not be able to put together the "big picture." And, if the superintendent left in any case, he—superintendents across the industry were universally male—would not leave the company in the lurch. For the project manager would be prepared to replace him.[72]

For their part, Wilson, Edwards, and Boyd grew Webcor Builders into a sizable and successful company. In 2009 it ranked thirtieth among general contractors in the *ENR 400*. It employed 800 people and had $1.3 billion in outstanding contract volume: It was almost three times larger than the Pankow firm on both counts. Like Charlie Pankow, the principals of Webcor Builders focused on commercial construction. They delivered projects using design-build, thereby contributing to the diffusion of the methodology. Focused geographically on California, Webcor made its name by building the headquarters of some of Silicon Valley's leading firms, including Adobe, Electronic Arts, Oracle, Siebel Systems, and Sun Microsystems. It also built Symantec's campus in Culver City. In 1994 Webcor merged with A. J. Ball Construction, the eponymous firm headed by Andrew Ball. In 2000 Edwards and Boyd retired, leaving Ball in charge of a firm that had grown its revenues from $60 million in 1994 to more than $600 million in 1999. In 2007, in the context of a deepening recession, Ball sold a controlling interest in the company to Obayashi, a Japanese construction company.[73]

Charles Pankow, Inc., at the End of Its First Decade: Poised for Growth?

CPI spent its early years in the "creativity" phase of the typology developed by Larry E. Greiner in a classic *Harvard Business Review* article on phases of "evolution," punctuated by stages of "revolution," in the growth of organizations.[74] Certainly, as Greiner found to be the case in the firms that he studied, Charlie Pankow and the men who followed him out of the Kiewit building division directed their creative energies toward developing and demonstrating their strengths in the marketplace. Still, the company was not much bigger in 1971 than the Arcadia-based building division had been in 1963. Communication within the firm remained "frequent and informal," as Greiner observed in other young, small organizations. The culture that had developed within the Kiewit building division persisted in the new firm. CPI remained organized only at the project level. The reputation of the company was based on the performance of small teams operating in platoon-like fashion at the building site. Charlie Pankow's men worked long hours at modest salaries, but could buy stock in the company. Charlie Pankow and Russ Osterman relied on their networks in the building community to sell work. Like Peter Kiewit, Charlie Pankow sought to be the best, rather than the biggest, builder of large concrete commercial projects. CPI had demonstrated its bona fides. Whether the company would grow in the manner of the organizations that Greiner observed, however, was open to question.

Charlie Pankow showed no interest in growth, if it meant implementing organization changes that might dilute his absolute authority over decision making, as he made clear at company's first annual meeting. He opened proceedings by announcing the departure of Lloyd Loetterle without explaining why one of company's founders had left. Apparently no one asked him to elaborate, either at the time or in subsequent years—itself a reflection of the trust that Pankow's men had in their leader and their willingness to defer to him. According to George Hutton, Loetterle's exit sheds light on Pankow's leadership style: Charlie Pankow was willing to engage in debate only to a point before he rendered a final, uncontestable decision. However vague, the story has the ring of truth about it. Without singling out Loetterle, Doug Craker states that some of the men who left Kiewit to join CPI disagreed with Pankow on the type of work that the company should undertake. That is, they were not necessarily as wedded to concrete construction as was Pankow. Some of the men—again unidentified—also wanted corporate governance to be more democratic or consensus driven. At the annual meeting, Charlie Pankow slammed the door on either of those possibilities. When queried on how the company would resolve disagreements on the direction of the business, Pankow reportedly replied, "We count stock."[75] Ironically, then, one of the most successful chapters in the company's history was written by its subsidiary in Hawaii, which functioned for two decades for the most part independently of Charlie Pankow.

CHAPTER 3

Pankow in Hawaii, 1965–1984

In May 1967, at the end of Charles Pankow, Inc.'s (CPI's) first project in Hawaii, an office building for the James Campbell Estate, Charlie Pankow flew to Honolulu to meet with George Hutton for a final briefing on what had been a difficult engagement. So the story goes, Hutton asked Pankow what he wanted him to do next. Pankow suggested to Hutton that he remain in Honolulu to see if he could develop additional business. Hutton asked Pankow how he should proceed. Pankow replied only that he had a plane to catch.[1]

Over the next two decades, George Hutton established Hawaii as the firm's most successful region. In fact, from 1970 to 1985, Charles Pankow Associates (CPA), a subsidiary, accounted for the entire increase in CPI's net worth. What was going through Charlie Pankow's mind when he left Hutton on the tarmac is unclear, but the decision seems to have been intuitive. Certainly, there was no business plan to establish a branch in Honolulu. The company would have to pay Hutton's salary and overheads associated with a new office. Since the local construction industry was booming, however, Pankow must have figured that it was worth covering these expenses to see if Hutton could line up additional work.[2]

Both George Hutton and Charlie Pankow enjoyed the fortune of good timing. In May 1967, Hawaii was in the midst of a construction boom. From 1966 to 1970, construction in the state increased in value from $370.9 million to $783.8 million, a level that, in real terms, would remain unsurpassed until the Japanese-fueled boom of the late 1980s. In the context of a chronic shortage of affordable housing, residential construction nearly doubled, to $234.6 million.

New multi-unit residential structures on Oahu, which would fill the bulk of CPA's order book, more than doubled, to $94 million. At the same time, non-residential construction rose in value from $82.4 million to $187.4 million. The robust demand for construction services may have been a factor in Pankow's decision to leave Hutton behind.[3]

Former president Dean Stephan believes that leaving Hutton at the airport with no instructions or objectives was the best thing that Charlie Pankow could have done for him. It reflected Pankow's role as a leader, rather than as a manager, of an effective group. He gave his men plenty of room to work as they saw fit. They either succeeded or were shown the door. Hutton succeeded beyond anyone's expectations, including his own. In doing so, he personified "the underlying thesis of the company," according to Stephan.[4]

Hutton ascribes a large share of this success to the culture that he brought with him to the Islands—the culture that had been forged within the Kiewit building division and transplanted in the Pankow firm. Hutton was perhaps the most enthusiastic disciple of Charlie Pankow's approach to concrete construction. He also had the highest regard for the founder as a leader. In this context, the alienation of their relationship in the 1980s would be especially dismaying and disappointing to him.

The James Campbell Building and the Revitalization of Downtown Honolulu

In the summer of 1965, Charlie Pankow called George Hutton at the MacArthur Broadway job site and instructed him to pick up the plans for a new project at the company's cramped "office" in Crystal Tower. Pankow charged Hutton with analyzing the documents with a view to producing a material takeoff, or MTO—an inventory of the amounts and weights of materials and their types that would be needed to build the design—in support of the company's bid. Only when he unrolled the plans the next morning did Hutton note that the project was located in Honolulu. His first reaction: This is an exercise in futility.[5]

The plans of architect Leo S. Wou called for a combined office and parking structure for the James Campbell Estate to replace the 2-story office block on Fort and Merchant Streets in downtown Honolulu. The Campbell Estate was one of Hawaii's leading developers, with some 70,000 acres of land in its portfolio, more than half of which Campbell had developed as sugarcane fields. The existing structure had served as the Estate's headquarters since it was built early in the twentieth century. (The Estate was established upon the death of James Campbell in 1900.) At one time, it had also housed the offices of the Hudson's Bay Company. As initially announced in October 1964, the project consisted of a 5-story building that would extend along Fort Street between Merchant and Queen Streets. Wou's design provided for parking below grade and on portions of floors two through five. The estimated cost of construction was $1.2 million. With CPI's involvement, Wou revised the design to include a sixth floor and additional parking. The new configuration nearly doubled the rentable office space, to more than 55,000 square feet.[6]

The James Campbell Building was part of the redevelopment of Honolulu's downtown, which Wou was instrumental in planning. In the idiom of urban renewal, downtown business interests concerned with blight and traffic congestion commissioned Wou and planning consultant David Y. C. Tom to prepare a master plan for the central business district. Wou had studied architecture at the University of Pennsylvania, where he received a bachelor's degree in 1950. From 1952 to 1954, Wou did graduate work in architecture and regional planning at Yale University, but he did not complete a degree. For a time he worked with Louis I. Kahn, who was then Chief Critic of Architectural Design and Professor of Architecture. In 1960 he established Leo S. Wou and Associates in Honolulu. Wou and Tom delivered "a flexible plan adaptable to the demands of progress," according to the boosters who commissioned it. They proposed to alleviate traffic congestion with ring roads, tunnels under Honolulu Harbor, and parking garages placed on the edge of the central business district, which would allow major streets to be converted to pedestrian use. Borrowing from noted urban planner Victor Gruen's revitalization plan for Kalamazoo, Michigan, which closed two blocks of its downtown in 1959 to create Burdick Mall, Wou proposed closing Fort Street to vehicular traffic. He allocated space for retail, entertainment, and cultural activities that would attract people throughout the day. His plan also responded to the interests of the financial and business communities in large-scale commercial development. He proposed razing blocks of 2- and 3-story structures built in the early 1900s to make way for modern office buildings. Both the City of Honolulu and the coalition of business interests that comprised the Downtown Improvement Association adopted the master plan as a basis "for a comprehensive evaluation of the central business district and environs which will direct further planning and action."[7]

The catalyst for Honolulu's commercial revival was the Financial Plaza of the Pacific, a cluster of three buildings on the block bounded by Bishop,

King, Fort, and Merchant Streets that boosters hailed as a "little Rockefeller Center." Initiated by Castle & Cooke, the largest company in Hawaii at the time, the project involved Wou as lead architect in a joint venture with Victor Gruen Associates and Bill Curlett, a property manager with Oceanic Properties. Curlett would become one of CPA's most important clients.

The Financial Plaza of the Pacific, which created an architectural and economic backdrop for the James Campbell Building, resulted from Castle & Cooke's search for new office space as the company expanded during the late 1950s. Its directors determined that the company's offices, housed in a 3-story structure at the corner of Merchant and Bishop Streets, were inadequate for the burgeoning enterprise. An imposing edifice when the company built it in 1924, the Castle & Cooke Building was one of several similar structures that Hawaii's so-called Big Five companies and their allied firms had erected near the harbor during the 1920s. Castle & Cooke organized Oceanic Properties as a real estate development subsidiary to explore options for the building, which it considered to be "obsolete" because it was "structured more appropriately as a bank than an office building," as Oceanic vice president Warren G. Haight explained. After it attempted, without success, to sell the building, Oceanic considered redeveloping a larger area that included the headquarters site.[8]

In December 1961, Alfred Boeke, architect, planner, and Oceanic vice president, retained Victor Gruen Associates to study Honolulu's central business district and make recommendations regarding its redevelopment. Gruen's reputation as an urban planner, built on the accolades that he had received for his proposals to revitalize the central business districts of Fort Worth, Texas, Rochester, New York, and the aforementioned Kalamazoo, Michigan, was reaching its peak. Numerous cities across the country were asking Gruen to plan the resurrections of their central business districts. (In 1966 the City of Honolulu would retain Gruen's services.) The Viennese-born Gruen, a savvy entrepreneur and marketer, never turned down an opportunity to remake one of America's downtowns, using the principles that he had perfected for the suburban shopping center. After Boeke guided him on a tour of the area around the Castle & Cooke Building, Gruen produced a four-block redevelopment plan that included transforming Fort Street into a pedestrian mall—a standard feature of Gruen's downtown plans. Oceanic also determined that there was substantial pent-up demand for Class A office space in the central business district.[9]

The delivery of Gruen's findings coincided with the publication by the Downtown Improvement Association of Wou's master plan. Gruen's team reviewed the plan and then met with Wou to discuss their ideas for Castle & Cooke in the context of Wou's plan. At the end of the meeting, Gruen's planners invited Wou to become their joint venture partner in the design of the project that materialized as the Financial Plaza of the Pacific.[10]

Because of the complexities of ownership, Oceanic reduced the project to the block that included Castle & Cooke's offices. The six owners of this property began to plan what would become the Financial Plaza of the Pacific. In addition to Castle & Cooke, they ultimately included the Bank of Hawaii, two savings and loan associations, and two family estates. The parties agreed to pursue the project as an office condominium, which enabled them to secure financing on favorable terms and therefore construct a project that was almost 40 percent larger than if they had proceeded individually. The $22 million project was the largest of its kind in America to date. Announced in 1964, it consisted of high-rises for Castle & Cooke, the American Savings & Loan Association, and the Bank of Hawaii. Each structure made a distinctive architectural statement. Wou integrated the buildings visually with exteriors executed in precast concrete etched with a permanent acid finish, tinted bronze glass, and dark bronze anodized aluminum; a terraced brick plaza; and a second-story walkway. The Financial Plaza of the Pacific offered 390,000 square feet of usable floor area, but the three buildings occupied just over half of the site. Victor Gruen Associates designed the open spaces, including a fountain plaza and an elevated promenade.[11]

"There was no real developer per se" under the condominium arrangement, according to Curlett, who joined Oceanic from Los Angeles–based Janss Corporation in 1964 to become project manager, succeeding Haight (who had replaced Boeke a year earlier).[12] Rather, each party was represented on a board of directors, which met monthly. Oceanic Properties managed the project under the direction of the board. Haight and Curlett retained the architects, engineers, and contractors, and participated in meetings with them on plans, specifications, budgets, and schedules. Curlett supervised construction, ensuring that contractors adhered to designs, stayed on schedule, and controlled costs. The process prepared Curlett, who had extensive experience in construction management and finance, to work with George Hutton in a design-build setting.[13]

Notwithstanding its reduced scope, the project, together with Wou's master plan, spurred other business owners to redevelop their downtown parcels. Alan S. Davis, chairman of the trustees for the James Campbell Estate, acknowledged the link between his project and the Financial Plaza of the Pacific, located across the street. As a member of the Downtown Improvement Association, the Estate had sponsored Wou's master planning exercise. It saw the James Campbell Building as an important part of the revitalization effort. Enthused Davis: "Our new building, together with the Financial Plaza and other new and remodeled projects, are all evidence of downtown vitality."[14]

The Estate selected T. Y. Lin, Kukla, Yang & Associates as structural engineers. Through Y. C. Yang and Felix Kukla, both of whom he now knew well, Charlie Pankow became acquainted with Wou. The latter's belief that architects "cannot remain in the traditional role of design alone but must be initial members of a development team" neatly dovetailed with Pankow's design-build approach. Wou would design several multi-unit residential projects that the Pankow firm would construct in Hawaii and the San Francisco Bay Area. The relationship between architect and contractor, however, was often contentious.[15]

The James Campbell Estate Building: George Hutton's Most Difficult Project

In September 1965, the month in which MacArthur Broadway Center held its grand opening, Charlie Pankow and Russ Osterman flew to Honolulu for another opening—that of the bids for the James Campbell Building. CPI was the third highest bidder. No bid met the owner's budget, however: a result that supported one of the main arguments of design-bid-build's critics. The outcome opened the door for Pankow to approach owner and architect about redesigning the project and negotiate a contract based on a guaranteed construction cost. After holding talks with Pankow and Osterman, Wou and representatives of the Campbell Estate, including property manager Wade McVay and assistant property manager Oswald Stender, accepted Pankow's offer to "salvage" the job.[16]

The process by which the company secured the project suggests the extent to which design-bid-build remained standard practice in commercial construction. Charlie Pankow's response to the call for bids also suggests that he was a practical businessman, pursuing work where he could get it. He would educate the building community on the merits of design-build one owner at a time, showing how commercial projects that could not be built under the traditional approach could be resurrected with the contractor's participation in their redesign. Diffusion of design-build would be slow, and would not gain traction until the 1980s.

Hutton and Yang spent the next two months in Honolulu, redesigning the project. As was the case with MacArthur Broadway Center, the new proposal substituted precast structural panels for cast-in-place elements. It also incorporated a slipformed service core, poured-in-place post-tensioned beams, and precast, hollow concrete core deck planks with a lightweight concrete topping. Given the lead time required to deliver it, Hutton ordered the tower crane to be shipped from California before Charlie Pankow signed the construction contract. Hutton moved his family to Hawaii on the first day of December. In early February 1966, the City of Honolulu issued a permit for the $2.23 million structure and construction began.[17]

For George Hutton, the job was the most difficult of his career. Water intrusion posed a much larger problem than anyone anticipated, despite the building team knowing beforehand that the excavated site lay below sea level and therefore near the water table (fig. 29). Yet, according to Hutton, most of the challenges related to the lack of a skilled labor force with experience in high-rise construction. It was a lament of local builders generally. From 1960 to 1966, contractors erected 120 multi-story structures on Oahu, only eighteen of which were commercial buildings. Now Honolulu was in the midst of a boom, with almost three dozen high-rise projects underway or planned to break ground before the end of the year. While all of this activity may have boded well for the Pankow firm's future prospects in Hawaii, it meant that labor was at a premium. As a result, carpenters supplied by the union hall typically had worked only on the "single-wall" houses that were common throughout the city. With neither the resources nor the time to train carpenters to build slipforms or perform other tasks related to mechanized concrete construction, Hutton tapped men with whom he had worked at Kiewit to be his crane operator, general foreman, carpenter foreman, and cement mason foreman. Owing to cultural differences, a lack of affordable housing, and other, personal reasons, not all of these recruits made the transition to life on the Islands successfully, requiring Hutton to replace them. For instance, general foreman Harvey Vocke had a rough style that included a generous use of foul language. One afternoon, Hutton had to intervene to prevent his entire carpenter crew from walking off the job. In addition to his supervisory role, Hutton was responsible for detailing and, after workers manufactured them, inspecting the 2,700 precast elements that the design prescribed, as he had

Figure 29. Excavating the James Campbell Building site, downtown Honolulu, 11 March 1966. Looking makai (toward the ocean), with Aloha Tower in the background. (Reproduced by permission of Williams Photography.)

no project engineer. And he had to work with architect Wou, a relationship that Hutton characterizes as "tumultuous." From Hutton's perspective, one source of tension was Wou's proclivity to insert himself directly into issues at the building site rather than work through Hutton's supervisors.[18]

Several months into the project, Charlie Pankow and Ralph Tice, who was now the company's general superintendent, or operations manager, visited the site (fig. 30). It was clear from their discussions with Hutton that the company needed to assign additional experienced people to the job. The project was on course to lose money, but Pankow did not want it to fail. After all, it was the company's first effort in Hawaii. Upon his return to Altadena, Pankow sent Tony Giron and Bob Crawford to Hawaii as field superintendent and general laborer foreman, respectively. (Field superintendents supervised so-called direct work, selected and scheduled equipment, and coordinated subcontracted work.[19]) Crawford was a Mississippi native whom Giron had hired to work on the Las Flores project at Camp Pendleton. He would soon acquire the nickname "Big Daddy," more so for his ability to lead other foreman and the building trades workers than for his towering profile (fig. 31). The author of a company newsletter article called Crawford a "man of action

Figure 30. James Campbell Building site, 1 July 1966. Looking makua (toward the mountains). Note the shoring of the foundation with precast concrete panels. (Reproduced by permission of Williams Photography.)

[who] makes things happen with innovation and efficiency." Giron also recruited the crane operator who had worked on the First and C Building to replace the man Hutton had hired. With these reinforcements in place, Pankow completed the James Campbell Building in May 1967 (fig. 32).[20]

Pankow lost perhaps $100,000 on the project—the only project to lose money during Hutton's tenure as head of the Hawaii office. As an organization, however, CPI benefited by sending its best people to support Hutton and minimize the financial losses that the project incurred. According to Doug Craker, who was CPI's assistant controller at the time: "Because of the way [the project] was handled in the learning process [*sic*] and what happened from that point on, it became a big success story for several years."[21] It was an exercise in reinforcing a culture that valued doing whatever it took to complete a project to the satisfaction of the owner. Charlie Pankow appreciated the difficulties that the project had posed and did not blame Hutton for the problems that the latter had encountered in completing it. The company as a whole had bit off more than it could chew, as it were. Hutton had been a competent, hardworking, and loyal member of Charlie Pankow's building team for five years. Pankow reciprocated that loyalty and service.

Figure 31. "Big Daddy" Bob Crawford on an unidentified building site.
(From *CPI News* 1 [Fall 1983]: 2.)

Rather than demote or take other action against Hutton, Pankow gave him
the chance to establish the company in Hawaii. Within several years, Pankow in Hawaii would be a self-sustaining and, for all intents and purposes,
independent operation.

George Hutton Launches the Pankow Firm in Hawaii

After completing the James Campbell Building, George Hutton rented a small
office on the second floor of the Hawaii Education Building at 1649 Kalakaua
Avenue. It had room enough for two people: Hutton and his part-time assistant. Over the next several months, Hutton worked to secure a second project while the company's men who stayed in Hawaii lined up work with other
contractors, such as Hawaiian Dredging and Construction, the Dillingham
Corporation subsidiary that headed the joint venture constructing the Financial Plaza of the Pacific. Recalls Hutton: "I just started knocking on doors, in a
way, call up an architect if I'd try to find somebody that knew an architect or
something like that, and I'd go try to meet with them somewhere." Since his
only means of private transportation was a "beat up" pickup truck, Hutton
arranged to meet potential clients at restaurants or other locations to avoid
having to drive anyone to meetings.[22]

Figure 32. The James Campbell Building, completed in 1967. Leo S. Wou, architect; T. Y. Lin, Kukla, Yang & Associates, structural engineer. (Reproduced by permission of Williams Photography.)

For the most part, Hutton was left to his own devices. Charlie Pankow traveled to Hawaii every two or three months, spending a couple of days per visit. On these occasions and in telephone conversations, Pankow might issue "marching orders," as he did to all of his lieutenants, on how to conduct business according to the criteria of his business model. It was almost impossible to follow these instructions—always delivered verbally and often without explanation of the reasons underlying them—and have any chance of success. As Rik Kunnath says, "There would be no way you could simultaneously meet all of [Charlie Pankow's] preferences for the way you would go about getting work or doing this or doing that." Adhering to the requirements that Pankow set for projects—in terms of profit margins, for instance— would mean abandoning many prospects. Kunnath and Hutton concur that Pankow issued his "marching orders" as general guidelines to be followed, or not, depending on the circumstances. Above all, Charlie Pankow expected those who were in the position to do so to sell work. Doing so required a fair measure of courage on the part of subordinates to act independently and with confidence exercise their creativity, judgment, and wisdom. Hutton operated within the framework that Pankow had established for the business, but in a permissive environment that allowed him to make decisions, short of signing the contract with the owner.[23]

The independence that Hutton enjoyed owed in large part to the physical distance between Hawaii and California and the lack of communications technology to overcome it. Before the introduction of fax machines, e-mail, and the Internet, long-distance telephony was the primary means of communications. It was prohibitively expensive, ruling out routine communications. Doing business in a timely manner required that decisions be made at the local level. Of course, relations with suppliers and subcontractors were conducted wholly at the local level as well. And the Altadena office was unable to offer administrative support, owing not least to the lack of computers to facilitate the transfer of data and consolidation of accounts. Early on, staff in Hawaii used an Olivetti adding machine to estimate jobs. The Hawaii office developed and maintained its own accounting and payroll systems. As a result, by the mid-1970s, the Hawaii office was operating for all intents and purposes as an independent company.[24]

During his visits to Honolulu, Charlie Pankow introduced Hutton to the handful of local contacts that had grown out of his Bay Area relationships. It was Hutton's responsibility, of course, to develop these leads into clients. Nevertheless, it was through these contacts that Hutton landed many of his early projects. The first contact to result in a construction contract was K. Tim Yee, general manager of Hawaii Kai, Henry J. Kaiser's residential city that was being built in the shadow of Koko Head, 10 miles east of the urban limits of Honolulu.[25]

Henry J. Kaiser's Hawaii Kai: One of Charlie Pankow's Local Contacts Bears Fruit

In the last decade and a half of his life, Henry J. Kaiser focused his energies on the promotion of tourism and development in Hawaii. The noted industrialist, who graduated from road and dam building to shipbuilding and aluminum, automobile, and steel manufacturing, turned a holiday on Waikiki Beach in 1954 into a capstone career. Kaiser saw an investment opportunity in the scarcity of hotel accommodation that he encountered on his vacation, and stayed on to exploit it. He was convinced of Hawaii's potential as a tourist destination for middle-class residents on the Mainland, who were becoming more numerous and prosperous in the postwar years. A year later, he bought the Niumalu Hotel on the west end of Waikiki Beach for $1,262,500. Over the next six years, Kaiser and partner Fritz Burns, a successful developer with whom Kaiser had built homes for middle- and lower-middle-class markets on the Pacific Coast after World War II, turned a collection of 1- and 2-story cottages frequented by US servicemen into Hawaiian Village. After building a lagoon, Kaiser and Burns renovated the cottages into beach huts and constructed a 13-story hotel and Diamond Head Towers, a complex comprised of twin, 17-story structures. All told, Hawaiian Village offered 1,000 rooms to visitors. In 1961 Kaiser sold his half interest in the project to Conrad Hilton for $21.5 million. The sale provided capital that Kaiser invested in the development of Hawaii Kai. He envisioned a city of 75,000 residents, most of whom he planned to lure from the Mainland.[26]

Kaiser had thought about a Hawaii Kai–like development for almost a decade before the project materialized. When Kaiser formed Kaiser Community Homes with Burns in 1945, he saw Honolulu as a "likely spot" for one or more of the partnership's projects. After his vacation in Oahu, Kaiser broached the idea with Burns, but the partners did not pursue it. Five years later, Kaiser's publicist built a home along the west slope of Koko Head. Kaiser became enamored of the location after visiting the site. Soon thereafter he approached the Bishop Estate to obtain a site to develop.[27]

Established in 1884, the Estate was the legacy of Bernice Pauahi Bishop, the last of the Kamehameha ruling dynasty. It owned 578 square miles of land. Most of it lay undeveloped. At the time that Kaiser became interested in developing its holdings, the Estate was generating annual revenues of only $3 million on assets worth at least $250 million. Kaiser was interested in a seven-acre parcel and found the Estate's trustees eager to lease it to him. Indeed, upon concluding the deal, the latter lobbied Kaiser to expand his project. A receptive Kaiser soon reached an understanding with the trustees on the development of Hawaii Kai.[28]

In October 1959, Kaiser formed Kaiser Hawaii Kai Development Company and began dredging Kuapa Pond to construct a marina. In April 1961, Kaiser Hawaii Kai signed a development agreement with the Bishop Estate that covered 6,000 acres. Henry Kaiser hired K. Tim Yee to be his general manager. Yee, who had worked for the City and County of Honolulu and the FHA, supervised construction of a sewer system. He also supervised the expansion of the narrow two-lane road that provided the only land access to

Hawaii Kai into the paved, four-lane Kalanianole Highway. Kaiser paid for the construction of this infrastructure out of pocket.[29]

Initial sales of homes in Hawaii Kai were disappointing, reflecting the community's remote location. Aware of his mortality and therefore all the more anxious to push the project forward, Kaiser turned to outside developers to accelerate construction. By mid-1964, 28 developers were involved in Hawaii Kai. Residential construction tripled during 1964–1965. By the end of 1965, some 1,000 homes had been sold. At the time of Henry Kaiser's death, in 1967, the development of Hawaii Kai was far from complete, but many of the remaining planned housing units were under construction and the requisite infrastructure was in place to accommodate additional housing and recreational and commercial facilities.[30]

To meet the consumption needs of residents, Kaiser Hawaii Kai constructed a community shopping center at the marina. As completed in 1963, the Koko Kai Shopping Center consisted of 17 stores and parking for 178 vehicles. Plans for a second phase of construction, which would expand the center to include an office building, a bank, restaurants, three dozen or more shops, and additional parking, were on the drawing board at the time of Kaiser's death.[31]

As a result of Hutton's development of his relationship with Yee, CPA built the second phase of the shopping center, which became known as Koko Marina Trade Center. In September 1967, Hutton met with architect Dave Stringer and consultant Bill Bone, both of whom Kaiser Hawaii Kai had retained for the expansion project. The three men began discussing a 2-story office building. The project was not delivered through a design-build approach—reflecting the need of the fledging firm to accept work as it came—but CPA controlled its structural design. Hutton estimated that he could meet Yee's $275,000 budget and, on that basis, construction began in June 1968, one year after Hutton hit the pavement in search of work. Hutton recalled Bob Crawford from Hawaiian Dredging and hired a carpenter foreman, Doug Nishida. Together with Hutton, Crawford and Nishida supervised the laborers and carpenters. Pankow completed the building in six months. As he would not hire an operations manager until 1972, Hutton made a practice of visiting the job every day before turning his attention to selling work.[32]

Once workers completed the building, Hutton negotiated a $644,000 contract for the construction of a second 2-story building on the site for the Bank of Hawaii. George Hogan and Donald Chapman, designers of the landmark American Savings Bank Building on Kapiolani Avenue, across from the

Ala Moana Mall, were the project's architects. Popularly known as the Pan Am Building because of the former tenant's large, illuminated blue sign on the building, the 17-story tower was completed just before construction of the Hawaii Kai project got under way. The project established the local reputation of Hogan and Chapman as designers of high-rise office buildings. The Hawaii Kai project was a rather more modest commission. Work on the 15,500-square-foot building began in October 1969 and finished in mid-1970. About this time, CPA began work on the expansion of the shopping center, which it completed in 1971.[33]

The Last of Charlie Pankow's Contacts: CPA's First Condominium and Regional Shopping Center

Charlie Pankow's professional acquaintances with Louis W. Riggs and Stanley H. Froid led to another opportunity for Hutton. Riggs and Froid were president and vice-president, respectively, of Tudor Engineering Company. Ralph A. Tudor, an engineer who was renowned nationally for his contribution to the design and construction of the San Francisco-Oakland Bay Bridge, founded the heavy construction firm in San Francisco in 1950. Tudor cut its teeth on bridges, highways, and dams. In the early 1960s, the company formed a joint venture with Bechtel and Parsons-Brinkerhoff to build the San Francisco Bay Area Rapid Transit (BART) system. In 1967 the Hawaii Department of Transportation hired Tudor Engineering to advise the agency on harbor development, maintenance, operation, and management. To accommodate freight operations, which were increasing even as airline travel cut into the volume of visitors who arrived by ship, the State of Hawaii had acquired 85 acres for waterfront expansion and now planned to build a deepwater harbor. The firm also landed construction contracts to extend the Lunalilo Freeway (Interstate H-1) from Barbers Point to Koko Head, complete an 8.5-mile section of the highway through central Honolulu, and build twin tunnels as part of an upgrade of the Pali Highway that linked Honolulu and Kailua. For reasons that are not entirely clear, given its focus on heavy construction, the company also entered into a 75-year leasehold arrangement for a parcel on Ohua Street in Waikiki to construct a high-rise condominium. In September 1968, Riggs and Froid traveled to Honolulu to open a branch office to handle this project.[34]

Hawaii was not the first state to adopt a "horizontal property regime," or condominium law, but it became a leader in its use. Used for centuries in Europe and South America, condominiums did not become part of the housing stock in America until after World War II. Based on individual ownership of a single unit within a multi-unit structure, with common ownership of hallways, elevators, recreational facilities, and other public areas, the condominium idea constituted an improvement over the cooperative—the only type of community-owned housing practically available to the general population in America. That ownership could not be mortgaged was the cooperative's primary disadvantage, precluding all but the wealthiest individuals from buying shares in them. Under condominium law, owners hold units in fee and therefore can buy, sell, and mortgage them. Condominiums, especially the high-rise variety (five or more stories), would account for more than 40 percent of the projects that the Pankow firm completed in Hawaii under George Hutton's leadership. Heavily weighted toward multi-unit residential projects, CPA's order book stood apart from its Mainland counterpart's.[35]

Hawaii passed its condominium law on 10 July 1961. From 1962 through 1971, the year in which the Tudor Engineering completed its condominium, developers added 19,638 units to the state's inventory. High-rise structures accounted for just over 70 percent of the total units constructed. Some 80 percent of the units were constructed on Oahu. Here condominiums comprised the largest share of multi-story buildings in terms of value—more than 40 percent. And most of these units were classified as luxury apartments in terms of price and size. Not surprisingly, projects in Waikiki—which, along with hotels, were rapidly urbanizing the district—commanded the highest prices. This was the case with 250 Ohua Street—the address and name of the project that Hutton's men built for Riggs and Froid.[36]

A look at the project's financial terms shows why owners of projects under construction are under pressure to sell units, in the case of a condominium, or lease space, in the case of offices, stores, and apartments. In June 1969, Riggs and Froid secured a permit to build the 94-unit structure. The company initially estimated the cost of constructing the building, developing the land, and selling the units at about $4 million, but increased the figure to $4.7 million by the end of 1970. Under the terms of its construction loan, Tudor Engineering could borrow up to $3.6 million, of which $500,000 was due on 1 March 1971, with the balance due on 1 September 1972. Meanwhile, the company paid 10 percent interest on the loan's monthly payment. It sold units as CPA completed them. As it sold units, the company reduced the amount of the construction loan, either by the net proceeds of the sale or by 85 percent of the appraised value of the unit as of August 1970, which was $65,000. CPA completed construction at the end of October 1971, at which time more than half of the units remained on the market. In the next eleven months, sales staff sold more than 30 units, yet 15 units remained unsold as the construction loan came due. Tudor Engineering refinanced them and used the proceeds to pay off the balance of the loan. In a favorable market, the company soon sold the remaining units. Speculative owners would not always operate under such conditions in the boom and bust condominium market in Hawaii.[37]

Another of Charlie Pankow's contacts who became a CPA client was T. Yokono and Associates. Hutton was introduced to "Yoko" during the construction of the James Campbell Building. At that time, the two men discussed a shopping center in Waipahu, a small city located on Pearl Harbor. The project failed to materialize. In 1968, however, Hutton negotiated a contract to build a shopping center in Hilo, on the so-called Big Island. Designed by Leo Wou, the Hilo Mall included 30 stores and parking for 1,000 vehicles. It also included a J. C. Penney department store, which Wou also designed. It was the first regional shopping center located on the so-called Neighbor Islands. Because of its location and Hutton's limited staff, CPA managed construction, subcontracting work that the company usually performed. Hutton did not supply the project manager. Rather, Don Kimball, who had worked for Eichler Homes before joining Pankow in 1965, relocated from San Francisco for the project. Hutton followed up this engagement with an agreement to build a 2-story retail and office building for Yoko in Honolulu, which CPA completed in 1971.[38]

Building a self-sustaining business would require Hutton to do more than develop leads that originated in California, however. Charlie Pankow may have known many people with an interest in commercial real estate development, but collectively they did not do enough business in Hawaii to enable Hutton to compete locally on a long-term basis. As Hutton puts it, "You can only carry that communication so far. Then it has to start feeding itself. It has to start generating on its own."[39]

The Victoria Ward Estate: George Hutton Develops His First Project

George Hutton's efforts to sell work on his afternoons away from building sites paid off when he convinced architect Donald W. Cutting that CPA should build his office complex in Honolulu's Kakaako district for Cutting's client, Victoria Ward, Ltd., the development entity of the Victoria Ward Estate.

The Estate owned 65 acres of land in the district, which encompassed some 600 acres between downtown and the Ala Moana Canal, which bordered the Waikiki district, making it the second largest private landowner in the Kakaako district, after Kamehameha Schools. The Estate was the legacy of Victoria Ward, wife of Curtis Perry Ward, who was the proprietor of a livery and dray business. The Wards acquired the land in 1875; at one time they owned more than 100 acres in the area. Over time, Kakaako developed as a light manufacturing district, but the Estate remained interested in developing its holdings for commercial and retail uses.[40]

In 1967 Edward C. Hustace, an heir of the Estate and property manager for Victoria Ward, Ltd., retained Cutting and his partners, Steven W. G. Au and Roger S. Smith, to design Ward Plaza at the intersection of Ward Avenue and Ala Moana Boulevard. It was the firm's first commission. As Hustace explained, the Ward Estate "wanted to have a first-class modern office building while maintaining an esthetically [*sic*] pleasing low-rise atmosphere."[41]

Cutting and his colleagues designed the speculative project as a four-level complex, consisting of three units, within a single structure, offering 43,000 square feet of office space on three of the levels. A parking level located partially below grade provided space for 140 vehicles. Street-level surface parking offered another 23 stalls. The architects arranged the units around large courtyards and used walkways to connect their upper levels. Offices opened onto either courtyards or walkways. As no office faced another, the architects could specify floor-to-ceiling plate glass for their fronts without compromising the privacy of tenants. The local American Institute of Architects (AIA) chapter recognized the design as Hawaii's best for 1970.[42]

Once Hutton convinced Cutting to involve CPA as the contractor, the architect persuaded Hustace to allow Hutton to submit a bid. Hutton prepared an estimate of $1,415,000 to execute the project using post-tensioned concrete panels and beams precast on-site. The figure fell within the Estate's budget. In November 1967, Hustace and Hutton signed a construction contract, five months after Cutting presented his first project model. CPA began construction early in 1968 and completed the project a year later. Principal initial tenants included the Bank of Hawaii, Prudential Life Insurance, and W. S. Myers Advertising Agency.[43]

As had been the case with the Hawaii Kai office building, Hutton supervised the project from early in the morning until just after lunch, and then turned to selling work. During construction of Ward Plaza, Hutton secured the contract to build the Hilo Mall.[44]

Hutton also found time to locate a new office for Pankow in Hawaii, in a Ward Estate–owned office building on Auahi Street. Its Kakaako district location made it more convenient for Hutton to see potential clients and for the latter to visit him, for it offered office space that was more in keeping with the identity of the firm that Hutton wanted to project.[45]

Meeting the Demand for Subsidized Housing

CPA completed almost four dozen projects under George Hutton (see appendix B). By 1971, the Hawaii office was generating most of its own work. Hutton might follow up a lead that Charlie Pankow sent to him, but he no longer relied on the latter's contacts to sustain his business. Four public housing projects on Oahu played an integral role in CPA's transition from takeoff to self-sustained growth.

In the late 1960s, Honolulu was struggling to supply housing for its low- and moderate-income families and seniors. The city, which never "at any time" in its history offered adequate housing for these groups, according to William G. Among, the state's social services director, was experiencing the worst housing problem in America, owing to in-migration from the Mainland and Pacific Basin; the highest construction costs in America, which were outpacing gains in income; the scarcity (and therefore prohibitively high cost) of land; "inadequate and unrealistic" federal cost guidelines; and the focus of developers on the lucrative hotel and resort sectors.[46] As much as 75 percent of the population was excluded from the housing market, according to Lieutenant Governor Thomas P. Gill, whose Interim Committee on Housing investigated the state's housing costs.[47] As State Senator Vincent Yano put it, the housing situation at the middle and low ends of the market constituted "an urgent and monstrous problem."[48]

A glance at the numbers suggests that residential construction kept pace with the increase in Honolulu's population growth. During the 1960s, Honolulu grew by 26 percent. As of 1970, its population stood at 630,528. Overall, Hawaii was home to 769,913 residents. From 1961 to 1970, annual residential construction authorizations in the state almost tripled, from $79.3 million to $234.6 million. Honolulu accounted for more than 80 percent of these authorizations in dollar terms. From 1961 to 1969, private builders constructed an average of 9,527 housing units annually. The state's housing inventory reached 209,598 units.[49]

Yet the average price of a single-family home, townhouse, or condominium exceeded the capacity of most people to service a mortgage on the unit. At $35,054, Hawaii's median price for an owner-occupied single- or multi-family housing unit in 1970 was unmatched nationally. In the estimation of Yoshio Yanagwa, the director of state's housing authority, builders needed to construct 1,900 low-rent units annually just to keep pace with growing demand in that segment of the market. The problem was most acute in Honolulu.[50]

A number of factors contributed to the high cost of construction in Hawaii. Concrete was 46 percent more expensive in Honolulu than in San Francisco. The demand for shipping military equipment and supplies associated with the Vietnam War was crowding out private shipping, doubling the time required to deliver materials from Pacific Coast ports. The aforementioned shortage of skilled labor also increased costs to owners. Developers' profits from housing projects were the subject of debate among officials, developers, builders, architects, and other interested parties, but there was little doubt that it was more profitable for developers to cater to higher-income residents and tourists than to build low-income housing.[51]

Responses to the growth in tourism diverted capital from low- and moderate-income housing because hotel and resort developments (as was the case with luxury high-rise condominiums) provided far greater profits. From 1961 to 1970, Hawaii enjoyed an almost sixfold annual increase in overnight visitors, from 296,517 to 1,746,970. To accommodate growth in the sector, which boosters claimed was the linchpin of the state's economy, planners authorized a massive increase in hotel construction during the decade. The focus of developers' attention on the tourism industry was, Lieutenant Governor Gill explained, "forcing up the price of construction for everything else, from the prospective home buyer to the government."[52]

Hotel and resort development generated employment for lower-income wage earners, who needed such housing. Columnist Thomas H. Creighton cogently noted, "the largest private industries in the State are low-paying ones." Overall, one-third of civilian jobs in the state directly or indirectly depended on the hospitality sector. Lieutenant Governor Gill wanted developers "who create a need for jobs in areas where they have inadequate housing for their prospective employees" to supply, or arrange for, housing for their employees as part of their development projects.[53]

Ever since it had authorized the FHA to subsidize multi-family housing through Section 207 of the National Housing Act, as amended in 1937,

Congress had expanded the scope of the FHA in meeting the demand for it. Since World War II, it had authorized the agency to participate in cooperative and condominium housing, housing for low-to-moderate-income families, housing for seniors, and urban renewal projects (see chapter 2). From 1961 to 1973, the number of subsidized housing units grew from 460,000 to more than 1.5 million nationally. Ironically, subsidized housing production peaked just as Honolulu's housing problem was reaching crisis proportions. From 1968 to 1973, federally subsidized housing accounted for more than 16 percent of housing starts in America.[54]

The local high cost of housing inhibited the federal government's role in expanding low-income and "elderly" housing on the Islands. As Housing Authority Director Yanagwa told state legislators, housing projects in Hawaii were not being built because FHA regulations set national limits on their development costs. For instance, a 40-unit public housing project proposed for Honokaa, on the "Big Island," received a low bid of $887,000. Under federal guidelines, the project's cost was limited to $514,000. Similarly, "elderly" housing projects planned by the nonprofit Hawaii Council for Housing Action (HCHA) cost $16,500 or more per unit. Yet federal officials capped per unit costs at $15,000. As a result, a 110-unit project proposed for Maui received no bids. As Director Yanagwa complained, "Officials in Washington don't understand the costs of construction here."[55]

There was certainly no lack of ideas on lowering housing costs. Lieutenant Governor Gill's report offered a litany of recommendations, including mobilizing a never-used state law, passed in 1961, to involve the state in condemning and developing public lands; requiring developers to donate lands for parks or nonprofit housing; developing training systems for the building trades; establishing a clearinghouse for gathering and disseminating housing data; and eliminating noncompetitive building industry practices. In a speech to the Hawaii Home Builders Association, Gill called for what now might be called "smart" development, combining "row or multiple unit dwellings" with more open space and common recreation areas. Honolulu Mayor Frank Fasi joined Gill, Honolulu Planning Commissioner Arthur A. Rutledge, and other public officials in calling for more coordination among public agencies, and between municipal government and private developers. State Senator Yano sought to increase state support for low-to-moderate-income housing and introduced Senate Bill 262 to achieve it through a $40 million bond issue. Richard Neill, executive director of the nonprofit HCHA, recommended direct government subsidies, property tax exemptions, mortgage extensions, and streamlined bureaucratic procedures to reduce financing costs, and research into better techniques to lower construction costs.[56]

Implementing these recommendations would take time and considerable resources. In the HCHA, George Hutton identified an organization that was interested in the construction program that he had to offer as a means of providing housing for low- and moderate-income families and seniors within federal guidelines.

The HCHA was founded in 1966 by religious, business, and labor leaders to promote and assist the development of housing for "gap group" families and seniors who did not qualify for public housing, but who also could not afford market-rate mortgages or rents. HCHA staff assisted in project planning and development. It set out to realize the construction of as many as 1,000 low-market rental units in three years. Its long-term goal was to develop as many as 7,000 units. Forty percent of these units would be set aside for seniors. The HCHA sought financing under Sections 221(d)(3) (see chapter 2) and 231 of the National Housing Act. Through Section 231, the FHA insured mortgages associated with new or rehabilitated rental projects of eight or more units designed for seniors or disabled individuals. For the reasons already cited, the HCHA did not enjoy much success in financing projects through these programs.[57]

The public face of the nonprofit organization was the aforementioned Richard Neill. A native of Rochester, New York, Neill set out to work on Wall Street after earning a degree in business administration from the University of Rochester, but soon changed his mind and returned to university to study for the ministry. After earning a divinity degree, he worked with migrant workers in New York and in community programs in Vermont before he became minister of Kahiluonalani Church in Pearl City, a Honolulu suburb, in 1961. Before working with Hutton, Neill spent three years during which, as he told the State Senate Committee on Public Health, Welfare, and Housing, "Housing has occupied my entire efforts, both physically and mentally, seven days a week, even in my sleep I dream housing." Yet, he testified, he had encountered "one or two bottlenecks for every forward step." The obstacles included inadequate funding, bureaucracy, and builders who "have placed profit and return ahead of service and need."[58]

Between 1969 and 1971, CPA constructed three projects for the HCHA in Honolulu's Kauluwela Redevelopment District: Kauluwela Co-op, a 126-unit tower for low-to-moderate-income families, designed by Leo Wou; Kauluwela Elderly, designed by Frank Slavsky (fig. 33); and Kauluwela Low-Rise, also designed by Wou (fig. 34). CPA participated in these projects on a "design-assist" basis, whereby the developer hired the architect and engineer, both of whom, in the case of the two high-rise projects, spent many months in development before the HCHA and Hutton discussed a possible relationship. In the case of both high-rise projects, however, Hutton's team certainly influenced the manner of construction, as CPA slipformed the service cores of both buildings. According to Neill, slipforming the Kauluwela Co-op marked the first time that a contractor had used the technique in Hawaii on a high-rise residential project. (The 250 Ohua Street condominium, which CPA was constructing at the same time, was poured in place.[59]) Most importantly, deploying the technique enabled the owner to build the project within FHA cost guidelines. As Neill explained, by pouring concrete "twenty-four hours a day, around-the-clock," the contractor was able to "cut construction time in half."[60]

Bounded by the Lunalilo Freeway (Interstate H-1), Vineyard and Liliha Streets, and the Nuuanu Stream, Kauluwela was one of several districts adjacent to downtown Honolulu that housing and redevelopment officials targeted for urban renewal. The HCHA projects were part of an $11.5 million program approved by the Honolulu Redevelopment Agency that included the housing projects, an elementary school and park, and a commercial complex. As the HCHA projects constituted replacement housing, families and individuals who lost their homes to slum clearance had first priority to purchase units. Each one of three housing projects was financed under Section 236, a new housing program established in 1968.[61]

The Section 236 program replaced Section 221(b)(3) financing. According to Neill, its terms offered the means of overcoming the financial hurdles that HCHA staff encountered in other programs. Eligible owner-occupants could borrow at interest rates as low as 1 percent. Eligible renters would pay a maximum of 25 percent of their income—rather than a fixed amount—on rent, which meant that the HCHA "would be able to produce lower rents for lower income people." The organization could also use federal rent supplements in a way that allowed the HCHA to lease units to the state's housing authority. From the developer's point of view, Section 236 enabled the HCHA to deal with local FHA officials, who presumably would be more flexible and timely in their decision making. It also allowed a developer to negotiate contracts with builders rather than put its projects out to bid.[62]

Figure 33. Kauluwela Elderly housing project, with tower crane. Completed on 8 December 1971. Frank Slavsky Associates, architect. (Photo J. Young, taken 30 August 1971. Courtesy of Charles Pankow Builders, Ltd., Pasadena, California.)

George Hutton negotiated lump-sum contracts for each of the projects. In doing so, he persuaded the FHA to change its policy, at least locally. In 1969 Hutton met David Cheng, who was consulting Neill. They discussed the cooperative high-rise project, on which HCHA staff had already devoted more than a year to "very tenacious negotiations," as Neill put it, with redevelopment and federal officials and other parties. (At this point, the Kauluwela Elderly project was farther along in terms of design, but it was on hold while the HCHA awaited a decision by the Honolulu Board of Zoning Appeals on a variance to the city's setback ordinance that it had filed. As architect Slavsky explained, a variance of 3 feet from the back of the building to the Nuuanu Stream was needed to permit "more controlled ventilation" in the high-rise structure.) Hutton and Cheng soon agreed on the terms of a $2.26 million contract to build the Kauluwela Co-op. Under FHA regulations, however, the fee that a contractor could earn on such a project was limited to a fixed percentage of its construction costs, which the contractor had to certify. Under these terms, the contractor had no financial incentive to find more economical ways to build the project. Hutton discussed the project with Charlie Pankow, who insisted that Hutton

Figure 34. Kauluwela Low-Rise housing project. Leo S. Wou, architect. (Photo J. Young, taken 28 May 1971. Courtesy of Charles Pankow Builders, Ltd., Pasadena, California.)

negotiate a lump-sum contract. Hutton contacted Alvin Pang, who directed the local FHA office, about the possibility of building the cooperative high-rise on a lump-sum basis. Pang told him that the FHA had never considered such a contract. Yet Pang eventually approved the approach, and, on that basis, the project—and the others that Pankow built under Section 236—went forward. There is no available record of internal FHA discussions on the matter or any documentation that might allow an assessment of the possible impact of the decision on other federal housing projects. The following year, however, Pang stated that the government should be more concerned about adding to the housing stock than how much builders were profiting from federally funded projects. In fact, Pang believed that both government and financial institutions could do more "to see that the [building] industry has its proper share" of the resources that it needed to meet the pressing demand.[63]

Groundbreaking for the Kauluwela Co-op took place at the beginning of November 1969. It was the third project for the HCHA, which had completed projects in Pearl City, just north of the Pearl Harbor Naval Station, and on Maui. Plans for another ten projects, including the Elderly and Low-Rise projects that CPA would build and a 348-unit low-income project in Waimanalo that would be a joint venture with the Hawaii Housing Authority, were in the works. Together, housing under construction and on the drawing board comprised some 2,000 units for low-income and senior occupants. The HCHA's efforts were halted, however, by President Nixon's moratorium on all new federal housing program starts, which he declared in January 1973 to defuse controversy that had accompanied the expansion of federal housing programs. (Housing advocates complained about developers' profits. Developers complained about bureaucratic regulation and the project delays that they caused. At the same time, rampant inflation was saddling owner-occupants and renters with utility and maintenance costs that they could not absorb. Defaults on subsidized loans were rising accordingly.) Nixon established a National Housing Policy Review, whose recommendations set in motion a change in course in US housing policy toward a focus on family income. The Housing and Community Development Act of 1974 replaced Section 236 and other FHA programs with income supplements to qualifying households under Section 8 of the legislation. HCHA folded its operations in the wake of the change in policy.[64]

CPA used its experience with the HCHA projects to complete a fourth Section 236 project as a sponsor-builder before the program ended: Kilani Village, located in Wahiawa, a small town in the central valley of Oahu. By the time the complex opened, in 1972, Hutton's marketing approach—getting to know people and establishing a track record that demonstrated CPA's ability to perform—had paid off.[65] It was an approach that was especially suited to the local commercial market, characterized by a relatively small number of prominent owners and developers. Charles Pankow Associates was functioning for all intents and purposes independently of Charles Pankow, Inc.

The Keys to CPA's Success: Salvage Jobs and Development Stakes

After peaking in real terms in 1970, construction activity in Hawaii remained stable through 1975, even as inflation, which soared to 9.1 percent in the wake of the 1973 oil crisis, was responsible for the generation of successive record increases in the nominal value of construction. Yet, the value of construction completed in 1975 was only $498.9 million in constant 1960 dollars, falling short of the $540.9 million mark set in 1970. And then the local market collapsed, owing to a downturn in housing attributed to stagnating incomes and population gains, tightening credit conditions, and rampant materials and labor costs. Over three years, the sector's output fell by one-third. A 16.8 percent decline in private construction in 1974 anticipated the collapse. Some 8,000 of 28,000 construction workers lost their jobs. The industry rebounded from 1977 to 1980, but not enough to erase the losses in jobs and output suffered during the bust. Then recession whipsawed the market once again. This time, the local market responded in line with the national market to the Federal Reserve's policy of raising interest rates to quash inflation, which climbed to 13.5 percent in 1980. From 1980 to 1982, overall construction in Hawaii declined by 22.7 percent. The number of jobs in the industry, which had rebounded, but not to the level reached before the 1974–1975 recession, decreased from 23,950 to 17,850. The value of private construction plummeted even further. In 1981 alone, authorizations fell 31.5 percent: the largest decline in two decades. Recovery, such as it was, was fragile. In its annual construction review for 1985, the Bank of Hawaii foresaw a "prolonged phase of little or no growth," based on the trend of the previous three years.[66]

In a topsy-turvy industry characterized by "surprisingly little advance planning and only a negligible ability to forecast future needs," where, as one contractor noted, "builders are either fighting for survival or riding a crest of peak activity," these conditions resulted in the demise, reorganization, or withdrawal of major contractors from the market. George Hutton was determined not to overextend CPA, as firms such as Swinerton & Walberg, which departed the Islands in 1980, had done. While other contractors took on as much work as possible during boom times, Hutton accepted only those projects that he could put on a "solid financial basis." If this meant that CPA was engaged in only one or two projects at a time, so be it. Hutton did business with people who adhered to the same standards as he did, and who had similar goals. "I like to identify personally with the people I'm doing business with," he explained. Once CPA was a going concern, he insisted that owners involve his staff in the planning and design of their projects. And, as CPI teams were doing on the Mainland, Hutton's men took responsibility for concrete construction tasks such as precasting and slipforming. Executing projects as Charlie Pankow had been recommending since his Kiewit days "set us apart" from other contractors and increased the "comfort level" of clients, Hutton explained.[67]

This conservative approach put CPA in the financial position to salvage a number of jobs when the market collapsed, recovered weakly, and faltered again. With credit tightening—the prime interest rate rose from 6.35 percent in 1976 to 20 percent in the spring of 1980—financial institutions required developers to inject additional equity into their projects before they would advance construction loans. In these circumstances, a developer might easily go bankrupt, as was the case with Waikiki Hobron Associates, a limited partnership formed for purposes of constructing the Hobron condominium (fig. 35). At the same time, CPA was in position to participate financially in projects by lending developers funds or taking an equity position. Salvaging projects in Hawaii involved more than redesigning the project to meet the owner's budget, as was the case with the salvage jobs described earlier (though, of course, this aspect of CPA's participation was also involved). CPA completed its first such job, Kaimana Lanais condominium, in 1974. It completed other salvage projects between 1977 and 1984, including the Hobron. CPA's participation as more than just a contractor enabled these projects to go forward, creating work for the company and additional profits for CPA (and, of course, revenue for the development entity partners). Salvage jobs, involving a variety of investment and financing configurations, comprised more than one-third of the projects that CPA completed between 1977 and 1984. Pure development plays accounted for another half a dozen projects (see appendix B). As a merchant builder putting its money at risk, CPA stood apart from its counterpart in California.[68]

In a presentation at the company's 1976 annual meeting, George Hutton used the Waikiki Tusitala condominium (later renamed Waikiki Lanais) to illustrate the role that CPA played in financing deals that otherwise would have fallen apart. The project was located on a leasehold site, one block from the Ala Wai Canal, on the eastern end of the Waikiki district. It offered one- and two-bedroom units selling from $44,800 to $75,500 (fig. 36).[69]

CPA became involved in the project in September 1975 through attorney Ronald Yee, who was seeking to arrange construction financing for a troubled project. Yee was an officer of the Waikiki-Tusitala Development Corporation, which was developing the project as a joint venture with Michael M. Ross. Indeed, construction had started in May 1974. The contractor, Oceanic Construction, had driven piles for the foundation and poured in place concrete walls and tie beams for its excavation. But Ross-Tusitala Associates, as the development entity was known, had launched the project without adequate financing. Work stopped once payments to subcontractors ceased. Several hundred thousand dollars in mechanics liens were filed against the project. The property was scheduled for sale at auction when Yee contacted Hutton. That the attorney approached Hutton suggests the growing reputation of CPA a decade after Hutton began working on the James Campbell Building.[70]

As Hutton understood it, Ross-Tusitala Associates would secure the construction financing that it needed to restart the project. Hutton's staff estimated the job and concluded that CPA could build the project within the developer's

Figure 35. The Hobron in Waikiki, completed in 1983. Jo Paul Rognstad, architect; Richard M. Libbey, structural engineer. (Photo © Bill Hagstotz/Constructionimages.com.)

pro forma budget. With some 10,000 units already under construction, however, Honolulu's condominium market was oversaturated in the fall of 1975. Banks began to condition construction loans on evidence of substantial pre-sales activity. Ross-Tusitala Associates could not meet the requirement in a faltering market, and the financing fell through.[71]

At this point, Hutton stepped in with a standby commitment to pay off the construction lender, if required. In this way, Hutton explained, "we would put our own funds to work as long as the stand-by was bankable." Hutton secured a commitment from a lender in Los Angeles, backed by Crocker National Bank. With the latter's approval and acceptance, and its agreement to extend a $6.5 million construction loan, CPA bought the standby commitment for $130,000. Ross-Tusitala Associates guaranteed it with a note and a second mortgage on the property, though the latter was of "dubious value," according to Hutton, given the mechanics liens in place.[72]

A shake-up at Crocker threatened to derail the bank's commitment to the project and expose the Pankow firm to the loss of its investment. The vice president who had approved the commitment letter was a casualty of the re-

Figure 36. Waikiki Lanais, completed in 1977. Looking makua. The Ala Wai Canal flows behind the building. Peter His Associates, architect; Dimitrios Bratakos & Associates, structural engineer. (Photo © Bill Hagstotz/Constructionimages.com.)

organization. After he departed, the bank withdrew its commitment. Hutton protested. In the ensuing weeks, attorneys and other representatives from both companies met at the bank's offices in San Francisco. Upon review, both sides concluded that the loan application, as submitted by the developer, did not make a convincing case for the project's viability. And so Hutton and his team returned to Honolulu to revise it. They determined that "a substantial amount of upgrading" to the plans and specifications would be required, both to reconcile them with the appraisal and to increase the salability of the units. Hutton and his staff revised the plans to include lanais for all of the units, air-conditioning, improved finishes, and other amenities. Notes Al Fink, who was involved in the redesign: "The value of [the upgrades] was much, much more than the cost . . . and that was part of [our] philosophy, always . . . to create value." To fund the upgrades, CPA eliminated the lease premium and rents due to the developer during construction. Ross-Tusitala Associates was in no position to object, for the partners would lose the property were the project not built.[73]

Ultimately, making Waikiki Tusitala viable meant removing Ross-Tusitala Associates from the project. Waikiki Tusitala now needed a substantial infusion of equity. When the developer demurred, Hutton persuaded Oceanic Construction and its subcontractors to defer funds owed them until the project retired the construction loan. For its part, CPA subordinated "a sizable amount of money" that it could expect from the construction contract. Once Crocker's bankers accepted CPA's final budget, Hutton sent it to Ross-Tusitala Associates, which now determined that the project's return on investment did not justify its guaranteeing the project; Ross-Tusitala's partners pulled out of the deal.[74]

Figuring that this outcome "would probably happen," Hutton sought Crocker's support for the creation of a development entity, Waikiki Lanais Corporation, to complete the project. After the bank approved the new approach, Hutton formed a development team that included Jim Adams, a structural engineer who, like Richard Bradshaw, was an avid proponent of mechanizing concrete construction though techniques such as precasting and slipforming. Adams enjoyed working with Hutton's office because when CPA's people in the field found problems, they brought them to Adams's attention; contractor and engineer would cooperate to solve them in the field. According to Red Metcalf, Adams felt that other contractors often failed to understand the problems that they faced or hesitated to inform the engineer about them. The Waikiki Lanais team also included developer Ralph D. Cornuelle, who managed the business end of the project. It was the first of several CPA development efforts that Cornuelle would manage. CPA also entered into a sales and management agreement with a local realty firm. Of course, CPA was the contractor. Crocker remained in place as lender and hired a construction manager to approve spending. Once in place, the building team reviewed the design and made additional upgrades, as the budget allowed. It also certified the work already undertaken by Oceanic Construction and otherwise prepared the site for the resumption of construction.[75]

With the third party financial agreements in place, CPA closed the construction loan with Crocker's bankers on 2 August 1976. Reflecting the complexity of the arrangement, it took nine weeks to draft the closing documents—six to seven weeks longer than normal. Construction resumed two weeks later; the project was ready for occupancy in July 1977. Hutton told his audience at the annual meeting that Waikiki Lanais was "probably the classic example of a salvage job," though one that required far more time and effort than he might have imagined up front.[76]

Not all salvage jobs were as complicated as Waikiki Lanais. For instance, CPA rescued the developers of Wilder at Piikoi, a leasehold condominium project named for the streets that the 30-story building faced, with a $650,000 loan that provided the developers with the equity that they needed to fund the construction loan. In return, CPA received a fee of $650,000. To ensure that the construction of Kawaiahao Plaza proceeded, CPA acquired the land. The $10 million project was to be a low-rise office and parking affair; similar in many ways to Ward Plaza, but with more space devoted to gardens and other landscaping. It was to be built on a site next to the historic Mission House in the Kakaako district. Developer James Trask held an option on the property, but knew that he would not be able to start construction before the option expired. He approached Hutton, who agreed to purchase the property. CPA held the site for six months until Trask and his partners were able to go forward with the project. CPA earned a larger construction fee. Regardless of the circumstances preventing the start of a project, Hutton generally found a way to make it work financially if owner, architect, and engineer were prepared to collaborate with his men in a design-build arrangement. As a result of its salvage work, CPA raised its profile among local developers, even as other contractors were retrenching or dropping out of the market altogether.[77]

Building the CPA Team

As CPA grew, George Hutton recruited additional men to manage the work. Once he landed the contract to construct the Kauluwela Co-op, he tracked down Red Metcalf to head the slipform operation. Metcalf was ready for a move. He had spent the previous five years with Kiewit's heavy division on various projects in California's Central Valley and faced the prospect of returning to Los Angeles, which he did not fancy. Now Hutton was able to fulfill Charlie Pankow's promise to Metcalf in 1963 to promote him to superintendent in due course if he left Kiewit to join CPI. In short order, Metcalf supervised the Kauluwela Co-op slipform operation, replaced the superintendent on 250 Ohua, and, after completing the latter project, supervised construction of Kauluwela Elderly. Until Charlie Pankow transferred him to the Mainland in 1985, Metcalf was involved in almost every major project in Hawaii.[78]

Al Fink joined Metcalf as a project engineer on the Kauluwela Elderly building site in June 1971. A graduate in civil engineering from Case Western Reserve University, Fink answered a help-wanted advertisement that Hutton had placed in the newspaper after he fired an underperforming project engineer. After working three years for San Francisco–based Bechtel Corporation, the free-spirited Fink had come to Hawaii "just for the heck of it" to earn an MBA from the University of Hawaii. He had nearly completed an accelerated program. After meeting Fink for lunch and giving him a tour of the job site, Hutton hired Fink, but was concerned that the "kind of hippie-looking" engineer might not reflect positively on the company's image. He instructed Metcalf to have Fink cut his hair when the latter reported for work. During the 1970s and 1980s, Fink took the path that engineering graduates typically traveled within the company, from field engineer to project engineer to superintendent to project sponsor—an off-site management position akin to film producer, involving business development, owner relations, estimating, design coordination, contract administration and cost reporting. Fink was responsible for numerous slipform operations, often in tandem with Metcalf (fig. 37).[79]

A mix of transfers from the Mainland and recruits composed the balance of Hutton's core team of field managers. Jack Grieger was a superintendent from 1974 to 1983. One of the original employees of CPI, he had worked as a carpenter foreman and field superintendent under Ralph Tice and Alan Murk. Grieger came to Hawaii after completing his first job as a superintendent, the parking structure associated with San José Plaza. An avid endurance athlete, Grieger became "one of Honolulu's famous marathon runners," according to the company newsletter. Tony Giron split his time as a superintendent between Mainland and Hawaii jobs after he helped Hutton complete the James Campbell Building. Don Kimball stayed in Hawaii after supervising the construction of the Hilo Mall. He helped to develop the Kauluwela housing projects and then managed subcontractors on half a dozen projects before becoming CPA's marketing director in 1984. Kimball also recruited Jim Thain, a native of Hawaii, to work with him in Hilo. Thereafter Thain became one of Hutton's most valuable superintendents (fig. 38). Jack Parker and Bill Deuchar, two men recruited by Hutton, rounded out the core group of superintendents. Hutton recruited Parker from Kiewit as a carpenter foreman while the latter was on vacation in Hawaii. Three years later, at age 27, Parker was put in charge of his first project. In 1975 Hutton hired away Bill Deuchar from E. E. Black, a well-established Hawaii contractor.[80]

Figure 37. Al Fink on the building site for the Pearl Two parking garage. (From the private collection of Albert W. Fink.)

Until 1972, Hutton had no general superintendent (operations manager), the role that Bob Carlson and, subsequently, Ralph Tice and Alan Murk played on the Mainland during the first two decades of CPI's existence. To fill the position, he hired William A. Heine, who had just left the US Navy as a lieutenant commander. "A very organized" and a "very solid, smart guy," Heine provided project management and subcontractor administration, leaving superintendents to organize project teams, schedule work, and direct activities in the field. "A stickler for process," Heine also wrote the first procedure manual for the company: an effort to standardize the way in which Pankow did business at the project level. Heine became the main recruiter of engineers to the Hawaii office and took on the responsibility of training them. As Kim Lum, who joined the company as a project engineer in 1980, explains, Heine's company-wide mission was to "make sure we were all doing things the same way so that we could move from job to job, work with different people, but still have common process."[81]

Like Charlie Pankow, George Hutton relied on talented self-starters to work innovatively and productively in the absence of managerial hierarchies to deliver projects on time and within budget to the satisfaction of the owner. Hutton believed culture was a "necessary ingredient for success." An admirer of General George S. Patton, Hutton sought to instill in

Figure 38. Jim Thain and Red Metcalf watch crews install precast columns for the Pearl One parking garage. (From the private collection of Albert W. Fink.)

his men the esprit de corps that had animated the Kiewit building division. In so doing, he aimed to create an identity for CPA as both a leading local contractor and a successful group within the company. Perhaps the most memorable expression of the latter occurred when the members of the Hawaii office arrived at an annual meeting wearing red, green, and purple aloha jackets. Annual meetings provided the forum for Pankow employees who were scattered among various job sites to get together and share their experiences of the previous twelve months. Though it may have struck Charlie Pankow as a statement of separation, it was not George Hutton's intention. Rather, in his view, donning the jackets was a statement of pride in the accomplishments of CPA.[82]

Geographic isolation helped to shape the company's culture in Hawaii. For many employees, working on an island that was located some 2,500 miles from California meant starting anew, severed from family and community networks. Opportunities to "get away," especially before the deregulation of air fares during the Carter administration, were limited. More so than was the case with staff based in CPI's offices in San Francisco and Altadena, people who worked for Pankow in Hawaii engaged with one another socially outside of working hours. Red Metcalf and his wife played the role of unofficial social directors, hosting Fourth of July parties and other events at their home. Playing on Pankow-sponsored softball and basketball teams also helped to weave CPA employees into a tightly knit group.[83]

The following case examples encapsulate how Hutton's teams executed projects in ways that established CPA's reputation as one of Hawaii's leading builders.

CPA in Action: Setting Slipforming Records on Pearl One and Two

As a contractor, CPA became renowned for its three-day slipform cycles. As Kim Lum notes, its crews could finish a floor every three days while competitors "were getting five days every floor . . . if they were lucky." It was what "everybody talked about," inside and outside the company.[84] CPA's aggressive completion schedules, based on slipform construction, saved owners money on interim financing and allowed them to rent their units. Impressed by CPA's performance at the building site, some of Hawaii's leading developers became repeat clients. Pankow in Hawaii thrived under George Hutton during the unsettled market conditions of the 1970s and early 1980s. The Pearl condominium projects demonstrated to the building community the determination of CPA project teams "to go for quality but to keep a job moving as fast and efficiently as possible," Al Fink explained.[85]

In November 1972, George Hutton met with Bill Curlett of Oceanic Properties to discuss the possibility of developing a multi-unit residential housing project for middle-income occupants on a site near the Pearlridge Shopping Center, overlooking Pearl Harbor. Since the completion of the Financial Plaza of the Pacific, Oceanic had completed residential projects that had targeted various income levels, including Mililani Town, the first "new town" in Hawaii; Queen Emma, "a new concept in 'walk to work living'"; Sea Ranch, an exclusive colony on California's north coast; and Hamilton, a "new town" community south of San José. Based on a feasibility study, Oceanic estimated that 950 one- and two-bedroom units could be developed on the Pearlridge site and formed Pearlridge Land Development as the vehicle for doing so.[86]

Curlett and Hutton assembled the design team. Strictly speaking, the Pearl projects adopted a design-assist approach, whereby the developer, rather than the contractor, hired the architect and the mechanical and electrical engineers. Hutton insisted, however, that Curlett hire an architect who was amenable to Pankow's design-build approach. Curlett selected Frank Slavsky, who had designed Kauluwela Elderly and, like Jim Adams, found working with Pankow to his liking. Hutton also insisted on hiring the structural engineer and retaining a veto over Curlett's selection of the other engineers. Once Curlett agreed to these conditions, Hutton tapped the local office of T. Y. Lin as the project's structural engineer.[87]

The design team produced preliminary plans that proved, upon CPA's review, to be uneconomical, given construction costs, the sales prices of the units, and the returns that Curlett expected. The plans called for six high-rise structures and three parking garages. The building team reconfigured the project as a complex comprising three towers. Hutton's staff also reassessed the developer's proposed amenities on both cost and marketability criteria. Hutton selected an electrical and mechanical subcontractor who had offered cost effective ideas on past CPA projects. Curlett's team accepted the revised preliminary plans, and on that basis Hutton negotiated a lump-sum construction contract that allowed Pankow to change its methods, if doing so did not compromise the quality of the project.[88]

CPA's men worked with Slavsky to ensure that the working drawings accommodated their favored techniques of mechanized construction. The garages would be framed using precast shear and retaining walls, reinforced concrete beams and columns, and post-tensioned girders. To speed construction, the CPA team set up a casting yard within the footprint of the garage (fig. 39). The frame for the towers would be fabricated from 8-inch, single-curtain reinforced concrete walls. The design also called for a mat foundation and the use of flying forms to install 6-inch floor slabs, reinforced with welded wire. As Dean Stephan explains, "Flying forms were a new and innovative way to efficiently form flat plate slabs." They consisted of prefabricated, reusable forms that crews could lower, roll out, and "fly" to the next level by means of a crane, once the concrete had set (fig. 40). Their use cut the number of workers nearly in half and saved time that would otherwise be devoted to building forms, placing and finishing concrete, and stripping forms after it cured. The savings in labor costs more than offset the cost of forms and cranes. The contractor coordinated the design of the slipform and non–load-bearing precast panels with T. Y. Lin's engineer (fig. 41). The idea, explains Hutton, was to "obtain the most efficient structural design possible and one that conformed almost exactly to our construction requirements."[89]

Owing to market conditions, only two of the three high-rises were built. Pankow began construction of the Pearl One condominium in November

Figure 39. Precast concrete elements stacked for use in the Pearl One parking garage. (Courtesy of Charles Pankow Builders, Ltd., Pasadena, California.)

1973 and delivered the building in January 1975 (fig. 42). Construction of Pearl Two began on 16 September 1974 and finished almost exactly one year later—some six months ahead of schedule. For the "construction excellence" that it demonstrated on Pearl Two, CPA received the Associated General Contractors of America's (AGC's) Build America Award, its highest honor. The relative speed with which Pankow crews erected Pearl Two owed to the experience they gained in constructing Pearl One, which was similar in design, if slightly lower in height. Keeping the Pearl Two project on schedule, however, required far more "ingenuity and never-say-die spirit" than CPA detailed in its submission to the association's prize jury.[90]

A problem that project engineer Al Fink had not experienced, and would never encounter again, threatened the slipform schedule. About the time that the slipform reached the tenth floor, the project team discovered that the compressive strength of the concrete on the first floor was substantially below the prescribed design strength. (The time interval is explained by the three days required to slip an 8½-foot floor and the 28 days needed for the concrete to cure to its maximum compressive strength.) With the encouragement of Bill Heine and superintendent Metcalf, Fink had set an aggressive schedule "that everybody said was absolutely impossible." Bill Fulton, a millwork installer, had bet Metcalf $25 that Pankow would not meet it. Now Pankow faced the prospect of having to tear down the building.[91]

Figure 40. Tower crane atop the Pearl Two condominium, with "flying form" for installation of concrete floor slab. Completed on 12 September 1975. Frank Slavsky Associates, architect; T. Y. Lin, Hawaii, structural engineer. (From the private collection of Albert W. Fink.)

Figure 41. Slipform operation associated with the Pearl One condominium. (Courtesy of Charles Pankow Builders, Ltd., Pasadena, California.)

Charles Pankow Associates
Project : The Pearl No.1
Location : 500 Kosuke Loop
Date : 11-20-74

Figure 42. Construction of the Pearl One condominium nears completion, 20 November 1974. Construction finished on 18 January 1975. Slipform construction of the Pearl Two condominium proceeds in the background. Frank Slavsky Associates, architect; T. Y. Lin, Hawaii, structural engineer. (Courtesy of Charles Pankow Builders, Ltd., Pasadena, California.)

After conferring with Heine, Fink and Metcalf brought the problem to the attention of developer Curlett, architect Slavsky, and Pacific Concrete and Rock, the supplier. Working with the structural engineer, they proposed to repair the compromised floor as construction of the upper floors proceeded apace. Oceanic Properties agreed to reduce the dimensions of each room by 1 inch to compensate for the inferior strength of the concrete. Local building officials analyzed, and approved, the proposed corrective measures, which involved shoring up walls at risk and removing low-strength concrete. The project team confronted the "most massive problems" in the history of the company to perform the repair work without jeopardizing the original construction schedule, according to Fink. The repairs added almost $100,000 to the cost of the project, including labor and materials, for which the ready-mix company was liable. But with "everybody working together harmoniously to get this done," slipform operations finished on the exact day that Fink had originally scheduled. The structural engineer certified that CPA had completed the repairs accord-

Figure 43. Red Metcalf collects his $25 bet in nickels with the completion of the slipform operation of the Pearl Two condominium on schedule. (From the private collection of Albert W. Fink.)

ing to his specifications. Metcalf collected on his bet, which Bill Fulton paid in nickels (fig. 43).[92]

CPA in Action: L. Robert Allen's Mixed-Use Projects

According to developer L. Robert Allen, Century Center and Executive Centre were "innovative and sumptuous," yet "cost-effective" office, retail, and condominium projects that met an unfilled, specialized need of small businesses for 500 to 700 square feet of nonexpandable space. At a time when the real estate industry was experiencing "rapid change" that was having an impact on all aspects of development, from design, materials, and construction to sales and marketing to financing, Allen's projects commanded record sales prices. Similar projects elsewhere inspired Allen to research the local market for projects that mixed Class A office space and "elegant living quarters." The configuration of the projects and the savings achieved by architect, engineer, and contractor working under design-build enabled Allen to offer buyers features and amenities whose cost "would have been prohibitive, had [they] purchased them individually." Hence, Allen's sales people were able to sell 90 percent of Century Center's units within 12 months even though, at $150 per square foot, they carried the highest price tag ever offered in Honolulu. Allen was quick to note that cost effectiveness had nothing to do with austerity. Rather, it was a "function of being appropriate to and responsible for conditions and circumstances" of the market. Allen and his building teams learned from previous projects to question doing things simply because they had worked before. Thus, Century Center, completed in July 1978, and Century Square, its sister project, completed in 1982, informed the design and construction of the $125 million Executive Centre, completed in May 1984.[93]

Architect for all three projects was Jo Paul Rognstad. A prolific designer, his portfolio includes some 80 condominiums that provide some 11,000 units of residential and office space. Rognstad made his reputation by developing an "extremely efficient" high-rise tower that maximized the number of floors, given local ordinances that restricted the height of buildings to a range of 350 to 400 feet in most areas. These limits dated from the 1950s and stemmed from the desire on the part of local planners and civic leaders to preserve ocean and mountain views, particularly of Diamond Head. In fact, assesses structural engineer Steven M. Baldridge, "monotonous densification" characterizes Honolulu's urban landscape, with "long continuous 'walls' of buildings blocking out the natural beauty." The merits of the argument aside, Rognstad's towers are typically tall and thin—and square or slightly rectangular in plan (fig. 44). At less than 4,000 square feet per floor, floor footprints are relatively small. Floor-to-floor heights are 8½ feet or less. Rognstad used thin, post-tensioned floor slabs—6 inches thick for Century Center and 4½ inches thick for Executive Centre—to squeeze as many floors as possible into a tower. (Post-tensioning floors was not a popular idea when it was introduced in the late 1960s. T. Y. Lin's "load balancing" system, which made calculations easier, and research and testing that showed that the amount of steel could be reduced by half, expanded its use.[94]) Century Center and Executive Centre featured 41-story towers that offered 375 units and 469 units, respectively. Rognstad's prototypical design made their configurations economically feasible.[95]

Allen provided incentives for CPA to build both projects as efficiently as possible. He had a financial interest in doing so. His $94 million construction loan on Executive Centre, for instance, was the largest recorded in Hawaii to date. And so the $58.1 million construction contract with CPA included a "sizable" bonus and penalty clause. CPA utilized slipforms to erect floors on three-day cycles and deployed flying forms to install floor slabs (fig. 45). Hutton's men incorporated a number of modifications to these techniques to accelerate the construction to Rognstad's designs.[96]

On Century Center, for instance, the architect clad the tower in a glass curtain wall. The contractor deployed a protective, 10-foot-wide plywood deck, cantilevered from the building perimeter, to permit workers to install the curtain wall below the canopy while others poured concrete several floors above it. Workers were able begin finishing interiors once a floor was enclosed. By deploying this approach, crews were able to complete the lower 19 floors about the time that the slipform operation reached the roof. Hutton's

Figure 44. Century Center, Honolulu. Looking makua from Waikiki. Construction started on 10 January 1977. Completed on 31 July 1978. Jo Paul Rognstad, architect; D. I. Shin & Associates, structural engineer. (Photo © Bill Hagstotz/ Constructionimages.com.)

men also redesigned the parking structure to utilize a 6-inch, post-tensioned slab and column system to replace the original beam and precast plank design. The change reduced the cost of the structure by eliminating beams altogether. It also allowed crews to work on the garage while others constructed the tower. These techniques enabled workers to complete Century Center on time despite the loss of four months, owing to separate labor actions in late 1977 and in early 1978.[97]

Rognstad's design of Bishop Tower as the centerpiece of Executive Centre complicated slipform operations on that project. The architect envisioned the public and retail areas as open spaces. (Incorporating indoor pedestrian space into the design allowed Allen to secure floor area bonuses under the city's zoning ordinance.) To enhance the ocean and mountain views, and therefore the marketability, of office and residential units, Rognstad developed an 18-sided floor plan for the tower that radically departed from his prototypical design. The architect did not want interior shear walls that radiated from the service core on the upper floors to extend to the mat foundation. He

Figure 45. A massive 5,300-square-foot deck slips all of the vertical structural elements for Executive Centre. The tower crane carries "flying forms" used to install floors below the deck. Construction started on 7 December 1981. Completed on 31 May 1984. Jo Paul Rognstad, architect; D. I. Shin & Associates, structural engineer. (Courtesy of Charles Pankow Builders, Ltd., Pasadena, California.)

prescribed 32-foot concrete columns, extending from the mat through the basement and ground levels to the second floor, to support the shear walls. Floors two and three were transitional; floors four and above were typical. The construction team, managed by Al Fink, saved $100,000 and one month's time by slipping, rather than casting-in-place, the vertical elements from the basement to the fourth floor.[98]

Workers collectively drew on all of the experience that they had accumulated on the job sites of 14 high-rise CPA projects to slip the elements from the basement. Fink, Metcalf, and others who were involved on the project worked closely with structural engineer Danny Shin on the requirements for the operations. Detailing the wall transitions from the second to the fourth floor and developing a structural support system for the 5,300-square-foot slipform deck posed the most challenging problems. The amount of rebar in the columns prevented workers from embedding the jack rods that supported the slipform deck into the concrete to provide lateral support, as was typically the case on CPA jobs. On Executive Centre, this approach could be used only on the elevators and stairwells. Engineers placed the jack rods that they could not embed in concrete beneath the upper walls and prevented them from moving or buckling with custom-built "jack ladders," constructed with four-by-eight boards, J-bolts, and two-by-four "kickers." Carpenters fabricated panels for the massive slipform in the yard of a nearby project. "As soon as the mat was poured, we started setting up the slipform," Fink explained. Once the slipform reached the typical floors, crews worked on the three-day cycle that had become the CPA standard.[99]

Construction perhaps benefited most of all from the experience of the engineers, foremen, and workers assembled on the building site. CPA had recently completed the Windward Mall in Kaneohe and three high-rise condominiums developed by George Hutton and Charlie Pankow: Hale Kaheka, Honolulu Tower, and Craigside Towers, the penthouse of which Hutton has called home since 1982 (see appendix B).[100] Executive Centre demanded significant resources. Many of CPA's field and project engineers on these projects were assigned to the job. More importantly, according to Bill Bramschreiber, a 1981 civil engineering graduate from Purdue University who worked as a field engineer on Executive Centre, the journeymen carpenters and other trades workers on the site "were the best we had." They were able to achieve a level of coordination between erecting the structure, on the one hand, and finishing floors, on the other, that Bramschreiber thinks has not been duplicated—and believes could not be duplicated, given the building inspection procedures and other practices that prevail today. Executive Centre, Bramschreiber notes, "was really run like clockwork. [It] was probably one of the best-run jobs and most profitable jobs that we ever did."[101]

CPA in Action: The Esplanade

For CPA's project teams, finding better ways to build was not limited to precast, slipform, and flyform operations. To speed construction of the Esplanade, a "unique multi-level condominium project" on Kuapa Pond in Hawaii Kai, operations manager Lyle Sheppard, superintendent Metcalf, and project engineer Fink devised a system to prefabricate walls made of CMUs (concrete masonry units, or blocks) that was "about 20 times faster" than the standard practice of workers lifting CMUs into place one at a time and applying grout, according to Fink. The system used a conveyor belt on rollers to deliver CMUs to masons, who arranged them in a row and applied two lines of Threadline epoxy to prepare the next row. Workers deployed a movable scaffold sys-

Figure 46. The Esplanade, Hawaii Kai. Construction of the 208-unit condominium began on 27 October 1971. Completed on 28 September 1973. Developed by Oceanic Properties. Leo S. Wou, architect; Dimitrios Bratakos & Associates, structural engineer. (Photo © Bill Hagstotz/Constructionimages.com.)

tem to repeat this mechanized procedure until they had completed a wall section. Crews used a crane to lift the section into place on the reinforcing dowels that protruded from the slab poured on grade. To construct the nine floors above grade, crews used flying forms to install floor slabs and a crane to place the wall sections (fig. 46).[102]

An Unsettled Future

When George Hutton landed the Hilo Mall and J. C. Penney store projects, businessman Minoru Morimoto urged him and Charlie Pankow to devote their energies and resources to developing a greater Hawaii. It was a laudatory acknowledgement of the potential role of the builder as the midwife of economic and social progress, as civic boosters understood the term throughout much of the twentieth century.[103] Associated with Big Way Supermarkets at the time and, later, president of the Hawaii Food Industry Association, Morimoto was the "key person" in helping CPA to land the construction contract for the Hilo projects. Over the next 15 years, Hutton took his cue from Morimoto and built CPA into one of Hawaii's three largest contractors and the one with the highest reputation for project delivery and owner satisfaction (fig. 47).[104]

Figure 47. George Hutton and Charlie Pankow exchange 20-year awards at the company's annual meeting, held at the Huntington Sheraton Hotel, Pasadena. Owing to CPA's performance, the volume of CPI's business had soared over the previous five years. (From CPI News 1 [Fall 1983]: 5.)

If the completion of Executive Centre represented the capstone achievement of a remarkable decade-and-a-half-long run, it also marked the beginning of a lean period during which the Hawaii office would lose the independence that Hutton believes underpinned its success. As of 1984, the local commercial real estate market was saturated, with building permits running 51 percent lower than they had in 1980. Forecasters were predicting lower construction activity in an environment of slower economic growth statewide.[105] At the same time, Charlie Pankow was reorganizing his company. A prolonged downturn in business, improved communications, the advent of computers, and the reorganization of CPI would alter the way in which the Hawaii office operated and derail the careers of many of Hutton's employees. Chapter 5 relates these events. For now, the narrative turns to Pankow on the Mainland for the 15 years during which Pankow in Hawaii was enjoying its greatest success.

CHAPTER 4

Pankow on the Mainland, 1972–1984

A fter he joined the firm as a project sponsor in 1972, Dean Stephan rarely saw Charlie Pankow. Indeed, it seemed to him as if the company ran itself: "We were maybe like a platoon in the Marines. We were a very small group, very tight knit, all working for each other and covering each other's back, and we really didn't have a colonel around telling us what to do. That's very much the way it was. . . . Hands off. He let you achieve your maximum. He did not artificially constrain you. And it worked."[1] To be sure, on paper, the organization did not work. For instance, superintendents still reported to the operations manager, even after the company created the project sponsor as the person with responsibility for owner relations, contract management, and other tasks related to the execution of an engagement. And with no organization chart to define lines of authority, both project sponsor and operations manager were directly responsible to Charlie Pankow. The organization worked in practice, however, as an accretion of project teams, because Charlie Pankow, George Hutton, and Russ Osterman were able to identify self-motivated and resourceful individuals who could work cooperatively and effectively in small groups, and let them get on with the tasks of building. Even as it grew in the 1970s, CPI differed from the Kiewit building division mainly in the number of project teams that were operating at any given time.

After a decade in business, Charlie Pankow was leading his company based on the same principles and in the same style as he had headed his building division. It was around this time that some people inside CPI began to refer to the company's culture and how it conducted business as the "Pankow Way." Unlike

Bill Hewett and David Packard, or Lockheed's "Kelly" Johnson, Charlie Pankow never put his organizational values into writing. But everyone who had followed Pankow out of Kiewit knew what their leader expected of them and what they expected of one another. And, much like George Hutton was doing in Hawaii, it would be the responsibility of those based in Altadena and San Francisco—the two California offices—to ensure that a new generation of engineers and construction managers—many of them university graduates—learned ways of doing business that had become second nature to them.

Thomas D. Verti, who rose up through the ranks to become senior vice president, has outlined what the Pankow Way meant to engineers, foremen, and superintendents in the field, managers and support staff in the office, and executive managers (fig. 48). The client service focus of the Pankow Way is unmistakable. Charlie Pankow's business model, including negotiated, lump-sum contracts and the promise of no contractor-initiated change orders; the involvement of the contractor as part of the building team under design-build; and techniques of concrete construction automation, were means to the end of meeting the expectations of owners, who often had projects languishing on the drawing board when they contacted the Pankow firm.

The demand for CPI to recruit a new generation of professional staff and inculcate it in the Pankow Way came from Winmar Company, a client that took Pankow project teams across America building office towers and shopping centers.

Winmar: Lifeblood of the Mainland Business

In the wake of the sharp 1974–1975 recession, Don Nash, a senior bond underwriter at Travelers Indemnity, reviewed the state of the surety market. Conditions were terrible, owing to the failure, default, and cash deficiencies of many of the firm's contractor clients. Executive managers of the latter were quick to point to adverse external conditions—inflation, high interest rates, labor productivity—as the cause of their problems. Nash concluded, however, that the root cause of his clients' demise or debilitation lay in "management's inability to cope with the conditional factors caused by the economy." That is, "construction management did not have the wherewithal—the financial, organizational, or managerial talent to effectively respond." These firms "were accidents looking for a place to happen." At the same time, other contractors were surviving, even thriving, during the recession, because their managers controlled their operations through goal-oriented performance measurement. Russ Osterman included Nash's speech in the binder distributed to Pankow's

To executives and upper managers, the Pankow Way meant:

Aligning all of our incentives with the Owner's goals

Being willing to assume responsibility for design and construction for the Owner's overall benefit

Ensuring that we performed, so that we would build a long-term relationship with the Owner

Acting as a Master Builder, not just as a contractor

Preaching to staff that we were not looking to profit from change orders; change orders should only come from Owner-initiated changes in scope

That we were responsible for guiding and managing the design process for the Owner's overall benefit of time and money

That we would never "nickel and dime" the Owner, because we wanted the Owner to succeed

That we would prove through our performance that we were the Owner's agent and partner, and so we would keep the Owner's goals and expectations at the forefront

That our success depended directly on our helping the Owner to succeed

That we would be fair, but not naïve, in all of our dealings with the Owner, subcontractors, agents, and workers

That our success and profit would be a direct reflection on our innovation and performance

That excellence in everything we did was who we were

To superintendents and other members of the project team, the Pankow Way meant:

Always meeting the Owner's schedule

Always going the "extra mile" to perform and meet the Owner's expectations

Always solving problems for the Owner instead of making problems for him

Always having a "can do" attitude in all aspects of performance under the contract

Always doing whatever it took to satisfy the Owner, even if it meant going beyond the scope of work

Always keeping quality, safety, and schedule equal as company priorities

Always trying to catch problems early enough to solve them before they became expensive, even if they were caused by other members of the building team

Always considering the Owner's budget as if it were our own

Always finding solutions to problems at any level and never making excuses

Figure 48. The Pankow Way. It had "various meanings to various people," but "was our way of distinguishing ourselves [from] the rest of the contractors in our industry." (From Thomas D. Verti, letter to the author, 14 October 2011.)

off-site managers ahead of their 1976 annual meeting. In his cover letter, Osterman noted, "All of [Nash's] comments are worth digesting."[2]

The Pankow firm survived the turbulent 1970s. In fact, largely on the strength of its operations in Hawaii, it turned a profit every year. Yet the business on the Mainland endured several lean years, particularly after the 1974–1975 recession. Ultimately, Pankow on the Mainland rode out the decade primarily on the back of a single client, Winmar Company.

In fact, Winmar was Pankow's most significant client across almost two decades. In the estimation of Dean Stephan, "If any entity built the company, it was Winmar. They were a very consistent client, a terrific client." As CEO Richard M. Kunnath observes, "It would be almost hard to overstate its importance to the success of the firm." CPI constructed almost two dozen projects for Winmar, from the Pacific Northwest to New England. During the 1970s, Winmar was Charlie Pankow's primary client on the Mainland, accounting for 60 percent or more of its business. The growth of the Pankow firm in the 1980s diminished Winmar's share of the company's order book, but not the importance of the relationship. Many of the projects that CPI built for the company were steel-framed regional shopping centers, but they also included several of the most noteworthy high-rise, concrete office buildings that the contractor erected during these decades. For Charlie Pankow, Winmar was a model client. As Kunnath puts it: The company "represented that single client who just did classic design-build, lump-sum, quality work over many years."[3]

Winmar was founded in Seattle by Frank A. Orrico and others as a real estate development and management firm. It conducted business strictly on a fee basis. A graduate of University of Washington, Orrico served as an officer in the US Navy during World War II. He was Winmar's president throughout the period in which it did business with CPI. As early as 1954, Orrico became a partner in Larry Smith & Company, a leading real estate consultancy. Smith, who established his firm in 1939, specialized in market analysis. He and his partners worked closely with noted architects John Graham and Victor Gruen, and developers, such as David D. Bohannon, the Stoneson brothers, and J. L. Hudson, on the design of dozens of regional shopping centers. By the mid-1970s, Orrico was regarded as one of the foremost authorities on the building type, having either consulted on or developed more than 75 projects.[4]

Winmar continued to develop and manage properties after the firm became a subsidiary of Safeco Insurance in 1967, but did so exclusively for the benefit of the parent company's investment portfolio. Founded in 1923 as General Insurance Company of America, Safeco originally wrote property and casualty insurance. By the 1950s, it was an industry leader. In 1953 General organized Selective Auto and Fire Insurance Company of America as a subsidiary. It soon changed the name of the latter to Safeco Insurance Company of America. In 1957 General began selling life insurance. Within a decade, it had $1 billion of policies in force. Capital accumulation prompted General to diversify into other lines of business. The Winmar acquisition followed from Safeco's decision to invest in high-quality commercial real estate that generated streams of income. In 1968 General changed its name to Safeco.[5]

Robert Heisler initiated the Winmar-Pankow relationship. Heisler had worked with the May Department Stores Company for a number of years. His key contact in the company was Dick Brewer, who was responsible for construction activities in the western United States. About the time that Charlie Pankow was starting his own company, Frank Orrico was looking to hire a shopping center construction expert. Winmar's business model included advising clients on aligning their developments with local or regional retail markets and leasing space in those projects. As Heisler tells it, Winmar had a number of prospective projects that promised to generate generous leasing fees, including a Montgomery Ward department store that would anchor Sherwood Manor, a shopping center in Stockton, California. Yet Orrico's staff "couldn't get the proper construction numbers" to justify building them. Orrico initiated a search for a construction manager. He recruited Brewer through the managers he knew at the May Company. When Orrico flew to Los Angeles to interview Brewer, the two men had lunch with Heisler. Once Orrico hired Brewer as vice president of development, Heisler's Key Mechanical Industries became Winmar's preferred mechanical and HVAC subcontractor. When Brewer sought Heisler's advice on contractors, he recommended Charlie Pankow.[6]

The first project that CPI built for Winmar was the Montgomery Ward store in Stockton. It anchored the north end of the 500,000-square-foot, enclosed shopping center. The first Macy's department store in the region—and the ninth in California—anchored the south end of the property. Stone Brothers and Associates was the developer. Formed by Max and Merrill Stone in 1948, the company had focused initially on residential development. After building some 16,000 homes in the area, the brothers expanded into multifamily residential and commercial real estate. Sherwood Manor was their largest retail project. Winmar was responsible for leasing and construction management. On Heisler's recommendation, he and Brewer discussed the project with Charlie Pankow over lunch. The latter responded with a design-build proposal that Brewer accepted. The project broke ground in 1964, making it the third project on the contractor's résumé (see appendix A). Pankow assigned Ralph Tice to be the project's superintendent. The store and shopping center opened the following year.[7]

With the backing of Safeco's $2 billion in assets, Winmar became "by virtually any measurement . . . one of the major real estate investment companies in the United States," Orrico enthused in his 1979 *Profile of the Company* letter. Winmar operated "on an investment level matched by few of its contemporaries." The Pankow firm benefited handsomely from this success. Like Peter Kiewit Sons', Winmar created divisions along geographical lines. In most of the regions, CPI constructed the "vast majority" of the projects. Still, the relationship was not an easy one. Explains Rik Kunnath: "The contract negotiations were certainly arm's-length." Frank Orrico remained a rather aloof figure in the relationship, leaving others to deal with the contractor. And Orrico's representatives at the negotiating table were tough, Brewer in particular. According to Russ Osterman, Dick Brewer and Charlie Pankow had such large egos that they could hardly stand to be in the same room with one another for any length of time. And so, as Osterman tells it, the task of negotiating with Brewer fell to him. At the same time, according to Brad Inman, who managed Pankow's Northern California operations from 1974 to 1988, Brewer "let us basically kind of run with a design-build approach on their projects." Both sides benefited from executing projects that "were really successful." Contributing to this success were CPI's capacities and capabilities, which the company expanded and upgraded at the urging of Brewer.[8]

Upgrading and Expanding Managerial and Engineering Capabilities

In the early 1970s, Dick Brewer complained to Russ Osterman that CPI needed to bolster its off-site managerial staff with talented individuals with degrees from top university programs. (The off-site managers were the people with whom Brewer, Orrico, and George Kohl, the Winmar president's "right-hand man," dealt.) To address the immediate need that Brewer identified, the company hired several experienced men as project sponsors.[9]

Three recruits came from the building division of South San Francisco–based Guy F. Atkinson Company, a contractor known above all, like Kiewit, for its highway and heavy construction projects. Bill Carpenter, the division's senior estimator, had joined CPI several years earlier. Between 1970 and 1972, he helped Charlie Pankow and Russ Osterman recruit Jon Brandin, Brad Inman, and Dean Stephan from Atkinson. All three had earned degrees in civil engineering from Stanford University. Inman had joined Atkinson out of school. From 1961 to 1966, Stephan had served as a commissioned officer in the US Navy Civil Engineering Corps, before joining Atkinson. At the time that Bill Carpenter queried their interest in meeting Pankow and Osterman, Inman and Stephan were engaged in the construction of the Dobie Center, a residential hall on the University of Texas's flagship campus, as superintendent and project engineer, respectively. Pankow, Osterman, and Carpenter flew to Austin to meet them. Inman and Stephan gave the CPI executives a tour of the project, which was one of the earliest to use flying forms, as a recent *ENR* article had noted. The design of the building also called for prefabricating brick curtain walls, which saved at least $50,000 and 2 months over conventional bricklaying methods on the $7.8 million job, according to *ENR*. Stephan had also worked on the Kaweah Delta District Hospital in Visalia, California, which *ENR* had cited for the structural vaults and foam forming that the Atkinson team had used on the project. Impressed by this résumé, Charlie Pankow invited Inman and Stephan to join CPI. Both men declined. The conversations continued, however. About a year after Pankow, Osterman, and Carpenter paid their visit, Atkinson closed the doors of its building division. At the time, Inman and Stephan were working on the preconstruction phase of a shopping center project. Atkinson had allowed them to continue the work, and also agreed to let them form their own construction entity to build light industrial projects on the side. After 18 months, Inman and Stephan concluded that this career path was a dead end. They returned to California with Atkinson's heavy and highway divisions, respectively. Within six months, both men sought out and joined CPI, deciding that there was more opportunity in buildings than in dams and roads. Inman worked on the construction of an office building in Lexington, Kentucky, for Winmar and then sponsored projects out of the San Francisco office for the next 14 years. Stephan joined CPI as a project sponsor. For 6 years he handled the Winmar account out of the Altadena office before becoming vice president of Pankow Construction Company, a CPI subsidiary (fig. 49).[10]

Jon Brandin did not join Atkinson immediately out of school, but by 1965 he was working for the firm. He was Inman's project engineer on a job

Figure 49. Dean E. Stephan. This photograph dates from his time as president of the company, 1993–1997. (Courtesy of Charles Pankow Builders, Ltd., Pasadena, California.)

in Daly City, California, and worked with Stephan on the Kaweah Delta District Hospital project. Brandin joined CPI about the same time as Inman and Stephan and spent most of his time on projects in the Pacific Northwest. In the mid-1970s, he headed a short-lived Seattle branch office before leaving CPI to work for Seattle-based Baugh Construction, so that he might continue to work where he wanted to live.[11]

At the time that Brandin, Inman, and Stephan joined CPI, Charlie Pankow also hired George Roberts, a University of California graduate in industrial engineering. Roberts was the brother of the partner in the prestigious San Francisco–based firm of Bohnert, Flowers, Roberts and McCarthy, which served as CPI's counsel. In 1971 he joined Carpenter and Inman in the San Francisco office as a senior project manager. He would leave eight years later to manage construction operations in Northern California for the Koll Company.[12]

Another significantly experienced hire was Rik Kunnath, who would eventually become CEO (fig. 50). He joined the firm in 1979. A graduate in civil engineering from the University of Detroit, Kunnath worked from 1971 to 1975 for Mobil Oil and from 1975 to 1979 for BASF as a construction engineer. In this capacity, Kunnath acted as a "service provider" with a limited career path because, as he explains, "You're not the business. The business is not what you do. . . . You can make some money, you can have some fun, but you're not going to be president." In joining CPI, he found a company that he thought he might one day run. Recruited by Bill Heine, Kunnath began his career as

Figure 50. Robert Law and Richard M. Kunnath at the 1981 annual meeting of Charles Pankow, Inc. (From the private collection of Albert W. Fink.)

superintendent on the construction of Windward Mall, in Kaneohe, Oahu (see appendix B). He then would spend five years as a project sponsor before Charlie Pankow would name him vice president of business development.[13]

The hiring of Brandin, Inman, Stephan, and Roberts met Dick Brewer's immediate demand that Pankow expand its stable of capable off-site managers. Looking to the future, Charlie Pankow also initiated a recruitment strategy that targeted engineers with newly minted degrees from leading university programs.

Hiring University Graduates, Steeping Them in the "Pankow Way"

In recruiting university graduates, Charlie Pankow and Russ Osterman sought men who would be eager to spend several years in the field before moving into management positions. Pankow also paid close attention to personality traits that he valued in himself and in the men whom he led in the Kiewit building division: character, integrity, sense of purpose, a drive for excellence, and work ethic. For he would expect new recruits to deliver projects under difficult circumstances. There was no formal training program to teach new employees the Pankow Way, but, by the end of the 1970s, the cohort hired during the decade knew from their experiences on building sites what the company expected of them.[14]

Charlie Pankow, George Hutton, Russ Osterman, and Bob Carlson, the operations manager, led by example, but new hires also learned the Pankow Way from peers who served informally as mentors. All university graduates were initially assigned to building sites as field engineers. "Career development was on-the-job training," Tom Verti explains. "Work hard and you'll get promoted" became the prevailing belief among new hires. Once an engineer reached the level of superintendent, he—the company hired no female engineering graduates prior to the 1990s—might remain at that level indefinitely or move into management. In either case, he carried the responsibility of inculcating the company's culture in those with whom he worked.[15]

The company approached the recruitment of engineers much as a professional football team might prepare for the draft. Led by Osterman, himself a standout lineman on a Rose Bowl–winning team, CPI sought, as Dean Stephan puts it, "the best players that [were] available, whether we need[ed] them or not, because we [would] find a use for them. . . . We had tremendous need for people [who] could perform." And, within a "highly unstructured and very, very fluid" organization, new recruits had the opportunity to grow into "very responsible positions within the company because nobody drew a box around them." Like myriad Silicon Valley entrepreneurs, Charlie Pankow explicitly avoided hierarchical structures and eliminated traditional boundaries between employer and employee.[16]

Most of the engineers hired directly out of university during this period graduated from Purdue University.[17] Alumni who rose to upper management positions include Robert Law, Dean Browning, and Joseph Sanders. Law and Browning were recruited at the same time. In early 1974, while students, both men attended a continuing education conference on campus. Charlie Pankow was the principal speaker. The topic of his talk was tilt-up construction. Armed with their résumés, Law and Browning spoke with Pankow individually at coffee breaks and sat on either side of him at a lunch. Pankow offered both of them jobs before he left town. Law and Browning joined the firm after graduating in May and August, respectively (fig. 50).[18]

Sanders was hired in 1979. During his senior year, Professor Donn E. Hancher, his adviser, arranged an interview for Sanders with Charlie Pankow, who was coming to campus to deliver the keynote address at the annual construction conference co-sponsored by the School of Civil Engineering, the Associated General Contractors of Indiana, and the Indiana Concrete Council. In his remarks, Pankow explained how his company used on-site precasting in the design and construction of concrete buildings. Sanders was "fascinated with [Pankow's] knowledge and understanding of building and how to put buildings together and how to control or manage the building process." The company newsletter would later call Sanders "a classic Pankow success story," characterized by, as he elaborates, one "working [one's way] successively up the responsibility ladder [and] taking on more and more challenges" within a culture animated by the archetypical master builder (fig. 51).[19]

To be sure, the company did not recruit exclusively from Charlie Pankow's alma mater. Tom Verti, a standout football player at the University of Washington, joined Pankow in 1971 after receiving a degree in architecture with a specialization in building technology and administration (fig. 52). Verti was attracted to the company because, after only eight years, it already "had a great reputation here on the West Coast for its creative abilities in performing innovative concrete construction." Working for Pankow also gave Verti an opportunity to work in the field, which he did for eight years before moving into management as a project sponsor, or, as he puts it, "I got demoted into the office." Verti's remark sheds light, both on the importance of the superintendent as the linchpin of the organization and the personal satisfaction and sense of accomplishment that engineering graduates got from work at the building site.[20]

Dick Walterhouse graduated from the University of Michigan in 1979 (fig. 53). A standout university baseball player, Walterhouse did not join Pankow immediately. Rather, he first played four years in the Pittsburgh Pirates minor league system. As the story goes, Russ Osterman, as an alumnus of the University of Michigan, wanted to recruit an engineer from the Ann Arbor campus to counterbalance the long string of hires from Purdue. At the annual Rose Bowl party that he held at his house, Osterman asked Michigan's assistant athletic director if he knew of any engineer-athletes who might make a "good fit" for the Pankow firm. The latter mentioned Walterhouse, who had won the Medal of Honor as the top student athlete in the Big Ten Conference in his senior year. Osterman first contacted Walterhouse during the baseball season and Walterhouse joined the firm in October 1979. For the next seven years, Walterhouse worked his way through the ranks, from field engineer to superintendent. In 1987 he began working as project sponsor in the San Francisco office.

Figure 51. Joseph Sanders. From 1979 to 1988, the Purdue civil engineering graduate rose through the ranks of field engineer, project engineer, and superintendent—the typical career path for university recruits. Thereafter, he held various leadership positions within the firm, including project sponsor, director of engineering, regional manager, senior vice president, and director of operations. (Photo taken in 2012. Courtesy of Charles Pankow Builders, Ltd., Pasadena, California.)

Figure 52. Thomas D. Verti at the 1981 annual meeting of Charles Pankow, Inc. (From the private collection of Albert W. Fink.)

Following his graduation from Stanford University in 1980, Kim Lum worked for Pankow that summer in Oahu as a field engineer, performing material takeoffs (MTOs) on the Windward Mall building site (fig. 54). He had been accepted into the master's program in construction management at Stanford and was supposed to begin his studies in the fall. Excited by how much he was learning on the job, Lum asked for a postponement. He put off the program another two years before dropping his plans to return to school altogether.[21]

Figure 53. Dick Walterhouse. The standout baseball player at the University of Michigan followed the typical career path of a university recruit, working his way through the ranks until he was promoted to project sponsor in 1987. (Courtesy of Charles Pankow Builders, Ltd., Pasadena, California.)

Figure 54. Kim Lum. The rare hire who left the company and returned, Lum has held a number of executive management positions in the firm. (Courtesy of Charles Pankow Builders, Ltd., Pasadena, California.)

Together with Rik Kunnath and CFO Kim Petersen, Sanders, Verti, Walterhouse, and Kim would form the leadership of the firm upon the death of Charlie Pankow (see chapter 7).

Tensions inherent in the traditionally dual role of the engineer as both professional and businessman "have been among the most important forces shaping the engineer's role on the job, in his professional relationships, and in the community at large," argues Edwin T. Layton.[22] His point of reference was the large corporation, where engineers who wished to advance their careers became managers beholden to the demands of commerce within rationalized organizational hierarchies. Such tensions do not appear to have surfaced within the Pankow firm, where everyone understood, and seem to have self-consciously embraced, the close link between the technical and the commercial in Charlie Pankow's business model. Like Pankow, who joined in 1954, many of the firm's engineers were members of the American Concrete Institute (ACI), an engineering society defined in terms in industry affiliation: a reflection of the duality of their professional lives. (Many engineers were also members of the American Society of Civil Engineers [ASCE], which Charlie Pankow joined in 1948.) Moreover, there were no organizational boundaries separating engineering and management. Functions, such as engineering, finance, marketing, and operations, were all of a piece, and focused sharply on client service at the project level. Those who worked as project sponsors, operations managers, estimators, and related positions remained closely involved in the technical aspects of building. There was no organizational chart. Job titles, such as they were, meant little. Engineers who were promoted out of the field and into the office did not manage, as professors of business administration understand the term.

At the same time, the company's second generation of professionals took responsibility for the formalization, rationalization, and computerization of internal systems, such as estimating and cost accounting, fundamentally changing the way in which CPI conducted business and foreshadowing the managerial structures that would be put in place, largely after the death of the founder. For instance, Dean Stephan and Bob Law implemented a project estimating and cost tracking system. When Stephan arrived from Atkinson—a company that had elaborate systems for pricing jobs—he was shocked to discover that CPI had no formal method of estimating jobs. The first time he participated in a meeting with Dick Brewer to negotiate the price of a contract, Russ Osterman, CPI's chief estimator, was essentially "writing numbers on the back of an envelope," based on the cost per square foot on the previous project, adjusted for inflation. Once Brewer and Osterman agreed to a price, to use Winmar as an example, operations manager Carlson would meet with the superintendent, once the job had started, to determine the cost control estimate, using technical reports from similar jobs. The technical reports recorded the history of a project's costs in dollars. The superintendent reported actual costs in weekly Labor Distribution Reports and monthly Job Cost Reports. Carlson would use the cost control estimate to determine variances after the project was completed. Stephan was "absolutely dumbfounded" by this approach to cost management (even though, as he notes, it worked because CPI was doing similar projects, and only three of them per year, at most.)[23]

To determine construction costs before the owner awarded the contract and enable CPI to handle more projects at a given time, Stephan initiated the development of systems to track costs in hours rather dollars (to account for inflation and costs that varied by location); standardize work items; establish guidelines for the preparation of outline specifications; and generate cost parameters with consistency, speed, and accuracy. Because managing the Winmar account was his top priority, Stephan prevailed upon Law to execute the task. With assistance from engineers who rotated through the office, Law developed a robust estimating system that supported the delivery of increasingly complex projects. In the four years before Stephan initiated the project, that is, from 1969 to 1973, actual direct and indirect construction costs on jobs ran an average of 11.5 percent above estimated costs. For the rest of the decade, actual and estimated costs were almost equal.[24]

Not all university hires performed as expected in the field. Those who did not thrive in this environment, however, soon departed. Attrition assisted acculturation in ensuring, as Law puts it, that "everybody was totally focused [on] what we were doing and how we were doing it and why we were doing it, how we could do it better . . . everybody was on the same page."[25]

Winmar's Architect: Welton Becket and Building Team Cooperation

Pankow's relationship with Winmar reconnected the men of the Kiewit building division with Welton Becket, designer of the developer's regional shopping centers and many of its office buildings. The Winmar-Becket relationship likely originated with Dick Brewer, as the May Company was a client of the design firm. And though Welton Becket was not the firm that the May Company retained for its shopping center projects, it had established its bona fides as a designer of regional malls. Welton Becket's work, as Richard Longstreth writes, "was not distinguished by artistic prowess so much as by efficient, nononsense resolution of complex programs in which budgetary constraints were paramount." Former Welton Becket architect Arthur Love echoes Longstreth, noting that clients "admired [Welton] Becket because of his business acumen as much as they did the style and design of the buildings that he produced. We had a reputation of producing efficient, well-thought-out, economical office buildings." This "no-nonsense" approach to design and construction may have frustrated the firm's designers who, as Love puts it, felt at times that they were unable "to push the design envelope as far as some of [their] major competitors," but would have impressed both Brewer and Charlie Pankow.[26]

Influenced by John Graham & Company's Northgate Center (1948–1950)—"the key prototype for work of the early 1950s," according to Longstreth—Welton Becket designed three major Northern California shopping centers in succession: Stonestown, in San Francisco (1950–1952); Hillsdale, in San Mateo (1952–1954); and Stanford, in Palo Alto (1954–1956). In Southern California, Becket designed two of the "large, fully integrated retail complexes" that opened in the Los Angeles metropolitan area in the wake of Lakewood Center (1950–1952), the first such center in the region. Each one was "oriented to a pedestrian mall and conveyed little semblance of a conventional retail district": Anaheim Plaza (1954–1957), developed by Broadway-Hale Stores, and Los Altos Shopping Center, Long Beach. By the time that Bob Heisler was introducing Charlie Pankow to Dick Brewer, Welton Becket had participated in the development of thirty or more shopping centers.[27]

Perhaps because of its similarities to the "total design" approach that they practiced (see chapter 1), Welton Becket's architects supported design-build as a project delivery method. For architect Love, who worked with CPI project teams out of Becket's Chicago office from its establishment in 1970 (with the acquisition of Childs & Smith) until 1981, working in a design-build configuration meant a return to his early days as a designer. As an undergraduate at the University of Illinois, he and several fellow students in the architecture program "design-built" homes for professors. And cooperating with the contractor during both design and construction made sense. As Love recalls:

> I was sort of excited about the whole notion that you could work directly with the builder and take advantage of their innovative ideas and incorporate them into the architectural design process and make it what I really felt, to quote Mr. Becket, was really more "total design," because it involved the construction aspects. It wasn't just the artistic end of design, but it was the reality, because without a good contractor and a successful construction, you really don't have a building.

According to Dean Stephan, architects who worked with the Pankow firm in a design-build relationship could be assured that their designs did not remain on the drawing board.[28]

Working with CPI permitted Welton Becket's architects to "push the design envelope." From Love's perspective, Pankow's engineers "were really very creative in their own right," notwithstanding their obligation to produce a building within budget. In the interest of using concrete "in more creative ways," they prodded Love and his colleagues "to reach out" architecturally in their designs. And they did. Reflects Love: "We rose to the challenge that they presented to us. . . . It was intriguing to be challenged and then, in turn, to challenge them. . . . We developed a relationship where the client would challenge both of us." Of course, Charlie Pankow's preference for concrete as a structural building material constrained design freedom, but no more so than if a client insisted on a steel-and-glass building, according to Love. Becket's designers and Pankow's engineers collaborated in a mutually productive relationship.[29]

Cooperation between Becket and Pankow, for instance, saved joint owners Winmar and Northwestern Mutual Life Insurance $2 million on the Citizens Fidelity Bank Building in Louisville, Kentucky (fig. 55). Maurice D. S. Johnson, chairman of the board of Citizens Fidelity, the state's largest bank, wanted to construct a building that would be the focus of attention of the city's commercial and financial sectors. State law prohibited banks from owning real estate, however, so the bank's directors and executives sought a developer who had demonstrated its ability to build projects of the highest quality and act as long-term landlord. Bank staff would occupy ten floors of the 30-story tower. In 1971 Winmar had completed Oxmoor Center, Kentucky's largest mall, on Oxmoor Farm, the 2,000-acre estate that bank director Thomas Walker Bullitt had inherited in 1957. Bullitt, a partner in the law firm of Wyatt, Tariant & Combs, and Winmar had developed the regional shopping center as a joint venture. The trust under which Bullitt had inherited the property, which had been in his family since the original owner, Colonel William Christian, had deeded it to his son-in-law, Alexander Scott Bullitt, as a wedding gift in 1786, prohibited him from selling the property. So Bullitt formed the Beargrass Corporation, named after the creek that runs through the estate, to handle the leasing of the land to Winmar. Bullitt likely played the key role in bringing the developer to the attention of his fellow directors. Impressed by Winmar's execution of the Oxmoor project, the bank approached Frank Orrico. Once directors approved

Figure 55. Citizens Fidelity Bank Building, Louisville, Kentucky, aerial view. Completed in 1972. Welton Becket & Associates, architect. (Courtesy of AECOM.)

the project, Winmar formed a joint venture with Northwestern Mutual Life Insurance to own the completed structure. "A strong advocate for modern architecture," Johnson hired Welton Becket as architect of the signature tower.[30]

As Charlie Pankow later explained in the *Journal of the American Concrete Institute*, Welton Becket and Winmar established "a rather unusual bidding procedure" for the project. They decided to design the building as a steel structure, but also to invite alternate bids in concrete. Given the bidding process and the project's high-profile status—the tower would be Louisville's tallest upon its completion—CPI formed a joint venture with a local contractor, Whittenberg Construction. Together they submitted a bid based on Becket's plans and specifications for a steel structure, but also an alternate proposal for a reinforced concrete structure with a slipformed service core. The contractors would precast beams, planks, and panels, and install the latter, with insulated window units, instead of the metal and glass curtain wall designed by Welton Becket (fig. 56). The gridded array of identical panels may have

Figure 56. Construction of the Citizens Fidelity Bank Building, 10 May 1971. Slipforming of central core proceeds two floors ahead of installation of precast concrete wall panels. (Photo Photography, Inc. of Louisville. Courtesy of Charles Pankow Builders, Ltd., Pasadena, California.)

produced a façade that one critic described as "particularly anonymous," but the window units markedly reduced heating and air conditioning costs. The low bid for a steel structure—submitted by the Pankow-Whittenberg team—exceeded the owners' budget, but their alternate proposal reduced construction costs by $2 million; the project went ahead on this basis.[31]

Welton Becket integrated the alternate proposal into the project to ensure that the project was completed on schedule. According to Charlie Pankow, the design firm "handled the transformation of the alternate reinforcing concrete design and exterior wall panels in a most efficient manner that

Figure 57. Workers install precast wall panels. The contractor substituted insulated window units for the metal and glass curtain wall in the initial design. (Courtesy of Charles Pankow Builders, Ltd., Pasadena, California.)

neither interrupted the start of the work nor delayed its progress." The architects completed working drawings in a sequence that enabled contractors to keep their crews fully occupied (fig. 57). Anticipating by some three decades the building industry's embrace of the concept of "integrated project delivery," a "high level" of cooperation and coordination existed among the owners, contractors, and architects, which "contributed substantially to the timely and trouble-free completion of the project." Pankow made the case for design-build: "Efficient coordination prevents change orders and disputes and contributes to the sense of satisfaction and enjoyment of everyone involved."[32]

CPI constructed other office towers for Winmar, including buildings in Bellevue and Renton, Washington; the First Security National Bank & Trust Building in Lexington, Kentucky; the Pacific First Federal Center in Portland, Oregon; and the 411 East Wisconsin Building in Milwaukee, Wisconsin (see appendix A). Together, these concrete building projects enabled CPI to perfect a system that, as Charlie Pankow wrote in *Concrete Construction* in 1980, has "worked very well for our company in constructing a variety of buildings." As he explained, the sequence of tasks in his construction program involved: slipforming the central service core, installing precast beams and floor planks, installing precast wall panels, casting in place exterior columns formed by wall units, and applying a lightweight concrete floor topping. The contractor demonstrated it that could deliver tall concrete structures using these techniques. Yet, in also erecting large shopping centers for Winmar, CPI showed that the value of design-build was not limited by the material used to frame the structure.[33]

Design-Build Applied to Retail Construction

Shopping centers were "the most pervasive new building type" in postwar America.[34] By the end of 1960, when almost 4,000 centers had been built, analysts fretted that the sector showed signs of maturing. Nevertheless, developers pressed on with new construction—with good reason, it turned out. As John F. Lynch, vice president for commercial and industrial sales at Eagan Real Estate—Winmar's co-developer of Penn-Can Mall in central New York State—asserted in 1978, "Shopping malls are the most attractive real estate investment in the country today." Lynch spoke at a possible tipping point. Saturation, recession, local development policy, and the increased cost of capital had slowed new development. Boom times for the industry resumed during the Reagan years, however. In all, developers added some 14,400 centers to the nation's inventory during the 1980s. By 1990, some 36,500 shopping centers in America were ringing up $681.4 billion in sales and employing 8.6 million people.[35]

Winmar's involvement in shopping center development, beginning in 1947, spanned this four-and-a-half-decade-long heyday in retail construction. The new regional malls that it developed after its acquisition by Safeco included the aforementioned Oxmoor Center and Penn-Can Mall, both designed by Welton Becket; Valley Fair Mall, in Salt Lake City; Washington Square, outside Portland; Bristol Town & Country, in Santa Ana, California; and Windward Mall (see appendices A and B). As the Penn-Can Mall example illustrates, Winmar sometimes worked with a joint venture partner. This was also the case with Windward Mall, which Winmar developed with partner Norris, Beggs and Simpson on behalf of the Bishop Estate. All but the smaller-scale Bristol Town & Country, a community shopping center comprising a mere 160,000 square feet, were enclosed structures that were typical of second-generation suburban malls. Ranging in size from 400,000 square feet to 1.1 million square feet, each of the regional shopping centers was one of the largest, if not the largest, in its region. From a marketing perspective, Winmar sought to distinguish these projects from their competition, resulting in several shopping center "firsts." Penn-Can Mall, for instance, was the first "multi-level, international fashion shopping center in Central New York."[36]

CPI was the contractor on all of these projects, except for the Valley Fair Mall. (It would be involved in its expansion and renovation, however.) On these projects, CPI was responsible for the concrete foundation work, and, if it was called for in the design, precasting and tilting up the exterior walls. It subcontracted most of the work, including the structural steel frame. Penn-Can Mall was an exception. Federal Insurance, CPI's bonding company, directed Pankow to execute the project as a joint venture with a local contractor, William E. Bouley Company. Superintendent Ken Hardee and project engineer Norm Husk supervised construction of the mall under a construction management agreement. All construction work was subcontracted, and all subcontracts were direct agreements between Winmar and the subcontractor. Bouley performed the concrete foundation work. The arrangement, involving two construction companies, was "a waste of time," according to Dean Stephan. "They didn't contribute anything. We ran [the project] like we did everything else." As examples of the innovation that Pankow brought to retail construction, however, interviewees pointed most often to the enclosure, expansion, and renovation of existing centers.[37]

With the stalling of new construction during the 1974–1975 recession, owners turned to redeveloping regional shopping centers built 15 or more years earlier. Such efforts dated from the early 1960s, when many first-generation malls were already considered to be obsolete. But the trend quickened as opportunities for new projects dwindled. Developers often acquired properties from the original owners. By the mid-1980s, owners also began to target their own second-generation malls for redevelopment. Both types of projects filled CPI's order book during this period. Expansion and renovation of first-generation centers included Capitol Court in Milwaukee and South Shore Plaza in Braintree, Massachusetts. Second-generation rehabilitation projects included Valley Fair Mall and Oxmoor Center.[38]

In undertaking these efforts, owners confronted demographic shifts, changes in consumer preferences, changes in physical access patterns (as a result of new highway construction, for instance), higher operating costs, new residential construction, and new competition, all of which contributed to faltering margins and loss of market share. Properties might be functionally or physically obsolete. First-generation centers, in particular, were visibly aging and suffering from outmoded appearances.[39]

Renovation and expansion addressed common areas, parking lots, landscaping and mall furniture, walkways, signage, and, of course, the buildings themselves. Redevelopers often added one or more anchor stores to the property, as well as specialty stores. Renovation typically involved the enclosure of centers that were built with open malls, often as a result of local building codes. Many of these ideas were not new. For instance, Allied Stores, owners of Northgate Center, enclosed that mall during 1962–1963. Still, as of the mid-1970s, many first-generation shopping centers operated with open-air common areas. Whatever the nature of the redevelopment effort, there was consensus within the industry on the need for centers to remain fully operational during construction. This imperative required owners to secure the cooperation of tenants. Contractors had to minimize interference with shoppers. The superintendents and engineers interviewed for this study cited the success in meeting the latter requirement as one of the most important achievements of their shopping center experiences. Two cases in point are Capitol Court and South Shore Plaza, which provided CPI with its first experiences in regional shopping center redevelopment.[40]

Capitol Court was Wisconsin's first regional mall when it opened on Milwaukee's northwest side on 28 August 1956. Developed on an "L" plan, owing to the triangular shape of the 61-acre site, and designed by John Graham & Company, the shopping center was originally anchored by a three-level branch of Schuster's department store, the developer of the $20 million project, and a two-level branch of T. A. Chapman's, an upscale local retailer. In October 1956, a two-level J. C. Penney's store opened for business. Wide overhangs and colonnades shielded shoppers from rain and snow. A U-shaped tunnel facilitated vehicular access to a basement that accommodated sales, deliveries, and storage. When Capitol Court opened, there were plans to construct three additional blocks. Two of the blocks were completed in September 1959, adding 60,000 square feet of space and 17 stores. The expanded mall offered 805,000 square feet of leasable area and 85 stores and services. In 1962 Gimbels acquired Schuster's. (Founded in Philadelphia in 1842, Gimbels had moved its headquarters to Milwaukee in 1887. Its 8-story flagship store, built in 1901–1902, extended along Wisconsin Avenue and faced the Milwaukee River.)[41]

Designed by Victor Gruen Associates, South Shore Plaza opened in 1961 with a branch store of Filene's and 30 specialty shops. (Founded in 1881, Boston-based William Filene's Sons had been part of Federated Department Stores since 1929 and was considered to be the region's leading retailer.) In 1962 the owners added R. H. Stearns as a second anchor, increasing its leasable area to 757,000 square feet. In 1968 the owners added a branch of Boston-based Jordan Marsh. (Founded in 1841, the company had been part of the Allied Stores empire since 1928.)[42]

Winmar invested some $13 million to enclose and renovate both of these properties. As it involved the addition of two new anchors and a parking garage, South Shore Plaza was the more complicated of the two projects. Welton Becket's design called for enclosing 75,000 square feet of common area, extending the mall, and constructing a truck tunnel to accommodate a new department store anchor, a branch of New York–based Lord & Taylor. A second phase called for the construction of a reinforced concrete parking structure and a new Sears store. While the latter was under construction, Winmar decided to expand the Filene's store. CPI performed this work under a change order. The Capitol Court project, designed by Los Angeles–based Charles Kober Associates, involved the enclosure of almost 72,000 square feet of exposed common area and the addition of 60,000 square feet of leasable space through both the construction of the third block originally planned for the center and the development of 15,000 square feet of new space in the existing mall. The developers later added a 15,000-square-foot food court to the new block. With this expansion, Capitol Court offered 107 specialty stores.[43]

Winmar insisted that the contractor complete these projects without interrupting business. A consultant retained by Winmar to study the South Shore Plaza project concluded, however, that this requirement could not be met. Winmar reported this finding to Charlie Pankow and Dean Stephan, who indicated that they might be able to develop a solution. As Robert Law, who began his involvement with South Shore Plaza as project engineer and ended it as superintendent, reflects, "We looked at it and we figured out logistically how it would be possible."[44] And so CPI secured the work. A year later, CPI's engineers also figured out how to enclose and renovate Capitol Court without interrupting retail business.

CPI's enclosure of South Shore Plaza began in August 1976. Under the direction of superintendent Norm Husk, Law, field superintendent Tom Rouhier, and carpenter foreman Dick Williams, crews finished the work involved in enclosing the shopping center, including roofing, heating, lighting, and interior finishes, by Thanksgiving weekend (fig. 58). Work then stopped for the Christmas season. To ensure that subcontractors performed their work expeditiously, subcontracts spelled out CPI's expectations in what the project's technical report deemed a "somewhat unusual environment." Subcontractors were responsible for considering all existing conditions prior to starting their work: Subcontracts provided that no payments would be made for work necessitated by omissions, problems, or discrepancies on the part of subcontractors. Because the mall would remain open during construction, subcontractors also agreed to schedule work to minimize tenant disruption.[45]

A year after it began work on South Shore Plaza, CPI applied this approach to the redevelopment of Capitol Court. The contract called for work to begin in August 1977 and finish 10 months later. CPI agreed to install a new roof and entrances and much of the new lighting and heating systems by Thanksgiving weekend, at which point work would pause until the end of holiday season. Work would resume with slab leveling and the tasks of the finishing trades.[46]

CPI worked slightly ahead of schedule. Crews erected the walls of the new block and finished much of the slab-leveling operation before the holiday break. Winmar allowed a reduced crew to work through the holiday period, enabling the finishing trades to "get off to a fast start" when work resumed on 3 January 1978. By late May, the project was essentially finished.

Figure 58. South Shore Plaza, Braintree, Massachusetts, under construction, January 1977. The enclosure of the regional shopping center was completed in time for the holiday shopping season. Welton Becket & Associates, architect; The Engineers Collaborative, structural engineer. (Courtesy of Charles Pankow Builders, Ltd., Pasadena, California.)

Crews then constructed the food court under a change order. Newly christened Capitol Court Mall, the shopping center opened with a week-long celebration, beginning on 23 August.[47]

Viewed through the lens of the company's culture, which valued the construction of concrete towers, redeveloping regional shopping centers did not carry the cachet of erecting office buildings and condominiums. As Rik Kunnath explains: "If it's not a high-rise building and it's not pouring 200,000 yards of concrete with a crane and a lot of other things, then it's not a real project." Whereas erecting a concrete tower might have required the services of some 250 carpenters, constructing a regional shopping center demanded perhaps only 25. Nevertheless, shopping center projects were, "dollar-wise, the equivalent or larger" of commercial high-rise projects and "very complicated, [with] lots of different parts and pieces."[48] As suggested by Federal Insurance's insistence that Pankow form a joint venture to build Penn-Can Mall, these projects entailed rather more execution risk than Charlie Pankow may have appreciated. And these redevelopment efforts, complicated though they may have been to complete, were prologue to even more intricate and demanding projects. As the next chapter relates, the addition of second levels to malls would involve some of the most innovative work in the history of the company.

Generating Construction Work through Captive Development Projects

Entering the commercial market as speculative developers was the other principal means by which Charlie Pankow and Russ Osterman filled CPI's order book during the 1970s and early 1980s. In a memorandum distributed to the company's off-site managers in May 1976, Osterman insisted that there were opportunities for designers and builders who could create a usable and economically attractive project, regardless of macroeconomic conditions. The commercial building boom of the early 1970s, during which annual levels of office construction peaked at 110 million square feet, had been followed by a severe recession, marked by a steady decline in the number of office buildings constructed.[49] Nevertheless, tenants still sought to lease space in Class A facilities at competitive rates, Osterman averred. He called on his colleagues to lead a "bottom line response" by tailoring the required finished product to the owner's budget and marketing the company's ability to make projects viable through "creative management, combined with, and including, performance ability and responsibility." During the 1970s, CPI had limited success on the Mainland in selling this "bottom line," design-build–based approach to clients other than Winmar. In part, this surely owed to the restrictions embedded in Charlie Pankow's business model. Osterman and Pankow, however, demonstrated the essential soundness of the former's analysis by developing millions of square feet of office space for their own account.[50]

During the 1970s and early 1980s, Pankow and Osterman developed half a dozen or more projects on the Pacific Coast. At the same time, Hutton and Pankow developed seven projects in Hawaii, beginning with the Nuuanu Brookside condominium in the Punchbowl district of Honolulu. (All but one of these projects, a timeshare complex on Kauai, was a high-rise condominium. See appendix B.) No particular strategy drove these development plays. Indeed, the first opportunities on the Mainland arose from the needs of one client, rather than through the impetus of Osterman and Pankow. Nevertheless, the projects entailed little financial risk to the developer in the way in which they were structured. To be sure, Osterman and Pankow invested millions of dollars to acquire sites for development, but land could be sold in a reasonable amount of time if construction did not materialize. At the same time, they typically were able to secure financing for the entire project, as developers often were able to do during this period. The captive projects were lucrative. They generated more wealth for Pankow, Osterman, and Hutton—the principal partners in each of the projects—than the company's construction activities. Generating construction work was always the principal reason for undertaking these projects, however. As such, with the exception of the Kauai timeshare, all of the projects were either office buildings or condominiums that leveraged the contractor's expertise in precasting, slipforming and other concrete construction techniques. They were typically turnkey projects, meaning that they included the finish work for tenants or occupants.[51]

The relationship between Osterman and Pankow on the Mainland developments paralleled Hutton's and Pankow's in Hawaii. As the person who spearheaded new business development for CPI, Osterman was often the general partner of the development entities created in conjunction with the projects and always put together the deals. Charlie Pankow typically participated as a limited partner. Yet, as Kunnath notes, "Regardless of who founded and managed the opportunities [Pankow] was *always* [in fact] the controlling partner."[52]

The entry of Pankow and Osterman into the development field began in 1969 with discussions that led to the construction of San José Plaza, a two-building complex in downtown San José, California, for the Pacific Telephone & Telegraph Company (PT&T) (fig. 59). Owing to an unsatisfactory arrangement with its landowner, PT&T was seeking new quarters for its staff, but did not wish to own the building. (The telephone company had no requirement for a structure to project corporate identity. Since 1925, PT&T had been headquartered on New Montgomery Street in San Francisco in its iconic, moderne skyscraper, designed by Timothy Pflueger.[53]) Charlie Pankow likely learned of the project through Derk Hunter, whose firm, Demmon-Hunter was the leasing agent. Both architect Allan M. Walter and the contractor were part of the team that had designed and built both 1625 The Alameda Building and Joe West Hall on the campus of San José State College (see chapter 2). Osterman negotiated a leaseback arrangement with the company. He and Pankow then formed a development entity to own the building, which was designed

Figure 59. San José Plaza, downtown San José, California. Construction of the first office building began in March 1970. It was ready for occupancy on 1 January 1972. Construction of the second office building began in March 1972. Completed in March 1973. Allan M. Walter & Associates, architect; Richard R. Bradshaw, structural engineer, San José Plaza I; L. F. Robinson & Associates, structural engineer, San José Plaza II. (Photo Peter S. Carter Photography. Courtesy of Charles Pankow Builders, Ltd., Pasadena, California.)

to meet the requirements of PT&T and other tenants assembled by Hunter. Construction began in March 1970. Pankow set up a casting yard in an adjacent lot, which it leased from the City of San José, since the building's footprint occupied the entire site. Construction stopped in May to allow structural engineer Richard Bradshaw and Walter to make design changes to meet unspecified "major tenant requirements." Work resumed in August. Pankow's crews, supervised by Alan Murk, completed the shell in 11 months. Tenants were able to move in on New Year's Day of 1972. [54]

The telephone company's need to accommodate a growing staff led to PT&T and its parent, AT&T, becoming a repeat client for CPI. Before the San José Plaza building was completed in late 1971, PT&T managers approached Osterman and Pankow about constructing a second, similar building as part of a complex. Owing to a lack of space, Pankow set up a casting yard some

Figure 60. Casting yard, Milpitas, California. Created to support precast operations associated with the construction of the second office building at San José Plaza. (Courtesy of Charles Pankow Builders, Ltd., Pasadena, California.)

four miles from the job site, in the city of Milpitas, on land leased from the San Jose Steel Corporation (fig. 60). Construction of the second structure began in March 1972 and finished 12 months later, despite a strike that halted work for almost 2 months (fig. 61).[55]

Following the completion of San José Plaza, Pankow and Osterman developed two additional buildings for the telephone company in San Francisco's Yerba Buena Redevelopment Area. Comprising some 635,000 square feet, the two structures met the company's requirement for additional space. Their form followed the function of housing a portion of the corporation's sprawling white-collar workforce as anonymously as possible. Thus, the opaque AT&T building that materialized was, in the words of one observer, "an imposing concrete fortress designed to keep the public at bay and business operations as secure and secretive as possible."[56]

Figure 61. Workers pour elevated slab, San José Plaza II. Looking toward the Bank of America Building, the tallest building in San José from its completion in 1926 until 1987. The completed San José Plaza I office building appears on the left. (Photo Peter S. Carter Photography. Courtesy of Charles Pankow Builders, Ltd., Pasadena, California.)

The construction of PT&T's regional headquarters shows how CPI's expertise in design-build and concrete construction intersected with developer-tenant relations to make Charlie Pankow and Russ Osterman millions of dollars. As was the case with San José Plaza, PT&T managers entered into a long-term lease arrangement with an ownership entity created by Pankow and Osterman—in this case, Building Enterprise, Ltd. Late in the project, which broke ground the first week of March 1975, Pankow informed PT&T that his team would be able to complete the project 90 days ahead of the lease date. The telephone company agreed to advance the lease by 3 months, if CPI met this target date. Alan Murk, the project's superintendent, met the accelerated completion date with a second shift. The contractor delivered the project 11 months after it broke ground—despite delays of 2 months in the spring of 1975 because of inclement weather and poor soil conditions. In doing so, Building Enterprise, Ltd., saved some $4.5 million in payments on the $12.8 million construction loan and related costs.[57]

Other CPI teams may have surpassed the performance of Murk's crew on subsequent projects in terms of schedule, but none have achieved comparable cost efficiencies. Acting as both developer and contractor, Pankow assembled a building team that included architect John Bolles & Associates, designer of Clementina Towers (see chapter 2), and subcontractor KMI. While the developers and SFRA officials worked through a five-month process associated with site acquisition, street abandonment, and utility removal or relocation, beginning in October 1974, designers responded to Charlie Pankow's call to minimize the project's cost and time to completion, without sacrificing quality. At about $23 per square foot, the PT&T building cost some 30 percent less to build than Walnut Center, Pasadena, completed in 1985, and Westside Media Center, Los Angeles, completed in 2002: two projects cited by Robert Law—a project engineer on the PT&T project—as examples of cost-efficient construction (see appendix A).[58]

CPI tailored slipform operations to meet the requirements of the tenant and the site. Redevelopment officials limited the building's height to 80 feet. Designers set the building 6 feet below grade to allow seven floors to be constructed. Despite the low height of the structure, using a slipform to cast the service core was cost-effective, Murk insisted. To reduce the amount of sheet metal ductwork and increase the number of walls that could be used to frame the structure, designers incorporated air-conditioning shafts into the core. Developed by KMI's Robert Heisler for the San José Plaza buildings, this technique also saved time on the installation of mechanical systems. Pneumatic jacks lifted the slipform at a rate of about 1 foot per hour. Crews erected each 60,000-square-foot floor in 11 days (figs. 62, 63).[59]

More than 400 precast panels comprised a key element in the construction cycle. Cast with cavities on all four sides, the cavities on the sides of the unit formed three sides of the exterior columns that were poured in place between panels. The top cavity served as the form for the spandrel connecting the columns. The bottom cavity reduced the weight of the panel, cutting material and handling costs. The panels, along with more than 850 prestressed girders and beams, were manufactured at the Milpitas casting yard—50 miles from

Figure 62. Slipform operation, office building for Pacific Telephone & Telegraph, 4 August 1975. Elevated view, looking southwest. (Courtesy of Charles Pankow Builders, Ltd., Pasadena, California.)

the job site. Working with the architect and structural engineer, CPI's five-man project team designed structural bay widths with the reach and capacity of the two tower cranes used on the project, minimizing the number of precast elements needed to erect the building. Casting operations commenced with excavation to facilitate just-in-time delivery of precast pieces to the building site.[60]

The building team took other steps to control prices and "protect the rent structure for the tenant." With inflation topping 12 percent and the Nixon administration's wage and price control policy in tatters, it committed to orders for rebar, elevators, and electrical and mechanical equipment before construction began. Building Enterprise, Ltd., also applied for building permits in stages to enable crews to perform excavation and foundation work before designers finalized working drawings.[61]

Development projects of the 1970s not tied to the needs of the telephone company included Jefferson Plaza, a professional building in Spokane, Washington, and the Citizens Bank Building in Eugene, Oregon. Architects for both projects were predecessor firms of Seattle-based Callison. Olsen & Ratti, also based in Seattle, were structural engineers for both projects. Jefferson Plaza,

Figure 63. Pacific Telephone & Telegraph office building, Yerba Buena Redevelopment Area, San Francisco, nears completion, 17 February 1976. Elevated view, looking southwest. John Bolles & Associates, architect. (Courtesy of Charles Pankow Builders, Ltd., Pasadena, California.)

built in the city's medical district, was similar in scale to the second Borel Estate office building (figs. 27, 64). The bank project involved a 50-50 partnership with Lew Walker, a local developer. Citizens Associates, the development entity, acted as owner, developer, and leasing agent. (The bank occupied only 3 of 10 floors.) How the relationship with Walker developed is unclear, but Osterman was likely responsible for it. The high-rise project was more typical of the projects that the company had recently undertaken (fig. 65). Atypically, the developers subcontracted the precast operations on both projects. They saved money by reusing the forms for the precast concrete panels from the Jefferson Plaza project for the exterior walls of the typical floors (fig. 66). The façades of the buildings were similar in appearance, of course, an effect reinforced by the designer's specification of buff-colored, lightweight, and sandblasted concrete for the exposed surfaces of both structures. For a contractor that "reveled in the fact that our buildings were different," the reuse of the precast

Figure 64. Jefferson Plaza, Spokane, Washington. Completed in 1974. Callison Associates, architect; Olsen & Ratti, structural engineer. (Photo Paula M. Siok, taken November 2010.)

molds caused some embarrassment for the firm. As Dean Stephan reflected, "We rued the day that we used the same molds on [those two buildings] because that created a bad reputation that we just did cookie-cutter buildings and that's how we produced them [for such attractive prices]."[62]

With the completion, in rapid succession, of four high-rise condominiums in Honolulu between 1979 and 1982 (see appendix B), Southern California remained the only region without a development project. That changed with the arrival of Jon T. Eicholtz in the Altadena office. A graduate of the University of Kansas, Eicholtz spent a decade in Hawaii before joining Pankow in 1978. He had been the project manager for Swinerton & Walberg on the construction of the 1,800-room Sheraton Waikiki, a complex project that finished on time and under budget. Six months later, Eicholtz took an executive vice president position with Pacific Construction. A wholly owned subsidiary of the Hawaii Corporation, Pacific was one of the largest contractors in the Pacific Basin. In 1972 Eicholtz replaced Bill Pruyn as president. Later he was named CEO. In 1978 the Hawaii Corporation, which was interested in many lines of business, including real estate, car dealerships, and savings and loan associations, went bankrupt. Pacific Construction emerged from the bank-

Figure 65. Slipform operation, Citizens Bank Building, Eugene, Oregon, January 1974. (Photo James Studio. Courtesy of Charles Pankow Builders, Ltd., Pasadena, California.)

ruptcy proceedings as a new entity. The Bank of Hawaii held the stock of the company until a buyer could be found. During this period, Eicholtz became chairman of the board, too. He agreed to stay on for six months, if a buyer were found. Artec Construction, a subsidiary of the interests of Clint W. Murchison Jr., son of the notorious Texas oil baron and the original owner of the Dallas Cowboys, acquired the firm. After the ordeal of keeping a company afloat under conditions imposed by the bankruptcy court, Eicholtz elected to resign once he had helped with the transition to new ownership.[63]

Eicholtz joined Pankow in Hawaii soon thereafter. During his tenure with Pacific Construction, he had befriended George Hutton. Both men were active in the local chapter of the Associated General Contractors of America. Eicholtz served a term as its president. He was also president of a construction industry legislative group, which enhanced his fairly high profile locally. Eicholtz had met Charlie Pankow as well, but the relationship had remained

Figure 66. Citizens Bank Building, Eugene, Oregon, ready for occupancy in September 1974. The developers reused the forms for the precast concrete panels from the Jefferson Plaza project. Callison, Erickson & Hobble, architect; Olsen & Ratti, structural engineer. (Photo Northwestern Photographics. Courtesy of Charles Pankow Builders, Ltd., Pasadena, California.)

a cursory one. When Eicholtz resigned from Pacific, Hutton approached him about joining the company. Pankow flew to Honolulu and offered him a job in business development. Eicholtz accepted and, for reasons that remain unclear, even to him, began working out of the Honolulu office. It soon became apparent, however, that there was no need for Eicholtz to develop business in Hawaii. George Hutton was well entrenched in the relatively small development community and was generating as much work as CPA could handle. As a result, Pankow asked Eicholtz to relocate to Southern California.[64]

Initially Eicholtz spent "95 percent of his time" on affordable housing proposals to the California Housing Finance Agency. CalHFA had been established in 1975 as the state's affordable housing bank to make low-interest loans to qualified buyers. US housing officials often worked through such state agencies to implement the new Section 8 Rental Voucher Program, which aimed to provide low-income families with the opportunity to rent units in privately owned apartment buildings. In the lean years of the mid-1970s, Charlie Pankow and Russ Osterman considered federally subsidized housing to be a business opportunity, if projects could be done on a lump-sum basis. Nothing came to fruition, however. Competitors often had large staffs that were devoted to this work. Eicholtz was acting alone—a significant disadvantage, given the time and effort needed to prepare proposals that met the expectations of housing officials. The opportunities that he pursued "were just very difficult to create, primarily because of the bureaucracy of dealing with any federal agency. . . . It was just a nightmare."[65]

By the spring of 1979, Eicholtz was working on the first of two major projects that materialized in Southern California during his eight years with the Pankow firm: a condominium in the Wilshire Corridor in West Los Angeles. It would be the first such project for Pankow on the Mainland. Charlie Pankow's participation in the project had more in common with George Hutton's participation in projects in Hawaii than with his prior efforts with Russ Osterman.

Until the late 1960s, Wilshire Boulevard between Westwood Village and the Los Angeles Country Club was lined with apartments. Developers seeking greater returns converted many of them into condominiums. When they exhausted this option by the mid-1970s, they built high-rise luxury condominiums. In all, developers erected more than 30 buildings, up to 27 stories in height, transforming the nine-block-long Wilshire Corridor. Eicholtz's project, named for its address, 10560 Wilshire, rose 22 stories above an underground parking structure.[66]

Working through a broker, Eicholtz assembled four lots that provided the requisite footprint. Charlie Pankow used CPI funds to acquire the land. Eicholtz then hired Maxwell Starkman Associates to develop preliminary drawings and outline specifications. After securing the entitlements that he needed to construct the project, Eicholtz produced a development proposal, including a pro forma budget and sales projections, as a tool to secure funding. He approached possible developers and lenders at a meeting, in Dallas, of the Urban Land Institute, of which he was a member. There he lined up a Greek developer and Citibank as investors. Together they provided 100 percent of the equity in the project, reimbursed CPI for the land, and gave Pankow Development Corporation (PDC)—an entity that Charlie Pankow had recently established, with Osterman as its president—a small stake in the project. The investors also agreed to let the construction contract to CPI.[67]

Eicholtz's development efforts also bore fruit with Catalina Landing, a 740,000-square foot, multiple-building project that extended along the Long Beach, California, waterfront on land owned by Crowley Maritime, the operator of a ferry service to Catalina Island from an aging terminal. Characterized in the company newsletter as "one of California's most exciting and unique office building developments," the complex included four office buildings, a three-level parking structure, a rehabilitated ferry terminal, a shopping arcade, and an "interconnecting two-and-a-half-acre elevated architecturally treated promenade" (fig. 67). What stood out, from the newsletter's perspective, was the "imaginative and functional use of architecturally sculpted precast concrete" (fig. 68). Workers under the supervision of Jack Grieger produced more than 2,000 structural columns, beams, and girders, and architectural wall panels, spandrels, and parapet panels for the project in a three-and-a-half-acre yard. Crews cast the decorative pieces in fiberglass molds that Tony Giron and others designed, detailed, and fabricated at a yard in Irwindale, a sparsely populated city in the San Gabriel Valley, and shipped to the job site.[68]

The project was a partnership of Crowley Development Corporation (CDC) and PDC. The former was an arm of Crowley Maritime, a transportation and logistics firm founded in 1892 in San Francisco. CDC contacted Charlie Pankow about remodeling the ferry terminal at Long Beach and developing the land around it. Pankow passed along the contact to Eicholtz, who worked with CDC staff on the project. Once they established the feasibility of the larger project, CDC and PDC formed Catalina Landing Associates to develop it.[69]

Charlie Pankow's last development project was 2101 Webster, one of three high-rise office buildings constructed in Oakland's Lake Merritt district during the mid-1980s boom. The steel-framed building is distinguished by the sculptured panels made from Sierra Mountain Granite aggregates and gray, tinted glass windows that, together, create a façade that "sparkles majestically in the Oakland skyline," as the company's newsletter enthused (fig. 69). Perhaps as dramatically, the 2-story lobby featured a Narcissus Quagliata–designed, 52-by-12-foot, stained-glass mural backlit by more than 600 feet

Figure 67. Catalina Landing, Long Beach, California, aerial view, August 1984. Hope Consulting Group, architect; Robert Englekirk Consulting Engineers, structural engineer. (Photo Warren Aerial Photography. From *CP News* 2 [Summer/Fall 1984]: 4.)

of fluorescent tubing. Those who built the high-rise point with pride to the way in which the design maximized floor efficiency. Mike Townsend, one of many business development professionals who rotated through the San Francisco office in these years, brought the opportunity to Charlie Pankow. Because Townsend lacked construction experience, Pankow asked Eicholtz to help launch the project. (Illustrating the oddities of lines of communication within CPI, Townsend reported to Pankow even as he worked under Eicholtz's direction.) The project broke ground in December 1983. The ceremony featured the presentation of a plaque from Mayor Lionel Wilson to Russ Osterman (who was not involved as a development partner on the project) for CPI's contribution to the development of Oakland over two decades, beginning with MacArthur Broadway Center (fig. 70).[70]

Development projects on the Mainland may have not been large in number, given the two decade period over which they occurred, but they helped to carry the company through the lean years of the 1970s and into the 1980s, when more office space was constructed in America than had been built cumulatively in the history of the country.[71] While Catalina Landing and 2101 Webster were under construction, the Pankow firm was also engaged in other large office projects for other developers on a design-build basis, including

Figure 68. Suspended promenade constructed of precast concrete columns, beams, and handrail panels cast on-site, with lightweight concrete paving. Flanked by parking garage on the left and office building on the right. (Photo Warren Aerial Photography. From *CP News* 3 [Spring/Summer 1985]: 4.)

Grand Financial Plaza in downtown Los Angeles, Crocker Plaza in Long Beach, and Walnut Center in Pasadena (see appendix A). CPI was now in the midst of a period of growth that was limited only by the capacity of the organization to recruit and train engineers in the Pankow Way.

Working with Subcontractors and the Trades

Constructing projects for Winmar and for Charlie Pankow and Russ Osterman as developers spread CPI project teams across the country, requiring them to work with unfamiliar subcontractors. In contrast to George Hutton and his staff, who became well-acquainted with all of their subcontractors within a rather closed community, CPI's sponsors and superintendents faced

Figure 69. 2101 Webster office building, Lake Merritt district, Oakland, California. Looking east toward the Oakland Hills. Completed in 1985. Bolles Associates, architect; Robinson, Meir & Juilly, structural engineer. (Photo Jack Duran. Courtesy of Charles Pankow Builders, Ltd., Pasadena, California.)

the challenge of identifying new firms that could perform to their expectations on almost every project that they managed outside of California. CPI may have stood out among contractors in the amount of work that it self-performed, which reduced its reliance on others for the execution of a project. Yet "buying out" subcontractors still cost some 60 percent of the gross amount of a typical construction contract. The financial consequences of working with subcontractors who performed below standards or even went out of business during construction could be significant.

Figure 70. Groundbreaking ceremony for the 2101 Webster office building, 16 December 1983. Oakland Mayor Lionel Wilson presents plaque to Russell J. Osterman and Pankow Development's Mike Townsend. Oakland City Council member Fred Cooper looks on. (From *CPI News* 3 [Winter 1984]: 4.)

With the company lacking in managerial hierarchies along either product or geographical lines, each project effectively operated as its own company, corporate accounting aside. Superintendents were given a budget to buy out subcontractors and line up suppliers. Project sponsors generally identified and interviewed potential candidates. As Dean Stephan recalls, this process typically started with a call to the local ready-mix concrete firm for its recommendations, then "you'd go to the Yellow Pages, you'd sit in your hotel room and you'd interview people by the hour as they trundled through." As the company's superintendents and sponsors learned, the outcomes of searches might vary significantly by location.[72]

In their technical reports, project engineers often took note of the challenges that teams faced in searching for qualified subcontractors and suppliers in new places. On the Jefferson Plaza project, for instance, some of CPI's bids elicited no responses. For, as project engineer Ronald Smith noted, the construction market in Spokane, Washington, was not competitive. As a result, the company had to arrange the shipment of many tools and materials to the building site from Seattle. Moreover, subcontractors' schedules were "seldom met," despite superintendent Jon Benner's "ample" briefings to the managers of those firms "well in advance" of deadlines. Owing to the un-favorable bids submitted by subcontractors for brick work associated with the Capitol Court Mall project, superintendent Norm Husk recruited a team of bricklayers directly from Local 8 of the International Union of Bricklay-ers and Allied Craftsworkers. Most of the subcontracted work associated with the addition of the Sears store and the expansion of Filene's depart-ment store at South Shore Plaza was performed in a cooperative manner and on schedule. The expansion of the shopping center followed its suc-cessful enclosure, and so satisfactory relations with various subcontractors might have been expected. Still, the project team, led by Robert Law and Dick Walterhouse, experienced difficulties with selected subcontractors. The drywall subcontractor did an "acceptable" job, but its failure to ad-here to schedules added time to the project. Moreover, the subcontractor's "constant changing of foreman resulted in added work and instructions for [CPI] personnel." The proprietor of the masonry subcontractor also "was difficult to deal with." One of Robert Heisler's subsidiary firms, Key Engi-neering, designed the HVAC system, but subcontracted its installation to a firm that was "extremely hard to keep on the job." Walterhouse had to apply "constant pressure" to keep the masonry work on schedule. The structural steel frame subcontractor left many items unfinished, requiring its employ-ees to return to the site "on numerous occasions" to complete its work. The firm also changed its job foremen "continually," causing problems for Law and Walterhouse. These examples suggest the difficulties that a contrac-tor's engineers and superintendents might experience in controlling work at the building site. [73]

In at least one case, the failure of a subcontractor to perform redounded to the benefit of CPI as contractor. As it was accustomed to doing, CPI pro-posed that it produce all of the precast elements needed in the construction of the Citizens Fidelity Bank Building. Winmar allowed the contractor to fabricate the structural planks and beams, but balked on the architectural pieces. As former controller Doug Craker relates, they "felt that they hadn't seen enough of precast work from us to justify doing this job." Winmar in-structed CPI to subcontract the work to Dolt & Dew. The leading precast and prestressed concrete firm in the area, Dolt & Dew had assisted archi-tect Norman Sweet in the design and construction of Louisville's Catalyst and Chemical Building and had manufactured the 3,346 precast pieces used in the construction of the city's federal building, completed in 1969. Having produced about 30 percent of the panels for the Citizens Fidelity Bank Building, however, the firm declared bankruptcy. Per company policy, CPI had bonded the subcontractor. Now project sponsor Dean Stephan convinced the bonding company that it could perform the work more cheaply than any other precast manufacturer. Under the supervision of Ralph Tice, the project team set up a casting yard next to a commercial concrete batch plant, 10 miles from the job site. Pankow acquired Dolt & Dew's fiberglass forms, A-frame trailer units, and cement truck and relocated the equipment to the yard, where crews manufactured more than 1,000 architectural elements used to clad the tower.[74]

At the same time that it sought to minimize the number of workers that it needed on the job by deploying techniques to automate construction, Pankow as contractor made it standard practice to recruit workers from the union hall, regardless of project location, reasoning that it made good business sense to use workers who were experts in their respective crafts. Each union trades worker might cost more on an hourly basis than a nonunion laborer. Nevertheless, the quality and timeliness of union work proved to be cost effective. The difference between union and nonunion labor hit home for Doug Craker when he left CPI in 1980 (for family reasons) and took a job with Brownlee Construction in Knoxville, Tennessee. Craker found that the nonunion workers he encountered were more likely to be jacks-of-all-trades rather than masters of specific tasks and that the resources needed to supervise them more than erased any savings that might be gained by paying lower wages.[75]

Union trades people sought work on Pankow jobs in California. Explains William "Red" Ward, who worked on several CPI jobs as a carpenter before he was recruited into the firm, word circulated in the locals of the United Brotherhood of Carpenters and Joiners of America (UBC) that "if you could ever get a job out of the hall on a Pankow job, you want[ed] to do that." If you were working at another site "when Pankow fire[ed] up a job, you want[ed] to drag up from where you [were] at and go back to work for them." For the contractor's superintendents "always seemed to have good direction. You worked hard [and] you got rewarded for that by getting paid and keeping [*sic*] on the job." According to Ward, his fellow carpenters in the hall agreed with him that the jobs of other contractors were less organized. Adds long-time superintendent Mike Liddiard: "We were always respected by the unions because we had such a good safety record."[76]

Unions supplied Pankow with people who became employees and supervised projects, as Charlie Pankow encouraged superintendents to identify outstanding candidates on their job sites. Both Liddiard and Ward were carpenters who became superintendents. Most union workers who joined the Pankow firm chose not to progress beyond this level, for doing so meant working in the office rather than in the field. Ward became the rare Pankow hire from the union hall to advance to upper management when, in April 2000, he was named operations manager for Southern California. In doing so, he followed in the footsteps of Alan Murk, who began his career, as we have seen, with Kiewit.[77]

Worker availability was the primary factor that shaped Pankow's experiences with union labor as the company broke out of California during the 1970s. In the summer of 1977, for instance, superintendent Norm Husk found it difficult to recruit skilled trades workers for the Capitol Court project, owing to a high level of local construction activity. Husk advertised for carpenters in the local newspapers, but no more than 25 percent of the workers he was able to secure were journeyman mechanics or had comparable levels of skill and experience. He turned to a local agent, but most of the workers the latter supplied were unskilled and inexperienced. Unsurprisingly, turnover was high. On the Jefferson Plaza project, only two men, the superintendent and the general foreman, had experience with precasting and other techniques deployed regularly on Pankow projects. As the project's technical report observed, "It took quite a bit of time and hiring of men from the union halls before a dependable, conscientious crew was formed. The final crew was a blend of old hands skilled in their respective trades and young men eager to learn and willing to accept the challenge of a concise timetable." When CPI placed the project trailer on the job site of Pacific First Federal Center in November 1979, project manager Alan Murk and superintendent Norm Husk worried about the impact of a possible construction workers' strike on the availability of labor. The strike did not materialize and Murk and Husk were able to recruit "good crews" in all of the trades. They found, however, that the productivity of field hands was "considerably lower" than the level to which they had become accustomed in California. Moreover, the cement masons "were lacking in prowess except for two older hands that remained with the project until all concrete work was finished." As these examples illustrate, the labor market varied by time and place. As was the case with subcontractors, CPI project teams adapted to the conditions of local labor markets to execute projects to meet schedules and budgets.[78]

With the cooperation of local unions, projects for Winmar stayed on schedule. In fact, local work practices might work to the contractor's advantage. For instance, Norm Husk found that the trades in Milwaukee were more tolerant of crossing lines than those in California, compensating for a relatively slower work pace that owed to the practice of downing tools whenever it rained. Moreover, adhering to local union contracts did not reduce productivity, as Winmar's First Security National Bank and Trust Building project in Lexington, Kentucky, showed. Union agreements required the use of composite crews. Carpenters and laborers stripped forms for poured-in-place and precast work on a one-to-one basis. Carpenters and ironworkers installed precast elements on a one-to-one basis, too. Crafts workers performed all other operations within their trades. The CPI project team encountered no problems with this configuration of work. As the technical report noted, the pride of union trades workers in their work was an important source of productivity.[79]

To be sure, Charlie Pankow automated concrete construction tasks primarily to reduce the cost of labor, which, he averred, increased by as much as 1,000 percent between 1955 and 1980. Indeed, he observed as ACI president, "slipforming did not become a major factor in our industry until construction labor costs made it competitive." The same was true, Pankow added, of precasting, "to some extent."[80] Self-performing these techniques helped the contractor insulate budgets and schedules from subcontractors, too. But these techniques did not eliminate labor from the building site, and many of the workers who remained were craftsmen. It is likely that Pankow at least sympathized with the political efforts of the Construction Users Anti-Inflation Roundtable, formed in 1969 to address union-driven wages in the construction industry.[81] Nevertheless, executing his construction program demanded levels of skill and experience that carpenters and other trades workers hired out of the union hall provided. The success of the company's project teams was closely linked to their performance.

Pacheco Village: Learning the Value of Bonding Subcontractors, Avoiding Litigation

On one project, Murphy's Law seemed to hold. Everything that might have gone wrong in the contractor's relationships with subcontractors and suppliers did go wrong.

Charlie Pankow worried increasingly about litigation during the 1970s. Legal claims had been minor during CPI's first eight years in existence. Then they became "quite extensive," as he noted at the 1976 off-site managers' meeting, held during the first weekend in May at the Del Monte Hyatt House in Monterey, California. Charlie Pankow took great pride in his relations with owners and his company's ability to resolve any problems related to construction to the latter's satisfaction. Though their financial impact was ultimately low, several legal disputes with subcontractors and suppliers had arisen because of inadequate construction documents or notices to proceed on work, suggesting to Pankow the need for the company to tighten its controls on subcontracts and materials contracts. More disturbing was the increasing exposure to jury trials that contractors apparently faced generally, prompting Pankow to insist that subcontracts include adequate arbitration clauses. It was company policy to bond subcontractors where the amount of the contract was at least $5,000, but not all subcontractors were bondable, as it turned out. The experience with the ready-mix concrete company on the Pearl Two condominium suggested that it might be appropriate to bond suppliers as well. In this case, as George Hutton noted, Pacific Concrete and Rock had $500,000 in product liability coverage, whereas the loss on the project could well have been $5 million. One project focused the mind of Charlie Pankow on his company's exposure to financial loss more than any other: Pacheco Village.[82]

Pacheco Village was a 180-unit apartment complex in Novato, California, that architect Leo S. Wou and his partners developed for moderate-income families under Section 221(d)(4) of the National Housing Act. A complex of redwood-framed, low-rise buildings, Pacheco Village was not the type of project that CPI normally would have been interested in building. As Brad Inman understands it, the company undertook it "as a courtesy" to Wou, who had recently established a San Francisco office. At the time, CPA was constructing the Esplanade and had recently completed the Kauluwela Co-op, both of which Wou had designed (see chapter 3.) Doug Craker, the former controller, adds that Charlie Pankow accepted the job "because it was available," and he wanted to learn "how [a project] like that would work." (In similar fashion, CPI built several parking garages in these years to provide newly minted engineers with field experience [see appendix A].)[83]

The project was run out of the San Francisco office under the direction of Bill Carpenter, who recruited a project manager with experience in the building type. Construction began in the spring of 1972. Almost a year later the project was in trouble, as Inman learned firsthand on a visit. In January 1974, Pacheco Village Properties, Leo Wou's development entity, defaulted on interest payments owed to Advance Mortgage, the lender, alleging that the contractor had failed to meet the terms of the construction contract. The mortgage lender, in turn, stopped disbursing funds under its building loan agreement with Pacheco Village Properties, leaving CPI partially unpaid for its work. The parties failed to work out an overall settlement, but the mortgage lender and the developer acted to reinstate the building loan agreement to satisfy HUD (US Department of Housing and Urban Development) officials, who allowed the project to go to "final closing." This stalemate over payments was only the tip of the iceberg, however, as far as problems that plagued the project were concerned.[84]

CPI's difficulties began with project management and extended to subcontractors and suppliers. The project accounting system set up in the wake of the departure of Rosser Edwards and David Boyd broke down, as Carpenter's project manager instructed the office manager—the person who completed project-specific accounting tasks and was responsible for furnishing data to the accounting department in Altadena—to ignore procedures. The project manager apparently failed to perform due diligence on subcontractors and suppliers as well. Both Domingo-Allison, the supplier of the wood-framing materials, and Green Brothers, the framing subcontractor, went bankrupt during the project. Whereas the bankruptcy of Dolt & Dew proved to be a blessing in disguise, the bankruptcies of these firms added to Charlie Pankow's growing list of problems. With the project in trouble, Pankow had the project manager fired—he was never the bearer of bad news—and replaced him with Alan Murk.[85]

The defaults of the developer, a subcontractor, and a supplier left CPI financially exposed. Since Domingo-Allison was not bonded, CPI had to pay $170,000 to cover the cost of materials. Green Brothers was bonded by Balboa Insurance, but was unlicensed. Balboa completed the framing subcontract after Green Brothers defaulted and then sued to collect from CPI what it had paid to do so. Charlie Pankow resisted, contending that Green Brothers had no legal recourse, as an unlicensed contractor, for the collection of such payments. He also sued to recover the balance owed to CPI by the mortgage lender, but his lawyers failed to convince the court that the contractor was entitled to an equitable lien on the construction loan as a third party beneficiary of the building loan agreement and a fully performing contractor. Ultimately, Charlie Pankow sought to recover the contractor's losses from the developer. The matter went to arbitration. Eventually CPI was awarded

$750,000. The project left relations between Pankow and Wou in tatters; contractor and architect never worked together again. The project embarrassed Charlie Pankow. He asked Alan Murk to remove the company's sign from the building site.[86]

Among construction projects built by the Pankow companies, Pacheco Village apparently has no peer in causing "major disruption within the company," as Dean Stephan puts it. For Charlie Pankow, the experience demonstrated the need for better daily communications between field and office. He charged his superintendents with keeping a daily job diary and called on managers to document their phone calls. Pacheco Village also prompted the company to rewrite its procedures manual, scrutinize more closely subcontractors' credentials and capacities to perform, deal with suppliers through materials contracts rather than subcontracts, expedite legal work, reduce legal expenses, and ensure proper representation for the company.[87]

Working closely with Robert E. McCarthy, a partner in Bohnert, Flowers, Roberts and McCarthy, Charlie Pankow also took every action that he could to insulate his company from litigation. As Rik Kunnath observes, however, this legal apparatus would not really be tested, even as anxiety over litigation gripped the construction industry.[88] Kunnath suggests that, were he ever asked about it, Charlie Pankow might well have argued that this lack of litigation owed to the provisions that he insisted be introduced into the relevant contracts. For his part, Kunnath suggests that the company's commitment to quality construction and follow-up with owners on actual or potential problems likely have played greater roles in limiting the company's exposure to lawsuits, including those with subcontractors and materials suppliers.[89]

Prologue to Reorganization

In contrast to the localized developer networks and tight-knit community of employees and subcontractors that characterized the milieu within which George Hutton grew CPA, CPI operated from coast to coast, its professional staff dispersed and relatively isolated. (Hence, the annual meeting was an important exercise in team building and technology sharing.) Constructing projects for Winmar and the development entities created by Charlie Pankow and Russ Osterman sustained CPI through uncertain times and carried it into the more prosperous 1980s. Captive projects aside, working on a dedicated basis for a single owner stabilized operations. Close-knit project teams, composed of old hands and new hires, all of whom were inculcated in the Pankow

Way, executed these projects to Frank Orrico's and Dick Brewer's satisfaction. Winmar remained a client for two decades.

Notwithstanding its moderate growth, the company for all intents and purposes operated as an aggregation of project teams, with Charlie Pankow firmly in control. The lack of rationalized managerial structures did not inhibit the company from executing a greater number of projects at a given time than it did in its first decade. Whether it restricted the firm's growth is a moot point, as Charlie Pankow set no targets. As Kim Lum explains, he simply took "jobs he liked where he thought he was making enough money and [forgot] the rest of them."[90]

A more pressing concern was the corporate structure of CPI, which had become a holding company for a bewildering array of development, construction, and equipment entities. By the mid-1980s, the university recruits from the 1970s were moving into positions of greater responsibility. Yet the growth in the book value of CPI made it exceedingly difficult for any of them to acquire an ownership interest. Charlie Pankow was not prepared to commit to a leadership succession plan and had no plans to retire, but he wanted his company to outlast him. Perpetuating CPI under the existing structure was problematic, however. How might he reorganize the company to broaden ownership without sacrificing control of the firm? Ironically, the question would be answered by a reorganization that Charlie Pankow undertook primarily for reasons related to corporate income tax policy.

CHAPTER 5

Reorganization, Growth, and Recession, 1984–1991

A s Charles Pankow, Inc. (CPI), neared its twentieth anniversary, Charlie Pankow pondered its corporate structure. Together, he, George Hutton, and Russ Osterman owned some 85 percent of the company's stock. It was prohibitively expensive to acquire any ownership position, as the book value of the firm had increased appreciably over two decades. The men who had been hired in the 1970s were contributing to the success of the company, but could not share in profits that they were helping to generate. As a corporate entity, CPI had moved away from the employee-owned model that Charlie Pankow had in mind when he started the firm. As Dean Stephan explains: "It became very apparent that it was going to be impossible to perpetuate the company under this kind of structure. The ownership was too concentrated, and there was no way to disseminate it." Thus Charlie Pankow reorganized the company "to provide a workable mechanism for distributing ownership of the company." "The idea," elaborates Timothy P. Murphy, whom Pankow recruited as chief financial officer near the end of the process, "was to create a new company with a low capitalization base where people could buy in and get a reasonable ownership interest."[1]

Pankow's interest in sharing ownership in his company with a second generation of employees was not the primary factor driving his reorganization of the company, however. An accounting event that involved the consolidation of myriad development companies that previously had reported separately from Pankow Construction Company (PCC), the construction entity, resulted in the creation of Charles Pankow Builders (CPBI), incorporated in October 1984. Changes in the corporate and individual tax rates and the repeal of the General Utilities doctrine, enacted in the Tax Reform Act of 1986, spurred Pankow to reorganize the company a second time. Advised by Peat Marwick, one of the so-called Big Eight accounting firms, the assets of both CPI, which had continued to exist, and CPBI were shifted to new limited partnerships under a new parent partnership.[2]

When Charlie Pankow announced the second reorganization at the company's annual meeting in September 1986, Dean Stephan bet him that the net worth of the new entity, Charles Pankow Builders, Ltd. (CPBL), would grow to the size of the dissolved CPI in ten years. In other words, the value of the new firm would double in half the time that it took the original company to reach its value at the time of its liquidation. CPBL would miss achieving that goal by only a couple of months.[3] The path to success was not an altogether smooth one, however. The company enjoyed late-1980s construction booms in Hawaii and on the Mainland, but was buffeted, along with the entire industry, by the sharp recession that followed. Speculative building at the height of the boom produced national office vacancy rates of more than 20 percent—exceeding the worst rates of the Great Depression. The impact of the bust in commercial real estate was especially severe in Hawaii, as the Japanese investors who buoyed the market withdrew almost overnight.[4] The protracted downturn produced changes in the market for contractor services that exposed the limitations of Charlie Pankow's business model.

Retirements and resignations of key personnel in the wake of the reorganization left the business more vulnerable to these external shocks. For one, George Hutton's and Russ Osterman's development work, which insulated the business from the downturns of the 1970s and 1980s, ended with the contentious resignation of the former and the retirement of the latter. In addition, efforts to establish branch offices during this period either were stillborn or faltered.

The company continued to turn a profit, even if volumes sagged. During the bleak early 1990s, the company still had work—indeed, good projects—validating to some extent Charlie Pankow's observation, made in a May 1988 interview, that "our business seems to do better in recessionary times." Based on the experience of the 1970s and early 1980s, Charlie Pankow insisted that troughs in the construction cycle had "had no bearing whatsoever on our business."[5] Yet, as CEO Richard M. Kunnath suggests, Pankow was either not recognizing or not appreciating or unaware of the changes taking place in the marketplace that were making past performance a less reliable marker for future prospects. Charlie Pankow's unwavering belief in the imperviousness of his business model to general economic and sectoral shocks did not bode well for the long-term viability of the firm. At the dawn of the 1990s, the company needed new products to differentiate itself in the marketplace, but had none to offer.[6]

Reorganizing the Company, 1984–1986

Whenever superintendent Alan Murk was in town on a Monday morning, he and Charlie Pankow would talk for an hour or two before the latter flew to San Francisco for the balance of the working week. On one occasion, Pankow told Murk that he had "made arrangements for the company to exist in perpetuity." He did not confide in Murk, with whom he had been associated longer than anyone else in the company, the details of his plans. But Pankow assured Murk that the company would endure without him.[7]

Charlie Pankow wanted CPI to outlast him. Perpetuating the company, however, would require the major shareholders to relinquish their dominant equity positions in the firm. As the largest shareholder, of course, Pankow would make the greatest sacrifice in terms of foregone potential wealth. He was willing to do so, however, "in order to encourage and maintain and attract future leaders of the company that [would] be given the opportunity to share in that equity," explains Thomas D. Verti, who was named regional vice president for Southern California on the eve of the second reorganization.[8]

Charlie Pankow had no intention of transferring his equity interest in the company to his children. As Murphy explains, he felt that "if it's [a] father-to-son kind of a succession, that maybe you get the right guy, maybe you don't, but if it's wide open, then you're going to get the best people to succeed." Rather, according to Murphy and others, Pankow believed that employees should be afforded the opportunity to acquire and accumulate ownership positions in a company based strictly on merit. If so, this was ironic. For in the first two decades of the company's existence, Pankow had not awarded stock as part of an employee compensation program. Moreover, there was no system for awarding bonuses or compensation generally. In fact, Pankow rewarded people as he saw fit. At best, there may have been an implied relationship between performance and ownership interest in the firm. With a second generation of employees poised to move into leadership roles, concentration of ownership became an issue—but apparently not one that was sufficiently pressing to prompt Pankow to reorganize the company. In essence, the company continued as a sole proprietorship.[9]

In any case, to acquire an ownership interest in CPI, the children of Charlie and Doris Pankow would have had to work for the company. Charlie Pankow did not exclude his children from working for him. Indeed, his three sons, "Chip" (Charles J. Pankow III), Rick, and Steve, spent their summers between university semesters working at company building sites. Only the latter, who was 11 years old when his father started CPI, pursued a career in construction. Soon after Steve Pankow graduated from California Polytechnic State University, San Luis Obispo, in 1976 with a degree in construction engineering, he and his father met for dinner in San Francisco. Pankow *fils* floated the idea of working for the company. The father did not reject the idea out of hand, as he suggested in a *Los Angeles Times* interview, but advised his son that he would benefit more were he to begin his career with another contractor; he would gain valuable experience and, his father hoped, receive proper credit for his performance. The son would also gain a better appreciation of what it took to succeed in business. Were Steve Pankow to work for him, Charlie Pankow continued, he would never be recognized for his work, no matter what he accomplished. And, from the company's perspective, hiring his son "would have inhibited [its] ability to attract a first-string organization." In sum, Pankow thought it best for his son to come in through the front door of the company rather than through its back door, so to speak. Ultimately, Steve Pankow pursued his career in the industry independently of his father's firm.[10]

Estate planning that Charlie Pankow performed in 1976 constitutes the earliest indication of his thinking about the long-term viability of CPI. In that year, he restructured his ownership in CPI with the creation of a family trust and related entities. He did so in a manner that enabled him to maintain his majority interest, and therefore, control of CPI. At the same time, he arranged matters so that a large share of any future increases in value of his interest would flow to his children and grandchildren.[11]

Precipitating the reorganization of CPI were factors that had nothing to do with employee ownership or leadership succession, however. Rather, during CPI's fiscal year that ended on 31 May 1983, Charlie Pankow created a number of new entities on the advice of a tax partner at Main Hurdman. (Main Hurdman was a large "second-tier" accounting firm, that is, it was national in scale, but not a Big Eight firm. It was the successor firm by merger to Cranston and Hurdman, CPI's long-time auditors.) Historically, CPI had kept its development assets separate from its construction activities. Now, under the rules of GAAP, or Generally Accepted Accounting Practices—the norms and rules that establish the framework for financial accounting—CPI was forced to consolidate development companies that had previously reported their financial information separately. These entities, known as "Chaskows" (a concatenation of "Charles" and "Pankow") on the Mainland and "PDs," or Pankow Developments, in Hawaii, were vehicles through which PCC, the construction subsidiary, participated in the development projects of Charlie Pankow, George Hutton, and Russ Osterman. The Chaskow associated with the 2101 Webster Street project, for instance, held a minority, but controlling, interest in the property. GAAP now required CPI, as the holding company, to include a $70 million asset and related debt in its consolidated statements. As a result, the size of both the balance sheet and income statement for CPI increased "dramatically" for the fiscal year ending 31 May 1983.[12]

To the managers at Chubb, the parent company of Federal Insurance, Pankow's bonding company, CPI now looked more like a development entity than a construction firm. Explains Kim Petersen, who joined CPI from Coopers & Lybrand, one of the Big Eight firms, in 1983 as assistant controller, and has served as CPBL's CFO since 2000: "bonding companies do not [generally] underwrite development activities the same way they measure risk for a [construction] contractor." And since he placed a high value on his professional and personal relationship with Chubb's George McClellan, Charlie Pankow "was more than a little troubled that he had to travel East to explain this sudden change in the company's financial statements." After meeting with McClellan, Pankow fired Main Hurdman and hired Peat Marwick as the company's auditors and tax advisers.[13]

Charlie Pankow and his accounting staff immediately began planning for the creation of a new and simplified construction group, which resulted in the incorporation of Charles Pankow Builders, Inc. CPBI would be the recipient of all new work. CPI would not be dissolved. It would retain its assets, but wind down its operations as work in process was completed. Selected employees would be offered shares in the new firm. The reorganization thus would go some way in achieving the diversity of ownership in which Charlie Pankow was interested.[14]

Changes in the tax code associated with the Tax Reform Act of 1986 precipitated the second reorganization of the company. As he announced in a nationally televised speech on 28 May 1985, President Reagan proposed to make the US tax system fairer, simpler, and more efficient. Recent congressional debates and well-publicized studies had raised public awareness of the extent to which loopholes had distorted the tax code and the fact that many corporations paid little or no tax on high levels of income. To increase the public's belief that the system was socially just, the Reagan administration sought to lower marginal rates and broaden the tax base. It achieved these goals by wielding expertise, not sparing powerful interest groups, and deferring consideration of contentious issues, such as a national sales tax as a substitute for the income tax, to mobilize bipartisan congressional support.[15]

The Tax Reform Act, which President Reagan signed into law on 22 October 1986, prompted Pankow's second reorganization by making the corporation "significantly less attractive from a tax standpoint relative to partnerships than it was under prior law," as Stanford University economists Myron S. Sholes and Mark A. Wolfson argue. It did so by reducing individual tax rates across the board and lowering the top marginal rate on individual incomes from 50 percent to 28 percent. It reduced the top corporate tax rate from 48 percent to 34 percent. At the same time, the act increased the top tax rate on capital gains from 20 percent to 28 percent. The act also raised the shareholder-level tax—a "second round" of corporate tax assessed on dispositions of ownership interest in a corporation—by eliminating opportunities to avoid or postpone payment of the tax or "time the realization of the gain to coincide with a period in which the taxpayer faced a low rate." As partnerships do not pay tax—only partners as individuals do so—it is apparent why Charlie Pankow's tax advisers at Peat Marwick would have recommended that he change his firm's organization from corporation to partnership.[16]

The repeal in the act of the so-called General Utilities doctrine, effective 1 January 1987, ensured that the second reorganization occurred before the end of 1986. So-named for the New Deal–era decision of the US Supreme Court in *General Utilities & Operating Co. v. Helvering*, the doctrine related to the nonrecognition of gain by a corporation on the distribution or sale of appreciated property in a planned liquidation. The ruling enabled corporations to pass along to shareholders profits relating to the transaction without paying capital gains tax. Shareholders were responsible for paying tax on what amounted to a special dividend. By repealing the General Utilities doctrine, the 1986 act required corporations to treat distributions of property as sales at full market value. Now both the liquidating corporation and its shareholders would have to recognize the full gain or loss of such transactions. The tax code also closed a loophole that would have allowed corporations to circumvent the repeal of the doctrine by converting to an S Corporation—a type of corporation that for all intents and purposes is taxed as a partnership. In repealing the doctrine, policymakers limited the role of the tax system in promoting what they perceived to be a dangerously high level of mergers and acquisitions among domestic corporations. Its repeal reflected, as Scholes and Wolfson note, the "greater emphasis [in the 1986 act] on double taxation of income earned in corporate form."[17]

The reorganization accomplished in 1986 shifted the assets of CPI and CPBI to new limited partnerships under the umbrella of a parent limited partnership, Pankow Management Company (fig. 71). The limited partnership structure allowed future earnings to be taxed at the now more attractive individual rates. To preserve the limited liability protection afforded by a standard corporation, two corporate general partnerships were created: Pankow Holdings, Inc. (for two holding partnerships, Pankow Management Company and Charles Pankow, Ltd.), and Pankow Corporation (for four operating partnerships, CPBL, Construction Specialties, Pankow Building Services, and PDC). Existing shareholder interests were paid off—in just three years, owing to the success of CPBL during the construction boom of the ensuing years and to the elimination of the corporate tax liability (which boosted retained earnings). There were no longer shares of stock to be sold, of course. Rather, a company unit program became the vehicle by which employees participated in the ownership of the firm. As Kim Petersen explains, it was a deferred compensation program that was the equivalent of a "phantom stock program," whereby units were valued in the manner that had been used to calculate share prices. "The difference was [that] the company unit program didn't cost you anything to get in. Units were awarded to you based on length of service and your job and the value of the company and whatever." The unit program would remain in place until the company reorganized again in the wake of Charlie Pankow's death in January 2004.[18]

With the reorganization, CPA ceased to exist. Its accounting department was folded into CPBL. As the reorganizations coincided with a "bust" period in the construction cycle in Hawaii, key people who worked for George Hutton were transferred, either to the Altadena office or to Mainland projects. Many of those individuals would leave the firm soon thereafter.[19]

Most employees were aware that organizational change was afoot, but could only guess at the outcome. Rik Kunnath, who was a project sponsor at the time, recalls, "I certainly didn't understand what was propelling the change." Tom Verti adds, "The company had amassed a significant amount

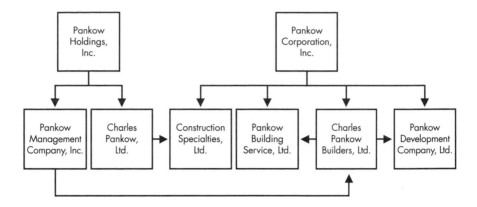

Figure 71. Partnerships following the reorganization of Charles Pankow, Inc., and Charles Pankow Builders, Inc., 1986. Arrows depict general or limited partnership interests. Charles Pankow, Ltd., was created as the repository of CPI notes. Pankow Holdings was dissolved after the holders of CPI notes were paid off. In 1989 Pankow Management, Inc., was created as the corporate general partner. (Modified from figures in binder, Offsite Managers' Meeting, Redwood City, California, 25–28 June 1987 [Charles Pankow Builders, Ltd., Pasadena, California].)

of equity [since 1963] and certainly there were rumors floating around within the company," as people speculated whether Charlie Pankow might sell or disband the company. As was the case with the men of the Kiewit building division, however, Verti and his colleagues placed their trust in their leader and hoped "that things [were] going to work out."[20]

For those who had significant ownership stakes in CPI, the reorganization was controversial. That Charlie Pankow acted unilaterally and did a poor job of selling it to those who were most affected by it made it even more so. For instance, George Hutton recalls that Pankow did not brief him adequately on the reorganization and never explained what he wanted to achieve. As a result, Kunnath explains, the reorganization became an "emotionally charged issue. . . . The people that used to own a lot [of stock] now didn't. People that owned very little now owned a lot, and so basically the deck got reshuffled. [Some] of the people at the top, who were making an awful lot of money on Tuesday, were no longer in a position to make very much money on Thursday. Charlie simply made the decision as to who was going to get what, to the angst and consternation of many people."[21]

Pankow might have done a better job explaining the reorganization and why he undertook it. Nevertheless, it was a smart move, Kunnath concludes, adding, "It was almost necessary to do what he did" to allow greater employee participation in the ownership of the firm. "[Many] people were helping to make the company successful, but very few people were really benefiting."²² As Verti explains, the outcome was "significantly better than our wildest dreams." Almost incredulously, he asks: "Who would have guessed that Charlie [would structure] it [so that] the future managers and leaders of the company [would receive] a much bigger share of the equity than in the original organization, and [that] he [would take] a much smaller proportion? Who would have assumed that's what he would do. . . . Who would do those things?"²³

The End of the Hutton Era in Hawaii

The reorganization of the Pankow companies coincided with what George Hutton calls the "emasculation" of the Hawaii office. First Hawaiian Bank economist Leroy Laney has observed that construction cycles in Hawaii during the first three decades of statehood were characterized by "three-to-four-year booms interspersed with two-to-three-year busts."²⁴ As an unsettled period, however, the mid-1980s did not fit well into either category. Rather, these years marked a halting and fragile recovery that produced uncertainty and uneven results in the commercial construction market, while the national market showed steady year-on-year improvement.²⁵ For Hutton's office, these years were particularly lean. Between the completion of the Hobron and Maile Court condominiums in Waikiki, in 1984, and Honolulu Park Place, in 1990, it was engaged in only two, relatively minor, projects, namely, the renovations of the Kapalua Bay Hotel and the Ala Moana Shopping Center (see appendix B). As a result, Charlie Pankow brought a large share of the underutilized workforce to the Mainland. By the time that Japanese investors fueled the "red hot" construction boom of the late 1980s, Hutton had been depleted of much of the staff that had coalesced around him over two decades.²⁶

The exodus began in the back office. For the better part of a decade, beginning in the late 1970s, the positions of CFO and controller became veritable revolving doors, for both organizational and personal reasons. Accounting became increasingly complex with the proliferation of development entities and the creation of subsidiaries for building services, construction, equipment, trucking, and so on. It was a function in which Charlie Pankow took a keen interest. He saw accounting and finance as sources of revenue as well as tools to deploy on behalf of minimizing corporate and individual tax liability. Many of the individuals whom he hired to handle the work during this tumultuous decade did not stay in their positions for very long. In some cases, the reasons were transparent; in others they remain murky. Of course, reorganizing the company heaped stress upon an already stressful environment. The record is unclear, but it appears that turnover owed in part to personality clashes that may or may not have been products of that environment. The reorganizations also set in motion a rationalization process that also accounts in part for the strikingly large number of individuals who held the position of CFO and controller from the late 1970s through the mid-1980s.

With the establishment of the CPA subsidiary, two controllers oversaw separate accounting systems. Jim Body, CPI's first controller, was transferred to Hawaii in the early 1970s to assume that position in Hutton's office. Conan "Doug" Craker, his assistant controller, became controller in Altadena. A few years later, Body left the firm to pursue executive recruiting. Hutton hired Robert Hewitt to replace him. Hewitt played the primary role in bringing the Pankow companies into the computer age, convincing a reluctant Charlie Pankow that investing in an IBM System/32 machine would be well worth the money. The Computer Age at Pankow began in Hawaii and soon spread to Altadena.[27]

Bob Hewitt became the first member of Hutton's team to transfer to Altadena, and its first casualty. Not long before Craker left his position as controller, in 1980, Charlie Pankow tapped Hewitt to become the company's third vice president of finance. Pankow and Russ Osterman had created the position in the early 1970s to meet the needs of the growing company. Hewitt remained at his new post for only a few months before Pankow let him go. As seems also to have been the case with Kay Jones, the company's first vice president of finance, Hewitt and Pankow did not enjoy good working relations.[28]

Jim Allyn, the man whom Hutton promoted from within his accounting department to replace Hewitt, was the next individual to leave CPA for the Mainland. At about the same time that Hewitt transferred to Altadena, Jim Body was rehired as the firm's CFO. As Kim Petersen puts it, "the accounting systems needed a lot of attention." They were "kind of disorganized." Body was recruited to fix the problem. He remained with the company until 1984, the year of the first reorganization. Charlie Pankow replaced him with Allyn, who remained on the job until early 1986, before the second reorganization was completed. As was the case with Hewitt, the reasons for Allyn's departure remain unclear. The timing of the departures of both Body and Allyn suggest that each case may have been tied to the accounting and tax "events" that were driving the reorganizations. Pankow hired a new CFO from the world of Big Eight accounting, most likely on the recommendation of Peat Marwick, the new auditors. His tenure was a brief six months. Two weeks before the effective date of the 1986 reorganization, Pankow hired Timothy P. Murphy from one of the Big Eight firms. With the reorganization in place and Hawaii's accounting department collapsed into the corporate system, the door to the CFO's office stopped revolving. Murphy would remain CFO until he retired in 2000.[29]

The dearth of work in Hawaii resulted in the departure of much of George Hutton's project staff. In contrast to the roller coaster market in Hawaii, California's commercial construction enjoyed a five-year boom from 1983 through 1987. At its peak in 1985, nonresidential building volume reached a record $14 billion.[30] (It was about this time that Winmar retreated from the development market. Hence, CPBL's order book became concentrated in California.) Building offices and expanding shopping centers in the Golden State created demand for Hutton's men. In short order, chief estimator Dick Ackerson, labor foreman Bob Crawford, operations manager Bill Heine, marketing director Don Kimball, superintendents Tony Giron, Red Metcalf, Jack Parker, and Jim Thain, plus a number of project and field engineers, were relocated to the Mainland. By 1986, few people remained in the Hawaii office.[31]

Organizational factors, including Pankow's business model and the role played by Hutton as developer, account in part for the lack of work in Hawaii at this time. As Kim Lum explains, neither Pankow nor Hutton were interested in taking on projects simply to keep people employed during troughs in the construction cycle: "If they didn't like the job, they weren't going to take it. And if they weren't going to make enough money on the job, they weren't going to take it, and it didn't matter whether we needed work or not." Indeed, Hutton saw in slow periods the opportunity to upgrade staff by letting go individuals who were performing below his expectations. As Lum perceived the situation, Pankow also had no strategic imperative to generate a growing, or even stable, volume of work: "When the work came along, they did it, they made a bunch of money, and then were content to cut back when things were slow, let people go." Moreover, since the business model called for negotiating work, CPA did not bid on it. As Lum observes, "I'm not sure they knew how to go get work if they had to get it. I mean, we just didn't bid work, so we didn't really know how to go about doing that." That CPA was viewed within the close-knit commercial real estate community more as a developer than as a contractor presented another obstacle. Because they viewed him as a competitor, developers were loathe to give Hutton work. The lack of work during the mid-1980s prompted many CPA staffers to look for employment elsewhere. Several of them followed Bill Heine, when the operations manager left CPBL to form his own company.[32]

Heine did not work long for Charlie Pankow after his transfer to the Altadena office. In his new role, Heine would have to work directly for Dean Stephan—a situation he apparently did not entertain as a long-term proposition. In fact, Heine left the firm within weeks of arriving in Southern California. An opportunity arose to build a project for a developer in Austin, Texas. Heine relocated to Austin and raised some money to start his own firm, American Constructors. He recruited several people from CPA to join him in Texas, including Jeff Bardell and Tim Cahalane, a carpenter by trade and a "very dedicated, very sharp guy" who distinguished himself at CPA as an expert in slipform operations.[33]

In 1987 Heine decided to establish a foothold in Hawaii, with Bardell heading the effort. Heine approached Kim Lum and queried his interest in joining Bardell. Recognizing that CPA's lack of work was hindering his career, Lum signed on. Together with Mike Betz, whom Bardell hired out of the University of Pittsburgh, his alma mater, Bardell and Lum built American Constructor's Hawaii operation into a $15 million business (in terms of annual revenues). After completing a costly and difficult project on the remote island of Lanai, however, Bardell decided, mainly for family reasons, to return to the Mainland. A year later, in April 1995, Lum rejoined the Pankow firm. A second unprofitable project on Lanai had spelled the end of American Constructor's Hawaii operation. Mike Betz also joined Pankow in Hawaii and soon would head Pankow Special Projects, which, as chapter 7 elaborates, engaged in the type and size of projects that Bardell, Betz, and Lum had completed for American Constructors.[34]

A Late-1980s Reversal of Fortune in Hawaii

Japanese investment revived the fortunes of Pankow in Hawaii. In 1988 representatives of C. Itoh and Company, Ltd., agreed to become development partners with Hutton and Pankow in Honolulu Park Place, Ltd., for the purpose of constructing a luxury condominium that would overlook Honolulu Harbor. (A leading *sogo shosha* engaged in general trading and commercial investment, C. Itoh traced its roots to 1858, when founder Chubei Itoh began trading in linen. In 1918 it began trading as a public company under the C.

Itoh name. In June 1992, it would adopt ITOCHU Corporation as its English-language name.) PDHP, Inc., a CPBL affiliate with Charlie Pankow, George Hutton, and Tim Murphy as its officers, served as the general partner. Mitsui Trust & Banking Company provided the construction financing.[35]

Honolulu Park Place was designed by Norman Lacayo, "the architect most in touch with what the upcoming affluent want," in the estimation of one real estate agent, who extolled the West Hollywood–native's work as "always stunning, exciting, and very popular." It featured curved shapes and fluent lines for which the architect was well known, a "distinctive" blue and white façade, reflective gray glass, and lanais with turquoise ceiling overhangs and curved aluminum railings finished in off-white silicone polyester (fig. 72). A deck above the parking garage offered tenants ample recreational space in a Japanese milieu. It included an elaborate koi pond, huts with wet bars and barbecues, a large entertainment pavilion, a putting green, a driving range, and tennis courts. Commented U. L. "Rick" Rainalter Jr., whose development company managed unit sales, the timing of the project was "perfect," as Honolulu Park Place was the first project of "the new market surge." Sited near Honolulu Tower, a residential high-rise that Lacayo designed, and CPA constructed, in 1982, the location of the condominium constituted "one of the last pockets of redevelopment in this particular area," according to Rainalter.[36]

The groundbreaking ceremony for the $125 million project took place on 21 March 1989. Owing to the depletion of the personnel of the Hawaii office, several men whom Charlie Pankow recently had brought to California returned to Honolulu to construct the condominium, including Red Metcalf and Jim Thain as superintendents and Bob Crawford as general labor foreman. By using the slipform system that "has been [the company's] forté [*sic*]," as the company newsletter put it, the contractor was able to "transform the curves that Lacayo likes to use into an efficient concrete reality" on a three-day cycle (fig. 73). The slipform was intricate and enormous—a footprint of some 14,000 square feet. In fact, it was the last time that the Pankow firm deployed slipform operations. Crews completed work in January 1991. Rainalter's team closed unit sales in March—not long before the commercial market crashed.[37]

The construction contract for the next ITOCHU-financed project that CPBL built, the $97 million Waikiki Landmark condominium, was the last one that George Hutton signed as head of Hawaii office. Charlie Pankow assigned Mike Liddiard to manage the project. Liddiard had demonstrated his bona fides on three complex projects: the YMCA in Oakland, Marathon Plaza,

Figure 72. Honolulu Park Place, downtown Honolulu, overlooking Honolulu Harbor. Project broke ground on 21 March 1989. Completed in January 1991. Lacayo Architects, architect; James Adams, structural engineer. (Photo © Bill Hagstotz/ Constructionimages.com.)

and the Resort at Squaw Peak (see appendix A). From Hutton's perspective, however, the assignment of Liddiard was further evidence that Charlie Pankow sought to exclude him from management and extinguish all remnants of CPA.[38]

Aloha Tower and the Departure of George Hutton

Tensions between George Hutton and Charlie Pankow had been simmering for the better part of a decade. Because the latter never explained his reasons for reorganizing the company and dissolving the CPA subsidiary, Hutton concluded that Pankow had acted, and was continuing to act, out of a personal desire to marginalize a rival who just happened to be the second largest shareholder in the company. The fact that many of the individuals who had spent their careers with the company solely under Hutton's wing either had left the firm or had been let go soon after their transfers to the Mainland reinforced Hutton's belief that Pankow was interested only in asserting his unilateral authority over what he considered to be *his* company. Why else, Hutton wondered, would Pankow kill the proverbial goose that was laying

Figure 73. Slipform operation at the 30th floor, Honolulu Park Place. Its use marked the final time that the company deployed the technique. (Photo © Bill Hagstotz/Constructionimages. com.)

the golden eggs that had provided him with much of his wealth? Tensions reached a boiling point when Pankow decided to run the University of Hawaii Arena project—included in next chapter's discussion on public sector design-build—out of the Altadena office. Hutton felt that he had been cut out of the loop completely. As he "just saw the handwriting on the wall," Hutton tendered his resignation, effective 1 September 1991.[39]

Hutton expected to continue to do business with Charlie Pankow, however. At his insistence, Aloha Tower Associates (ATA), a development team in which he became interested, designated CPBL to be the builder of the Aloha Tower Marketplace. As events played out, this project marked Hutton's complete break with the man who had hired him, brought him into his new company, and had allowed him to build the Pankow firm into one of Hawaii's most respected and successful contractors.[40]

At 184 feet, Aloha Tower was Hawaii's tallest building when it was completed in 1926. One of the state's most recognized landmarks, it traditionally served "as a welcome sign for shipboard tourists," as journalist Andrew Gomes writes. The first call to redevelop the surrounding area came in 1973, when a local developer proposed transforming state-owned warehouses into a foreign trade center. In 1979 Governor George Ariyoshi outlined an ambitious redevelopment program. Encouraged by the popularity and financial success of Faneuil Hall Marketplace, which had opened in August 1976, and inspired by the conversion of Baltimore's gritty inner harbor into Harborplace, which was then under construction, Ariyoshi announced his ideas on turning the industrial land around Aloha Tower into a similar, nostalgia-animated "festival marketplace," replete with shops, restaurants, and an international trade center. The governor commissioned designers to build a model and proposed that the State of Hawaii spend $8.5 million to catalyze private development, which he hoped would finance the effort.[41]

A legislative audit concluded that the project was flawed financially. Nevertheless, state lawmakers established the Aloha Tower Development Corporation (ATDC) for the purpose of attracting private investment to underwrite a revised plan. Two bids to secure backing for subsequent proposals were aborted in the mid-1980s. In 1983 the ATDC accepted a $100 million bid by the development arm of the Southern Pacific Railroad to build hotel, office, and retail space, and facilities for cruise ships. The developer pulled out the following year, however, in the wake of the merger of the Santa Fe and Southern Pacific Railroads. In 1985 the ATDC began working with Cordish Embry & Associates on a $200 million proposal. Cordish Embry had been involved in the development of the now-celebrated Harborplace, and so local expectations soared on the belief that the ATDC and the firm would finally produce a development blueprint that could be built. A legislative audit completed in 1987 dashed these hopes. It concluded that the financing for the project was "unrealistic and unworkable" and recommended that the State of Hawaii scrap both the plan and the agency. The plan died, but the beleaguered agency lived on.[42]

Notwithstanding the warning signs raised by these failed bids, Japanese investment revived interest in the redevelopment of Aloha Tower. In 1989 the ATDC called again for bids. Four parties responded. The ATDC awarded the project to ATA. Its $750 million proposal included a retail marketplace, two high-rise condominiums, an office tower, a hotel, a cruise ship terminal, and underground parking for 2,000 vehicles. In an agreement signed in 1990, the developer agreed to pay the state an estimated $4 billion in ground rents on a 65-year lease on 17 of the redevelopment site's 22 acres.[43]

Rick Rainalter headed ATA, which also included local developers Peter Smith, Robert Gerell, and Glenn Okada. James W. Rouse, the developer behind both Faneuil Hall Marketplace and Harborplace, also joined the group. As was the case with Honolulu Park Place, ITOCHU was a financial partner. Rainalter and the other local partners invited Hutton to join the partnership. He agreed on the condition that they designate CPBL as contractor in a design-build relationship and negotiate a lump-sum contract. They agreed. Rainalter and his partners had no interest in including Charlie Pankow as a development partner, however. Hutton subsequently acted as the group's managing director. On that basis, he entered in discussions with CPBL on the final terms of a construction contract early in 1993.[44]

By that time, ATA had scaled back the project considerably, as the Japanese-fueled construction boom of 1988–1991 had gone bust. For Japan's economy, its "lost decade" began with the steady decline of the stock market from its December 1989 peak. Investors did not immediately recognize the onset of financial crisis. Indeed, the domestic property bubble did not burst until 1991. Japanese financing—much of it locked in place—sustained Hawaii's building sector through the end of that year, when it registered a record $4.3 billion in revenues. Once it began, however, the sector's decline was precipitous and unrelenting. In 1992 revenues fell 10 percent. Private building permits dropped 18 percent. In the first half of 1993, permits sank another 30 percent. With the US and Hawaiian economies slumping, it was clear that Honolulu's commercial sector was significantly overbuilt. Citing the retrenchment of its Japanese financiers, owing to the uncertainties created by the Persian Gulf War and US financial conditions, and the weakening local condominium and office market, Rainalter informed the ATDC that he and his partners could proceed only with the retail portion of the project. ATA also requested, and received, a waiver on its scheduled $60 million lease down payment. Early in 1993, the project was dealt another blow when the ATA partners determined that the planned underground parking structure could not be built as designed. The initial redevelopment effort, reconfigured as Phase I, would consist of the $100 million, 190,000-square-foot Aloha Tower Marketplace, a cruise ship terminal, reduced parking, and the renovation of the tower itself. Construction would be fast-tracked on a 14-month schedule to facilitate an August 1994 opening.[45]

Two weeks before ATA expected the City of Honolulu to issue the building permit, Charlie Pankow sent a letter to George Hutton, informing him that CPBL would not build the Marketplace. He gave no explanation. Publicly, CPBL stated that it was voluntarily withdrawing from its role as designated contractor for "business reasons." The developer was similarly tight lipped. Rainalter stated only that the withdrawal was "in the best interests of the project." In June 1993, CPBL cited the uncertainties of the project in the recession and its—meaning, of course, Charlie Pankow's—decision to cut costs by no longer developing projects as reasons for not building the Marketplace. It was a cryptic explanation. For, as noted, neither Charlie Pankow nor the company had a development interest in the project. Given the decline in the local building sector, it seemed to people within the firm who were familiar with the project that it made sense to accept the work. Ironically, the ultimate contractor was, in effect, a Pankow spin-off company. U.S. Pacific Builders, founded in 1989 by former CPA project sponsor Bill Deuchar, former CPA estimator Dick Ackerson, Kirk Clagstone, and Steve Grimme—all of whom had worked most recently for California-based A. T. Curd Builders—employed many former CPA employees. From the perspective of a general contractor, the spring of 1993 was "a good time to price out a job and a good time to be negotiating a construction contract," as Hutton put it.[46]

According to Dean Browning, whom the company sent to Hawaii in the fall of 1990 to manage Aloha Tower, Charlie Pankow made the right decision, at least in hindsight, because "there were a lot of problems on that job" associated with payments to the contractor and subcontractors, change orders, and "a few other things." Browning surmises that Pankow "recognized that [the project] was trying to be built on a shoestring, and he decided that we shouldn't get involved."[47]

From George Hutton's perspective, however, Charlie Pankow's decision was strictly personal. From the moment that he informed Pankow that he had been invited to participate in the Aloha Tower redevelopment, Hutton felt that Pankow treated him "like I was a traitor." Still, until he sent his eleventh-hour letter, Pankow had raised no objections to CPBL constructing Aloha Tower Marketplace. As Hutton saw it, there was no reason for Pankow not to sign the construction contract, given the recession. Whatever Pankow's reasons for withdrawing CPBL as contractor so soon before the start of construction, the decision created problems for ATA's partners, who had to scramble to find a replacement. In Hutton's view, Pankow's timing constituted further evidence of the personal reasons that lay at the heart of his decision. For, according to Hutton, Pankow was well aware of the deadline to secure the building permit and the difficulties that the abrupt withdrawal of the contractor would create.[48]

Hutton's resignation, coupled with the retirement of Russ Osterman a couple of years earlier, spelled the end of development as a way of generating work for the construction company. Now, as Jon Eicholtz explains, "there was nobody around with the talent and the background to create development opportunities." To be sure, the market for construction finance changed in the wake of recession and the savings and loan crisis. Institutional lenders began to require developers to commit 30 percent or more of their own funds to their projects—something for which Charlie Pankow had no appetite, in part because he was growing more adverse to risk as he aged and accumulated wealth. Yet Pankow always had been reluctant to commit his own capital to a project, even if it was only "front" money that he would soon recover. As Rik Kunnath and Kim Petersen note, the captive projects of the 1970s and 1980s were completely leveraged. There was no risk capital. The development entities borrowed money on a nonrecourse basis. As such, they carried significant "upside" potential, as long as the developer completed the project within budget—and as "captive" projects, Pankow, of course, had control over the contractor. Nevertheless, as Eicholtz and Osterman would demonstrate after their departures from CPBL, there were always development opportunities to be had, regardless of market conditions. Together they would enjoy considerable success in investing in, upgrading, and selling (often poorly managed) apartment buildings. With development no longer a source of work, CPBL would have to explore other means to sustain the business in a "challenging environment for commercial builders," as Kunnath noted in the company newsletter.[49]

Branching Out: A Series of Cautionary Tales

In the spring of 1990, CPBL opened a branch office in San Diego and installed Tom Verti as its manager. Given a "quite vibrant" local development and construction market—staff were engaged in preconstruction activities on several large projects, including the $200 million Four Seasons Resort Aviara in nearby Carlsbad—the decision to establish the office appeared to make business sense. The company newsletter hailed the opening of the office as "evidence of our steady growth and expanding capabilities." For the

San Diego market offered "the perfect atmosphere for us to promote the Pankow philosophy of strict cost and quality control through technical construction innovation." As was the case with all other attempts to branch out, with the exception of Hawaii, the San Diego gambit soon withered, however. Set to open in the summer of 1992, the Aviara project was not completed. The developer cancelled the contract halfway through construction, owing to its failure to secure credit in the context of the oil price shock precipitated by Iraq's invasion of Kuwait in the fall of 1990 and the recession that ensued. With the "launch pad" for the office removed, there was no longer enough work to justify the maintenance of the branch office. It was quietly wound down. Staff returned to Altadena.[50]

Other efforts to branch out also failed. The company acquired 35 acres of land along Interstate 85 in Gwinnett County, Georgia, north of downtown Atlanta. A local developer convinced Charlie Pankow of the site's potential as a suburban office park and proposed that he and the contractor develop the property as a joint venture. By demonstrating the value of its services to local developers and subcontractors, CPBL would establish itself in the market, justifying the opening of a branch office. Reflects Rik Kunnath, whom Charlie Pankow planned to send to Atlanta to manage the prospective office: "Basically, I think we got sold a piece of the Brooklyn Bridge." The idea for an office park at that location was premature by a decade or more. Explains Kunnath: "When we started to look at the ability to get money and attract tenants and really make economic sense out of an office development application at that time, none of it made sense. And we never got a loan. It never went anywhere. . . . So [the] project went away, [the] idea went away." About a decade later, Pankow sold the undeveloped property. Even then, the level of suburbanization did not justify the construction of an office complex. The buyer, a used-car dealer, established a CarMax Auto Superstore on the site. Likewise, plans to establish offices in Chicago and New York came to naught. A branch office was established in Seattle after CPI completed several projects for Winmar in the Pacific Northwest. It lasted but a short period of time. There, as in San Diego, Kunnath notes, "we had a lot more difficulty penetrating the market than I think we probably assumed we would."[51]

In each case, Charlie Pankow put the proverbial construction cart before the business development horse. According to Kunnath, there was "never a concept of doing a market analysis or a business plan." There was no systematic effort to determine whether a prospective business opportunity was viable. There was little appreciation of "what it takes to [develop] real in-depth market knowledge, having the right boots on the ground, the right [people] with the right contacts." Driven by a belief that business success must follow from proven engineering acumen, Pankow acted on tips or concluded that a single project signaled that the market was receptive to an ongoing presence. The attempts to open branch offices showed that "we [knew] nothing, really, about business." And outside of George Hutton and Russ Osterman, "we knew nothing about real estate development, regardless of what we thought. We wanted to build concrete buildings and [had] some mechanized systems that were really pretty great." Relying on his instincts as a builder, Pankow apparently figured that his "budgeted construction program" would sell itself. Kunnath concludes, his approach was "so casual and so unbusinesslike in the decision making and so unstructured in the thought process. [Looking] back [it's] just a complete head-shaker."[52]

Working in the shadow of Charlie Pankow and Russ Osterman would also have diminished the long-run viability of any branch office on the Mainland, according to Doug Craker and George Hutton. As Craker notes, Pankow and Osterman "were so powerful within people's focus that it was hard for anybody to go in and work on their behalf." Moreover, "if developers knew Charlie and Russ, they wanted to work with Charlie and Russ." The direct involvement of Charlie Pankow in the affairs of the office would also be stifling. Both Craker and Hutton point to the performance of the San Francisco office under Brad Inman. If Charlie Pankow hardly interacted at all with George Hutton or, for that matter, Dean Stephan in Altadena, he micromanaged Inman. Rather than consider Pankow's instructions as general guidelines that could be followed, or not, depending on the circumstances, as did Hutton, Inman felt obligated to follow Pankow's "marching orders": "He managed by discussing issues with you . . . and he was very diplomatic and always listening to whatever [you] had to say. . . . But he'd always then, either by suggestion or direction, set the line of march out very clear." This arrangement worked well enough in tasks related to project management. The problem, Kunnath explains, was that Inman did not sell work, as Pankow expected (even if, as Inman notes, this requirement was never part of any job description): "There would be no way you could simultaneously meet all of [Pankow's] preferences for the way you would go about getting work." In 1988 a frustrated Charlie Pankow removed Inman and promoted Kunnath in his place. To date, Hawaii has remained the firm's only successful branch office.[53]

Innovation in High-Rise Concrete Construction

The company continued to rely heavily on the construction of high-rise concrete office buildings, hotels, and condominiums. Charlie Pankow's business model did not change, and he had no plans to change it. When asked in 1988 what the company would be doing in 5 or 10 or 15 years hence, he replied: "Design-build in heavy commercial buildings for the private sector. That's it." From the mid-1980s until the early 1990s, the company's order book grew largely because of such projects. The company continued to deploy methods and techniques that it had used since 1963 to deliver projects. At the same time, project teams deployed new techniques or varied existing ones to execute increasingly sophisticated designs. The completion of signature high-rise projects in California and Hawaii demonstrated the company's capabilities as the "premier design-build company for what we do," Pankow explained.[54]

Because reinforcing requirements stipulated by increasingly stringent building codes for seismic zones eliminated the cost advantages of slipforming, the company began to use jump (or climbing) forms to erect service cores in high-rise structures. Walnut Center, an office building completed in 1983 in Pasadena for Kaiser Permanente, was the last project on the Mainland to deploy slipform operations. In contrast to the slipform, which moves upward incrementally while crews pour concrete, workers raise the jump form to the next level only after they pour the concrete for an entire floor and it hardens sufficiently to support the weight of the floor above it. After workers roll back the forms, they raise the entire form system to the next level by means of a crane or hydraulic jacks. Forms raised by means of the latter are said to be self-climbing. The contractor deployed self-climbing forms to build Chase Plaza in downtown Los Angeles—the first granite-clad structure that the company built—and the office building associated with Shoreline Square in Long Beach, a three-building complex that constituted the company's largest project to date. The structural designs for both projects met the requirements of the Uniform Building Code for Seismic Zone 4 with a concrete core that provided drift resistance and a lightweight steel moment-resistant frame that provided flexibility during earthquakes (figs. 74, 75). Developed in association with John A. Martin & Associates and Robert Englekirk Consulting Structural Engineers, both of Los Angeles, the designs constituted a new approach in seismic engineering. Using self-climbing forms was cost-effective for both projects and enabled the project teams to meet grueling, fast-track schedules. Their deployment provided better continuity of work, reduced labor

Figure 74. Chase Plaza under construction in downtown Los Angeles, August 1985. The use of the jump form to construct the service core marked a departure in the firm's methods of erecting high-rise structures. Note the steel and concrete composite structural system, designed for UBS Seismic Zone 4. Gin Won Associates and Gensler & Associates, architects; John A. Martin & Associates, structural engineer. (Photo Warren Aerial Photography. Courtesy of Charles Pankow Builders, Ltd., Pasadena, California.)

costs associated with pouring concrete, and permitted the construction of the service cores to occur independently of the tower crane, which could be used concurrently on tasks associated with the steel frames and decking (fig. 76).[55]

Pankow project teams did not confine their use of jump forms to structures designed for earthquake zones. In the case of the 30-story "contemporary minimalist" 411 East Wisconsin Building in Milwaukee, Wisconsin, for instance, the team used a jump form system because the erection of the core occurred during winter. As project engineer Kevin Smith explains, "A slipform isn't really easily weather-protected." In contrast, the jump form protected the concrete as it hardened overnight. "Usually, once [the concrete has] been cast overnight, then it's safe after that, as long as it's been kept from freezing in that first 500 psi or 24 hours." No pours were lost because of freezing on the proj-

Figure 75. Shoreline Square under construction in Long Beach, California, August 1987. As was the case with Chase Plaza, the contractor used jump forms to construct the service cores of the Sheraton Hotel on the left and the office building on the right. A 1,400-stall parking garage completed the complex, finished in the summer of 1988. Maxwell Starkman Associates, architect; Robert Englekirk Consulting Engineers, structural engineer. (Photo Warren Aerial Photography. Courtesy of Charles Pankow Builders, Ltd., Pasadena, California.)

ect. Weather prevented crews from performing concrete operations associated with the six-sided building's core, frame, and floors on only 19 days (fig. 77).[56]

CPBL teams adapted their concrete construction toolkit to types of projects that they had not encountered. Every project, of course, presents the building team with a unique set of problems to solve and challenges to address. The Alameda County Metro YMCA project, however, stands out as truly "unique" in the history of the company: "Unique in concept, development, design, and construction," as the company newsletter described it.[57]

The downtown Oakland project was a public-private effort that met the need of Webster Street Partners, the development entity created by Charlie Pankow for 2101 Webster Street, for additional parking for the building's

Figure 76. The use of self-climbing jump forms permits construction of the service cores of the hotel and office building associated with Shoreline Square to occur independently of the tower crane. (Photo Warren Aerial Photography. From *CP News* 5 [Fall 1987]: 1.)

tenants and of the YMCA for a new facility with a three-building, mixed-use complex. Russ Osterman concluded a condominium agreement with the YMCA of Alameda County whereby Webster Street Partners owned the parking structure and the latter owned their facility. PDC donated the land, valued at $1.5 million.[58]

The complex included a 4-story fitness center, a single-story office building, and a 700-stall parking structure that rose eight stories above underground and ground floors, where a 25-meter swimming pool, saunas, steam rooms, and locker rooms were located. Reflecting the dissimilar uses of the project's components, the building team included separate architects for the parking structure and building shell (John S. Bolles Associates, whose relationship with Pankow dated from Clementina Tower), the YMCA (MacKinlay, Winnacker, McNeil & Associates), and the swimming pool (the YMCA's

Figure 77. 411 East Wisconsin Building, downtown Milwaukee, Wisconsin, elevated view. The project used more than 45,000 cubic yards of concrete and 6,450 pieces of precast concrete. Crews drove the first of 710 piles on 7 March 1983. The tower was topped out four months ahead of schedule, on 3 August 1984, and the first four floors were certified for occupancy on 5 November 1984. Harry Weese & Associates, architect; Chris P. Stefanos Associates, structural engineer. (Reproduced by permission of Eric Oxendorf.)

national architects), as well as structural engineers Culp & Tanner, parking design consultant International Parking Design, and CPBL.[59]

The project's owners wanted a state-of-the art, long-lasting facility that required low maintenance, delivered within a stringent and fixed budget. Designers prescribed a parking structure with precast columns, prestressed girders, and mild steel, poured-in place decks, and an adjoining fitness center with a structural steel frame. A "complex seismic joint system" separating the two structures met the City of Oakland's requirements for mixed-use buildings, including lateral stiffness of structural systems. The concrete shear walls of the parking structure supported each building. After consulting the YMCA's architects, the building team opted to construct the swimming pool as a "fully cast-in-place concrete structure" and cast the structural slab in a single pour.[60]

Erecting the parking garage over the swimming pool posed perhaps the most significant challenge. As project superintendent Mike Liddiard explains, "[We had to] incorporate the precast system to go above the YMCA and then to coordinate all the mechanical that it takes to do the plumbing for the showers and the swimming pool." All of the structural elements for the parking structure were precast on-site, including 75-foot column segments, and "in sequence so that we could erect them," using a Pecoo SK400 tower crane and several mobile cranes. In fact, Liddiard recalls, "we had a record on how many columns we set in a day."[61]

▪ ▪ ▪

Since 2007, "Thinking Beyond the Building" has served as the company's tag line. As Al Fink explains, "[It means that] we're not just focused on one element. We're not just focused on the building. [We look] at the big picture, trying to integrate all the elements of it into this big picture, and solving whatever problem and impediment there might be to achieving the goal [of] building a [good] project." According to Fink, the company's project teams were "thinking beyond the building" long before the company adopted the tag line. In his view, the Waikiki Landmark project provides perhaps the best example to date of "Thinking Beyond the Building." As well he might: Fink was primarily responsible for securing this project after the developer failed to receive any bids that fell within his pro forma budget.[62]

To ensure that the structure that he built was worthy of its prime location at the "gateway" to the Waikiki district, developer Sukarman Sukamto, owner of the Bank of Honolulu, held a design contest. Honolulu-based Architects Hawaii submitted the winning entry: a 38-story complex, consisting of twin 320-foot-high triangular towers, clad in pre-glazed Italian stone and glass panels and topped by 5 stories that spanned the 65-foot gap between them. It met Sukamto's desire for a "bold statement," as architect Alex Weinstein put it. But constructing the design, as Sukamto learned when he put out his project to bid, would cost at least $10 million more than the developer and ITOCHU, his financing partner, were prepared to spend.[63]

In deciding to secure a contractor through bidding, the developer had rebuffed Fink's suggestion that he negotiate a lump-sum construction contract with George Hutton. Because of ITOCHU's participation in the project, the Hawaii office monitored the project through the bidding process. When it became clear that building the project as designed was not feasible, Fink approached Sukamto's representative, Colby Jones, and persuaded him to allow him and his colleagues to "go ahead and see what we can do." Jones agreed. [64]

Fink reviewed the project and reduced construction costs by more than $10 million without sacrificing any of its revenue stream. He achieved most of the savings by eliminating a basement that the designers had intended as space for tenants' storage lockers. Since the water table sits less than ten feet below street level, including the basement would have required a substantial "dewatering" operation and costly waterproofing work. During a contentious meeting that included the entire building team, Fink argued that each storage locker was costing the project approximately $50,000. The developer agreed that there was no good reason to include the basement. Its elimination enabled the structural engineers to simplify the foundation—which nevertheless included some 1,300 precast piles of different lengths, driven to depths of up to 300 feet, to minimize the difference in settlement between the towers. [65]

In subsequent meetings, Fink proposed dozens of so-called value engineering items to cut costs. For example, he showed that savings of some $250,000 could be achieved by reshaping each rounded granite-clad column as an octagonal, cornered cylinder. The architects resisted this and other recommendations, such as enclosing lanais with precast panels instead of curtain-wall units, as Sukamto already had approved them. Yet the latter had done so without assessing costs and benefits on a line-item basis. Now the developer acted to reduce the cost of the project without sacrificing its quality. Sukamto's acceptance of 35 of the changes that Fink proposed reduced the construction cost from $114 million to $97 million, including a $2.5 million reserve fund that was never tapped. [66]

Executing Waikiki Landmark's design demanded "a much tighter tolerance on construction" than was usually the case, structural engineer Gary Chock observed. Pankow's project team, led by superintendent Mike Liddiard and project engineer Jeff Doke, decided to build one tower and then align the second tower with it, to ensure that the two structures were "within a quarter of an inch [of] being perfect from the bottom to the top," as Liddiard explained. A similar approach had been taken with Marathon Plaza in San Francisco, which involved constructing two "non-towering" towers of 9 and 10 stories. Confronted by uneven subsurface conditions, Liddiard's team had built the cores of the two towers sequentially. Owing to variable soil conditions on the Landmark site, Liddiard's team also had "to allow for the possibility that one tower might settle differently than the other." As Chock explained, the two towers were designed to be similar structurally to minimize stress on the connection between them from wind and earthquakes. Both towers were designed to "move as an integrated structure with a total lateral sway of three inches." Further, as project sponsor Stuart Feldman noted, "the design allowed for the inner connections to accept a one-inch difference [in settlement] between the two towers."[67]

"Extensive" planning and coordination, involving the contractor, structural engineers Martin & Bravo (an affiliate of John A. Martin & Associates), crane operators, state safety officials, iron workers, and others, was required to complete the complex project successfully. Lifting each of four assembled, 64-foot-long and 10-foot-deep trusses—each weighing 32 tons—to the base of the tower connection at the thirty-third story posed the most difficult challenge. It was certainly a new experience for the contractor. Two cranes were used to ensure that the trusses remained level as they were raised at an angle, rather than "straight up between the towers," and provide the requisite tolerance for the operations, Chock explained (fig. 78). Those responsible for signaling the cranes climbed from story to story to verify that the cranes kept each truss level. Once the truss reached its prescribed height, the operator swung it into position. Workers set it on three-by-six-foot corbels—projections built into the wall to provide support—fitted with special pot bearings that permitted movement in any direction and welded it into place. The operation took two days to complete, but hundreds of hours to plan.[68]

Liddiard's team closely monitored the settlement, plumb, and alignment of the towers. Required to retain an outside contractor to measure the outside of the structure every fifth floor, Liddiard had his engineers "shoot" each floor "to keep the thing plumb," he explained. As a result, Feldman reported, "Pankow was able to deliver a vertical elevation difference from one tower to the other within one-quarter of an inch" (fig. 79). John Bravo, president of the structural engineering firm, complimented the contractor for its role in ensuring the quality of the project, about which he initially had "great anxiety." Project engineer Doke's response neatly encapsulated a key aspect of the company's culture, namely, that of doing whatever it takes to get the job done: "We've got to make the structure work and we've got to stand behind it when it's done." He added: "It's just good business."[69]

Figure 78. Two cranes were used to raise four trusses into position between the Waikiki Landmark towers. (Photo © Bill Hagstotz/Constructionimages.com.)

Expanding Regional Shopping Centers

The decades-long dominance of regional shopping centers in the retail sector began to falter in the second half of the 1980s. Rehabilitated community centers and Main Streets, so-called festival markets like Faneuil Hall Marketplace and Harborplace, and strip centers began to compete successfully against the suburban mall. Institutional investors took advantage of the market's uncertainty and the difficulties faced by original owners to acquire many of these properties. By 1986, for instance, both Equitable Life Assurance Society and Prudential Insurance Company owned more than 60 shopping centers; Rosenberg Real Estate Equity Funds owned 32. To maximize returns on their investments, real estate portfolio managers expanded and rehabilitated hundreds of properties to increase gross leasable areas and improve the quality of new and existing spaces so that owners might command higher rents. High land values, which in many cases owed to development that the regional shopping center itself had spurred, generally made the cost of expanding horizontally prohibitive. So major expansion projects often involved adding a second level. The conditions that made the regional shopping center "the darling of the investment community" provided CPBL with work for a

Figure 79. Careful planning, coordination, and monitoring enabled the building team to maintain a vertical elevation difference of no more than ¼ inch between the Waikiki Landmark towers. (Photo © Bill Hagstotz/Constructionimages.com.)

new generation of owners, including New York–based Corporate Property Investors, which became an important repeat client. It was America's largest private equity real estate investment trust (REIT), with 23 regional shopping centers and four office buildings in its portfolio. The vertical expansion of existing properties presented the firm's engineers and their building team partners with new opportunities to innovate on behalf of clients, as the Tyler Mall project illustrates.[70]

Tyler Mall, in Riverside, California, was one of 32 shopping centers built by Ernest W. Hahn.[71] It opened as an enclosed mall in 1970 with 810,000 square feet of leasable space, three department store anchors, including the Broadway, the May Company, and J. C. Penney, and 75 specialty stores. In 1980 Hahn sold its shopping centers to Trizec Corporation, a Calgary-based real estate holding company. CIGNA Investments, a subsidiary of the Connecticut-based insurance giant, acquired Tyler Mall. In conjunction with Newport Beach, California–based developer, Donahue Schriber, CIGNA agreed to undertake a major renovation and expansion. The decision resulted in, as CPBL regional manager Norm Husk described it, "one of the largest and most sophisticated mall expansions ever planned." The new mall, rechristened The Galleria at Tyler when it reopened in October 1991, included a new anchor, Nordstrom's, 2,000 additional parking spaces, and 192,000 square feet of new leasable space produced by the addition of a second level (fig. 80). Its design and construction illustrates how project teams continued to find solutions to difficult engineering problems, to paraphrase Charlie Pankow, in remarks that he made in a company training video produced shortly before his death.[72]

The second-level expansion involved building the floor slab between the 12-foot tenant ceilings and the 23-foot-high original roof. This solution was developed by Larry McLean, of Hayward, California–based McLean Steel—the structural steel subcontractor—in collaboration with CPBL. The second floor was suspended from a new metal deck roof, using hanging columns attached to the bottom chords of its "super trusses." An independent foundation comprised of more than 600 steel pipe piles supported the new roof (fig. 81). A structural steel system of lightweight open web joists and girders integrated more than 25,000 square feet of skylights into the new roof. A joist system supported the electrical, plumbing, and HVAC systems of existing tenants. Lowered through openings created in the existing roof, it also served as a temporary platform for the eventual removal of the existing roof and the installation of another joist girder system that formed the second-level floor. The solution, as Suzanne Dow Nakaki, an engineer on the project with structural engineer Robert Englekirk, puts it, "was absolutely fascinating." Suspending the floor reduced the numbers of columns needed to carry the load, minimizing interference with existing stores—one of the primary requirements of both owner and developer. It also met their demand that the expanded mall retain standard floor-to-floor heights for multi-level shopping centers.[73]

Because Tyler Mall had been built as an enclosed shopping center, constructing the second level on top of the existing columns would have yielded unacceptable floor-to-floor heights and a multi-level second floor. A second-

Figure 80. Tyler Mall, Riverside, California, nears completion, July 1991. Callison Architects and MNA California, architects; Robert Englekirk, structural engineer. (Photo Warren Aerial Photography. Courtesy of Charles Pankow Builders, Ltd., Pasadena, California.)

level floor above the industry norm of 19 feet would have impaired shoppers' views of stores on the second level from the ground floor. It also would have required 4- to 6-foot transitions—escalators, ramps, or stairs—from the second levels of the existing department stores, blocking sight lines between the new stores and anchors, "seriously disrupt[ing] shopper traffic patterns." Upgrading foundations and enhancing columns that lay within existing stores would have been "very disruptive and messy," requiring the closure of large areas during construction. CIGNA Investments and Donahue Schriber insisted that Tyler Mall remain open for the duration of the project (fig. 82). The requirements of owner, developer, and anchor tenants thus ruled out the least technically complex and least expensive option—building over the existing mall.[74]

Construction of the project, which took 18 months, was confined to the hours between one o'clock and ten o'clock in the morning. Ensuring that the project was completed on schedule within these hours required a high degree of cooperation and coordination. As Nakaki describes, the night crews "were lifting pieces of steel over these shops and sliding them in between the roof and the ceiling below. [Even though] there was room to do it in there [the]

Figure 81. Second-level addition to Tyler Mall under construction, November 1990. Aerial view, showing the structural steel roofing system made of lightweight "open-web" joists and girders. Once completed, crews removed the existing roof and metal deck and installed a joist girder system that formed the second floor of the expanded shopping center. (Photo Warren Aerial Photography. Courtesy of Charles Pankow Builders, Ltd., Pasadena, California.)

logistics of getting in there was incredible." The window for driving piles to support the new loads was even narrower because the mall was located in a residential neighborhood: from six o'clock to ten o'clock in the morning. Robert Law, who estimated the project for CPBL, explains: "We ended up suspending pile-driving apparatus over top of the existing mall with a huge crane, driving pile through the roof through the store into the ground, and then doing the pile cap on top of those." The logistics involved with this task were "incredibly complicated."[75]

For his part, Donahue Schriber's Ernie Weber hired extra security to reassure tenants that thefts would not occur during construction hours. Weber also hired staff to act as liaisons between the contractor and tenants, and additional maintenance workers to ensure that retail areas were cleaned up before the mall opened. The developer, which had spent four years bringing the projection to fruition, took these and other steps to keep

Figure 82. Tyler Mall remains open during construction. (Courtesy of Charles Pankow Builders, Ltd., Pasadena, California.)

tenants informed during construction and address their complaints whenever they arose.[76]

CPBL's execution of the vertical expansion of Tyler Mall led Corporate Property Investors to select the contractor for its $71 million vertical expansion and renovation of Roosevelt Field Mall in Garden City, New York. CPBL had previously completed two shopping center projects in Orange County, California, for the REIT. More or less concurrently, it had added a food court to the Westminster Mall and had expanded the Brea Mall with a Nordstrom's department store and two parking garages. Pankow's work on the Roosevelt Field project would be similar to that which it performed at Tyler Mall. In this case, however, floor-to-floor height was not an issue. Designed as an open-air mall by I. M. Pei and Associates, Roosevelt Field had been enclosed in 1968, but with an atypically low floor-to-ceiling height. As project sponsor Dave Dwyer explained, "The old roof [did] not affect the new construction and [was essentially] dead structure within the interstitial space between the first and second levels." Rather, he continued, "Our biggest challenge [was enhancing] the existing columns so that they [could] support the load of the new second level." Crews reinforced the existing structure with high-strength, poured-in-place concrete or steel plates. Even though this work affected almost all of the tenant spaces, work proceeded without disrupting business. In fact, during the year-long construction period that began in December 1991, the mall reported an increase in sales.[77]

Even before Pankow began to expand Roosevelt Field Mall, however, the regional shopping center as a commercial property was in dire straits because of overbuilding during the 1980s. Many malls were opened and many expansion projects were completed just as the economy plunged into recession. As *The Economist* put it, "Having bred too quickly, they began to cannibalize each other" in a struggle to survive against discount stores, such as Walmart, so-called category killers, such as Toys "R" Us, and factory outlet stores, as the middle-aged population that formed the malls' customer base dwindled. The situation confronting the seven regional shopping centers in Onondaga County, New York, including Winmar's Penn-Can Mall, illustrates the cost of oversaturation. One of the properties, Fayetteville Mall, underwent a $25 million facelift that, as Al Lipsy, a spokesperson for Rochester, New York–based Wilmorite, which owned six of the local malls as of 1992, put it, "unfortunately [was] completed at the conclusion of the [1980s] growth." Two of the malls opened between 1988 and 1990. Over the next two years, occupancy rates for all seven of the shopping centers fell below the 95 percent level that was the traditional retail industry benchmark for centers of their size. Penn-Can Mall's acute distress—its vacancy rate soared from 10 percent to 26 percent during 1991—prompted Wilmorite, which acquired the mall from Winmar in 1989, to convert it into a discount and outlet center. Wilmorite also repositioned Fairmount Fair, operating with a vacancy rate of 37 percent, as a strip mall. The term "dead mall" had not yet entered the lexicon, but the term aptly describes the swift and terminal decline that these two centers experienced and their five neighbors faced in prospect.[78]

In this environment, "only the best managed, best located, and best stored mall" could hope to thrive, observed Michael Beyard, director of commercial development research at the Urban Land Institute. Fortunately for CPBL, Corporate Property Investors would continue to upgrade its shopping center portfolio. Owing to the quality of its work and its ability to complete work on time and within budget with minimal disruption to tenants and shoppers, the REIT would retain the contractor on several major expansion and renovation projects. In so doing, it would help to carry CPBL through one of the most difficult periods that the construction industry faced in the twentieth century.[79]

Looking Forward

On the twenty-fifth anniversary of the company, Charlie Pankow called for "a continuing effort to improve our services." In addition to design-build project delivery and "sound construction practices," Pankow highlighted "quality assurance, loss prevention, good public relations and, with growing concern, litigation" as key areas on which to focus attention. With a nod toward the reorganization, he noted that the firm had been "structured for perpetuity." Employees therefore "not only had an unlimited opportunity but also an obligation to continue to improve the high standards of our design-build services."[80]

The message was a conservative one. Writing during a boom that was benefiting all three offices in California and Hawaii—states whose business cycles had occurred more and less countercyclically for two decades—Pankow saw no need to modify the business model that had proven successful for a quarter century. Indeed, exposure to litigation was his greatest concern, if one measures concern by the number of words that he devoted to the issue in this article. During the 1980s, company's volume of business grew significantly (see fig. 17). Yet it also marked a period of transition for the Pankow firm as an organization. The departures of George Hutton and Russ Osterman, and the retirement of operations manager Alan Murk in 1991, signaled to the generation of professionals hired in the 1970s that it was incumbent upon them to realize the "unlimited opportunity" to which Charlie Pankow referred in his message.

It was unclear, however, whether the business model that had sustained the company for 25 years would enable it to thrive, or even survive, over the next quarter century. Charlie Pankow had no doubt that it could. Yet the recession of the early 1990s did not simply mean a temporary, if deep, drop in contracts. It changed the market for commercial real estate development in ways that threatened to marginalize Pankow's model, even as owners increasingly chose design-build as a project delivery method. The new environment narrowed the niche within which Pankow's firm operated, if not threatened to extinguish it altogether.

CHAPTER 6

Promoting Design-Build and Funding Concrete Construction Research

This chapter steps aside from the chronological framework of the narrative to consider the diffusion of design-build and the role that Charlie Pankow and his firm played in spreading the practice. As design-build and certain concrete construction techniques were intertwined in Charlie Pankow's construction program, efforts by the firm to promote the former entailed advocating the use of precasting, slipforming, and related techniques. The Pankow firm stood apart from contractors as an underwriter of primary research, and so this chapter also examines its contribution to the development of a structural framing system that promised to improve the ability of buildings to survive seismic events intact.

Charlie Pankow maintained a professional and proprietary interest in incorporating design-build and advances in concrete construction into engineering education and research. Though much of the company's design-build and concrete construction know-how was developed at the building site, Pankow understood that the engineers and construction managers whom he recruited from university stood a better chance of meeting his expectations in the field if they possessed a solid technical foundation, especially if it dovetailed with the ways in which he conducted business. The chapter's first section examines Pankow's ongoing interest in Purdue engineering and shows how his efforts promoted the diffusion of design-build, in particular.

Throughout their careers as builders, Charlie Pankow and others in the firm promoted design-build and advances in concrete construction from positions of leadership in national associations and societies, the American Concrete Institute (ACI) in particular. The Southern California chapter that Pankow helped to found in 1957 was the first local chapter established within the national society. In 1980 he served as ACI president. Dean Stephan and Tom Verti subsequently led the organization. All three men used the monthly "President's memo" and other means that the position offered them to recommend what they believed to be better ways to build concrete commercial structures. Charlie Pankow's advocacy of design-build culminated in his playing a direct role, along with Stephan and Rik Kunnath, in founding the Design-Build Institute of America (DBIA).[1]

As a glance at the notes to this book would suggest, Charlie Pankow and others in the firm, Dean Stephan in particular, published articles that promoted design-build and automated concrete construction techniques. Signed essays appeared in trade journals such as *Concrete Construction* and *Journal of the American Concrete Institute*. Articles that detailed the construction of specific projects appeared in *Engineering News-Record*, *Urban Land*, and other publications. Pankow and Stephan also contributed chapters on on-site precasting and slipform construction to successive editions of the *Concrete Construction Handbook*.[2] In this way, the medium-sized contractor reached a wider audience.

No doubt, these efforts ultimately benefited the bottom line of the firm. At the same time, the choice of Pankow and others in the firm to publish articles in these outlets rather than simply write copy for proprietary brochures and other advertising media suggests that they had a professional interest, too, in spreading their ideas among colleagues, and not just pitching them to potential clients. Of course, convincing other contractors of the value of Charlie Pankow's construction program in professional venues might increase its acceptability in the marketplace, and therefore benefit the company's bottom line, too. At the same time, promoting ideas on contracting might just as easily increase competition. It would appear that Charlie Pankow and others in the firm calculated that it would be better to enlarge the size of the market for their favored practices than do business strictly on a proprietary basis.

Charlie Pankow's Interest in Engineering Education

The technical training that Dean Browning, Bob Law, Joe Sanders and others recruited into the Pankow firm during the 1970s received placed far more emphasis on theory and science than it did when Charlie Pankow studied at Purdue. A seismic shift in engineering education followed in the wake of the reorientation of research that occurred with increased federal funding of basic scientific research that began during World War II and accelerated with the launching of Sputnik in October 1957. As Bruce Seely argues, the war exposed engineers' inability to contribute to advanced engineering projects, which had implications for both their research and teaching. To compete successfully for their share of the millions of dollars appropriated by Congress for research that addressed the needs of the national security state during the Cold War, academic engineers and their students pursuing advanced degrees anchored their proposals in science. In this context, it was more likely, too, that the doctoral candidate who wished to pursue a career in academia would accept an assistant professorship without first gaining practical experience. The development of the computer also facilitated the shift in emphasis in research by enabling hypotheses to be tested with numerical procedures like the Finite Difference Method. Where research led, education followed: By 1970, the transformation of engineering curricula, as leading educators had advocated since early in the century, was essentially complete.[3]

Purdue Civil Engineering embraced the trend toward science-based research. Gone were drawing classes and courses such as Steam and Gas Power. Students now studied Mechanics of Fluids, Statically Indeterminate Structures, and Elements of Thermodynamics, for example. Whereas Charlie Pankow took Applied Materials and Engineering Materials, Browning, Law, and Sanders were required to take Mechanics of Materials and Materials Science. In place of Contracts and Specifications, civil engineering students took Construction Engineering and Management. Reflecting the Computer Age, students took Laboratory on Programming for Digital Computers rather than Mechanical Laboratory. Courses also updated their names to reflect the shift from the practical to the theoretical. City Planning, for instance, was now City Planning Philosophy and Theory. Gone, too, was the summer surveying camp that Charlie Pankow had attended between his sophomore and junior years: It had been eliminated in 1960.[4]

At the same time, Purdue Civil Engineering did not dispense with training in the practical application of engineering design. While students may no longer have spent nine weeks at summer surveying camp, they still took Engineering Surveys, a two-semester course, and undertook an engineering inspection trip (CE 490) in their senior year. The School also offered a five-year cooperative program that enabled a limited number of qualified students to acquire practical experience while they pursued their bachelor's degree.

In fact, in the late 1960s, Purdue Engineering began to direct an increasing amount of attention and resources to preparing undergraduate students for careers in industry.[5]

As dean of the Schools of Engineering at Purdue University from 1967–1971, Richard J. Grosh initiated efforts to align programs with the needs of employers. He asked industry leaders, including Charlie Pankow, to provide professional expertise and guidance, development opportunities for students, and financial support. As a corporate leader who was seeking to hire its graduates, Pankow took a keen interest in the civil engineering program of his alma mater.[6]

One outcome of Grosh's initiatives was the establishment, in 1976, of Construction Engineering and Management (CEM) as a division within the Schools of Engineering. Under its second head, John A. Havers (1979–1986), CEM established a mandatory three-year internship program that exposed students to work at the building site, project management, and estimating over successive summers. It was a program that Charlie Pankow applauded, and his company participated in it. He was concerned, however, that the CEM program weighted management too heavily at the expense of engineering. Pankow sought recruits with a solid technical foundation in mechanics, materials, structures, and so on: They could learn whatever they needed to learn about business from the company, he believed.[7]

Pankow also wanted recruits to know about design-build and so he recommended to the heads of the CE and CEM programs that design courses "include a strong dose of this method of project delivery," according to Vincent P. Drnevich, who was head of Civil Engineering from 1991 to 2000. Pankow's suggestion prompted Drnevich to learn about design-build and incorporate it into Civil Engineering Design Project, the capstone course in civil engineering, as the required method of project delivery. Pankow's influence is evident in the articles that Drnevich and his colleagues wrote about the course.[8]

Charlie Pankow also supported engineering education and research at Purdue with substantial gifts. These contributions established an endowed chair and a concrete laboratory that bears his name, and contributed materially to the construction of the Robert L. and Terry L. Bowen Laboratory for Large-Scale Engineering Research. In all, Pankow's deep and abiding relationship with his alma mater helped to spread his ideas on improving the delivery and construction of commercial building projects.[9]

Construction Management and Design-Build as Alternatives to Design-Bid-Build

The diffusion of design-build was slow, as may be expected for a methodology that may be characterized as a lost art in the commercial segment of the construction industry in postwar America. Indeed, as late as the mid-1980s, design-build was used to deliver just 3 percent of total US construction. The methodology apparently was not top of owners' minds as they grasped for solutions to the problems that plagued their projects during CPI's first two decades in existence. Indeed, during the inflationary 1970s, owners turned to construction management (CM) as an alternative to design-bid-build. As inflation and interest rates fell, owners cooled to CM and became increasingly receptive to design-build. To build on this momentum and propel its use among private sector owners, Pankow and other design-build and design firms organized the DBIA in the early 1990s.[10]

Before design-build gained in popularity, there was CM. As Dean Stephan tells it, owners increasingly came under pressure to control costs once the pent-up demand for commercial structures created by wartime restrictions was satiated. Until that point, "it really didn't make much difference what the buildings cost, because they were going to be leased up instantaneously because everybody needed space. [The] economy was growing, and you could afford [the] huge inefficiencies [of design-bid-build]." Stephan argues that "the cracks [in the traditional approach] started showing" once owners had to compete more vigorously for tenants. With the cost of materials, labor, and financing spiraling upward, owners adopted a much more proactive role in the building team in order to realize expected returns on their investments. They sought ways of configuring work to increase cooperation and compress schedules. Owners insisted that architects, engineers, and contractors adopt more of a team approach to ensure timely project delivery. Charlie Pankow, of course, had been enjoying some success in selling owners on the merits of design-build, which seemed to offer a tailor-made solution to the cost problems that owners faced. But he succeeded in carving out only a small, albeit profitable, niche for design-build as a project delivery system. At the same time, architectural and engineering firms (A&Es) and some contractors who specialized in industrial buildings promoted "one-stop" design and construction. But the commercial projects that they delivered on a lump-sum basis were often of poor quality. So owners who were willing to depart from the traditional approach turned to CM as the answer to the cost pressures that they faced.[11]

Under CM, a construction manager acts as the agent of the owner and co-ordinates the activities of the building team. Rather than sign a single contract with a general contractor, as would be the case under design-build, the owner lets multiple prime contracts with the designer and multiple trade contractors, often securing the latter through bidding (fig. 83). As CM emerged in the late 1960s, the construction manager typically collected a fee for acting exclusively as the owner's agent. A decade later, the construction manager was operating increasingly "at risk." Under this variation, the construction manager guaranteed a maximum price for the project and assumed some of the risk that the owner incurred when the construction manager simply worked for a fee. Under this arrangement, the construction manager consulted the owner during preconstruction and design, but assumed a role akin to the general contractor during construction. Owners might recruit CM experts to work in-house, as many of them did. (With the recessions of the 1970s, owners were quick to shed these resources.) In any case, owners expected their construction managers to advise them on scheduling and estimating. CM advocates touted the coordination of designer and builder as one of the system's major advantages over design-bid-build. Construction managers, they insisted, would ensure that designs could be built easily and economically and that builders adhered to the construction documents. In the context of high inflation and soaring interest rates, owners turned to CM because it promised to reduce the time needed to complete their projects. For design and construction now could overlap on a fast-track schedule, rather than occur sequentially, as was the case under design-bid-build.[12]

For the general contractor, CM was controversial, since the construction manager replaced him in the owner relationship. Under CM, the contractor was just another prime contractor. (Architects objected as well, for they saw the construction manager infringing on their traditional advisory role.) The Association of General Contractors insisted that the contractor should play the role of construction manager. Were this not the case, the AGC argued, the project delivery system would provide insufficient incentive for the building team to deploy cost-saving innovations or guarantee project costs. Agency CM, whereby the construction manager worked without guaranteeing a maximum project cost, was especially "fiscally irresponsible," according to Campbell Reed, director of the AGC's building division. In contrast to the contractor, the AGC argued, few architects and structural engineers were capable of estimating costs accurately. Further, they lacked the general

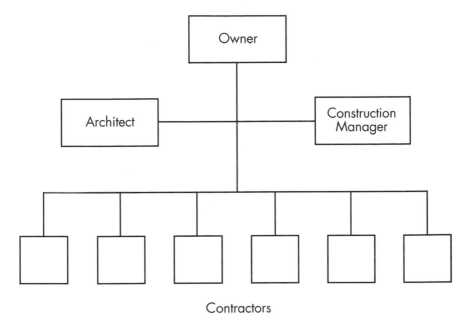

Figure 83. Construction Management. The owner signs contracts with the architect and all of the prime contractors, including the general contractor. The construction manager acts as the owner's agent and coordinates the activities of the building team.

knowledge about methods, materials, and labor costs that contractors possessed. The AGC agreed with CM's goal of eliminating inefficiencies associated with the owner hiring architects and engineers to design a project and contractors to build it on a separate, sequential basis, but was adamant that CM did not provide the solution.[13]

Contractors, of course, agreed with their trade association. For one, Saul Horowitz Jr., president of New York–based HRH Construction, insisted that CM did not represent "anything different in today's building market from the well-qualified general contractor." Hinting at an argument for design-build, Horowitz continued, the functions of the construction manager were no different than those "which have been traditionally handled by general contractors who were able to negotiate their jobs early." D. C. Peters, president of Pittsburgh-based Mellon-Stuart Company and a past president of the Master Builders Association of Western Pennsylvania, called CM "superfluous": "A reputable, capable general contractor needs no further supervision." As Tom Verti notes, CM was a "four-letter word" within the Pankow firm. CM was the "antithesis of what Charlie's philosophy was." There was no need for a construction manager, he felt, "because as the [master builder] we would do what a good construction manager would do, but we'd also take ownership in those decisions, and we'd back that up with guarantees, bonded lump-sum contracts." In contrast to the design-build contractor, adds Bob Law, construction managers had no incentive to drive innovation in the industry. The construction manager "doesn't really do any of the work—all they're doing is just managing other people doing the work—they don't have the experience to know what you can innovate." Indeed, they do not have the information required "to know what questions to ask or what things to attack or [even] what areas need innovation." CM is pernicious, because construction managers sell owners on the idea that, for the lowest fee possible, "they'll beat up all the subs and make it so the subs make just a little bit of money." Whereas Pankow as contractor "actively assume[d] the risk for construction costs," as an early CPI marketing brochure explained, construction managers avoided it. And "where there is no risk, there is no incentive." Concludes Dean Stephan: As an agent, the construction manager does not "have any ability to the control the cost [of the project], and really [has no] stake in the fire."[14]

The contractor community agreed with Charlie Pankow that the solution to the lack of coordination between designer and builder under the traditional approach lay not in CM but in involving the contractor in the entire delivery process. That is, they saw CM as an opportunity for the contractor to assume the role of construction manager and thereby coax owners toward accepting design-build as their preferred project delivery system. HRH Construction's Horowitz believed that, were the unsettled economic conditions prevailing in the early 1970s to persist, owners would be compelled to negotiate "a single general contract with well-organized and well-managed contractors." Resolving the adversarial relationship between builder and designer would also mean convergence in the form of joint ventures, he predicted. Arthur O'Neil, president of Chicago-based W. E. O'Neil Construction, believed that CM contracts that "welded into one team" the architect and contractor held the key to achieving the reconfiguration of design and construction delivery that would meet owners' need to control costs. Even contractors who were not prepared to discard bidding work in favor of negotiating design-build contracts saw the need for a contractor-led team approach. For instance, Lou Perini, chairman of Perini Corporation, believed that contractors had to become much more involved in "total construction management." In criticizing CM, these men built the case for design-build.[15]

The logical end of this reasoning, given owners' seizing on CM as panacea for cost escalation, was positioning the contractor as construction manager. And, in fact, many a contractor marketed their CM abilities to owners and developers. (A&Es did so as well. Among the cacophony of voices, there was no agreement on a working definition of CM.[16]) As Charles Kendall, an executive vice president with Geupal DeMars, an Indianapolis-based CM and contractor, reflected, "When contractors first got involved with CM in the early 1970s, everyone who was a general contractor just turned his sign around and put CM on the back, even if he had no management expertise."[17] The reputation of the contractor suffered, because not all firms were able to deliver on their promises. This was ironic. As Glenn J. Chell, president of Chicago-based contractor Chell and Anderson, noted, the construction manager owed much of his success in selling his services as a "composite miracle worker" to the fact that "the name, 'general contractor,' had been degraded by many so-called general contractors whose performance has provided much less—those for whom a contract is a hunting license for cheaper prices and inferior subcontractors."[18] This reputation persisted at least into the 1980s, hindering design-builders from convincing owners to give them sole responsibility on the building team.

Nevertheless, with the Federal Reserve Bank's merciless crushing of inflation during President Reagan's first term in office, CM fell out of favor. As Charles B. Thomsen, president of 3D/International, a Houston-based A&E firm, explained, "When you had 20 percent inflation, you could knock a year off the construction schedule for a high school and give the owner 20 percent more school." But now owners were no longer "in a hurry like they used to be. . . . The pressures that drove [them] to use [CM] aren't there to the extent they were."[19] Design-build advocates took the opportunity to counter critics who argued that their approach reduced the owner's control over the design process, created adversarial relations among members of the building team, and encouraged the contractor to discourage creative design and reduce the quality of the building (to achieve a low bid price or maximize profits on lump-sum contracts).[20] Or, as Dean Stephan puts it, it provided a chance to put to rest the widely held notion within the industry that, under design-build, "you had the fox guarding the hen house, the nasty old contractor's going to steal the owner blind by cheapening everything in sight."[21] Noted Richard Wilberg, marketing director for Opus Corporation, a Minneapolis-based design-builder, with owners taking more time to do their projects, they "want[ed] to be assured that the project [would] be done right, because escalating property values [were not] going to bail them out."[22]

On the thirty-fifth anniversary of his firm, Charlie Pankow took note of the "gradual acceptance" of design-build during the existence of the company. The mid- to late 1980s, however, were critical years. By the end of the decade, design-build was "no longer considered a process used almost exclusively for constructing industrial buildings" or the choice only of owners "willing to compromise quality to gain speed." And design-build firms were delivering commercial and institutional buildings that were "increasingly sophisticated in design." Public agencies were also becoming more receptive to the process. According to Stephan, design-build gained momentum during the late 1980s because both architects and engineers became more receptive to it and owners became more familiar with it.[23]

Market research by *Building Design & Construction (BD&C)*, a trade journal, traced the acceptance of design-build. A survey of 326 building owners by the magazine's research affiliate suggested that design-build delivered roughly 14 percent of total US construction in 1983. Between 1986 and 1988, the share of "diversified developers" who reported "some use" of design-build in the journal's annual survey of the 300 largest owners and developers increased from 26 percent to 32 percent. At the same time, the figure for developers of retail projects increased from 52 percent to 61 percent, while developers of hospitality projects reported an increase in their use of design-build from 17 percent to 26 percent.[24]

The editors of *BD&C* conceded that these survey figures only suggested the positive trend in the use of design-build—the "hard numbers" needed to fix the value of US construction delivered by design-build were unavailable. Yet anecdotal evidence reported by A&Es and contractors reinforced the survey responses and suggested that the Pankow firm was no longer virtually alone in the marketplace. The aforementioned Charles Kendall, for instance, expected his firm's design-build order book to double from 1989 to 1994, at which point it would comprise as much as 30 percent of its business. Smith, Hinchman & Grylls Associates, Inc., a Detroit-based A&E, noted that they had no design-build work at the end of the 1970s. Now it was a growing, if still small, part of their business.[25]

Descriptions of successful design-build projects in the trade press helped to promote the benefits of design-build that Charlie Pankow had been touting for years. Some of the private sector projects profiled in *BD&C* during the 1980s (excluding Pankow's work) included the headquarters for Federated Department Stores, Cincinnati; the headquarters for the Farm Credit Banks of Baltimore and Metro Plaza, Silver Spring, Maryland; the American Transtech headquarters and operations center and the Prudential Records Center, both in Jacksonville, Florida; a Frito-Lay food processing plant in Frankfort, Indiana; the Orchard Point Office Center, Rosemont, Illinois; First Citizens Bank Plaza, Charlotte, North Carolina; FMC's Technology Center, Minneapolis; and Casio's US headquarters in Dover, New Jersey. The showcasing of design-build on projects of varying types and sizes across the country by *BD&C* and other trade publications demonstrated the benefits of the approach to the members of the building team, owners in particular. It also showed that the market for design-build services was becoming more competitive with the participation of new entrants.[26]

The publication of design-build documents for owners, designers, and contractors by professional societies both reflected the growth of design-build and promoted its use. The AIA's change of position on design-build constituted a seminal moment in the growth of the practice. Fundamentally opposed to design-build from its inception in 1857, the AIA in its Code of Ethics prohibited its members from offering design-build services, citing the conflict of financial interest that it saw as inherent in the arrangement. In 1978 it relaxed the restriction on a probationary basis. In 1983 the AIA's directors approved a policy statement that gave support to qualifications-based selection in the procurement of design and construction services on public projects. The statement also declared competitive bidding to be not in the public interest. The following year, the AIA released initial drafts of proposed design-build contracts. In 1985 it approved a revised Code of Ethics that dropped the ban on design-build and published its first series of design-build contracts. After the 1983 Policy Statement sunset in 1988, the AIA formed a national task force to deliberate on design-build. In so doing, the society recognized its growing popularity. In August 1990, the task force produced an interim report that supported the adoption, in May 1991, of a new policy on public sector design-build. The AIA affirmed its approval of qualifications-based selection criteria in project procurement. Finally, in 1997, the AIA's directors issued a position statement on public sector project procurement that endorsed the use of design-build competitions, featuring a two-phase selection process, described below.[27]

Winning support for design-build from engineering societies was less protracted. In 1987 the Engineers Joint Contract Documents Committee, whose work the American Consulting Engineers Council, the American Society of Civil Engineers (ASCE), and the National Society of Professional Engineers all endorsed, began to develop standard design-build documents. At the same time, the American Consulting Engineers Council (ACEC) initiated action with a view to adopting a general policy on design-build. In the early 1990s, the group issued a design-build guide for owners and contractors. For its part, the ASCE reported on the advantages and disadvantages of design-build from the points of view of building team members. In 1995 the Engineers Joint Contract Documents Committee (EJCDC) issued a set of design-build contracts and guides—the culmination of the process it initiated eight years earlier. The institutional support of architects and engineers provided crucial support for the arguments of individual contractors like Charles Pankow Builders, Ltd. (CPBL), that design-build provided a superior approach to project delivery.[28]

Perhaps most important of all to the diffusion of design-build were the performances of contractors, making good on their argument that they were in the best position to guarantee a project's cost, schedule, and quality. In less than a decade, the share of nonresidential construction delivered with design-build tripled. Indeed, as one observer noted, by 1992 design-build had long been a "hot topic," at least as far as private sector construction was concerned.[29] Still, the traditional design-bid-build approach overwhelmingly remained the project delivery system of choice for owners. To fuel the growth of design-build, leading design-build contractors, including CPBL, joined forces to create the DBIA.

Promoting Design-Build through Institutional Means: The Design-Build Institute of America

In February 1993, Preston Haskell, founder, in 1965 in Jacksonville, Florida, of his eponymous design-build firm, gathered in Washington half a dozen executive managers of the leading national design-build firms. Dean Stephan represented the Pankow firm. At a dinner and a half-day meeting the next morning, the group discussed whether the industry was ready for another association, one that would promote the concept of design-build project delivery and the interests of the companies who did business on that basis. For a decade or more, Haskell and his like-minded colleagues had failed to persuade industry associations to establish design-build practice sections within their organizations. Those who joined Haskell in Washington met several more times before they agreed to form an association dedicated to promoting design-build. Eleven firms organized the DBIA. They selected Haskell as chairman and Jeffrey Beard, formerly head of regulatory affairs at the ASCE, as executive director.[30]

Rik Kunnath represented CPBL in the organization. After attending the initial meeting, Stephan bowed out, citing the time constraints of running the company's operations in Southern California. He selected Kunnath to attend future meetings. In 1995 Kunnath succeeded Haskell as chairman—an appointment that both reflected Kunnath's talents and recognized the firm's place in the industry. Subsequently he served as director. Kunnath recalled, "[It] turned out to be an incredibly rich opportunity for me. . . . I was highly identified with starting that organization, so it's given me a lot of personal credibility. [It also] turned out to be a great learning opportunity." In keeping with the Pankow Way, Stephan acted as mentor. Recognizing Kunnath's potential for top management, Stephan already had encouraged Kunnath to prepare for general contracting licensing exams in California, Florida, and Nevada. Kunnath says that Stephan's support "was very, very, very helpful" to his career.[31]

From the outset, the DBIA aimed "to place design-build on an equal footing with the 'traditional' project delivery method," as Beard put it. To accomplish this goal, the group acted on a number of fronts. It disseminated best practices to owners and practitioners through its publications, including the *Design-Build Manual of Practice* and *Design-Build Dateline* magazine, courses, and conferences. In 1998 it published its own set of contracts and forms for the building team. It developed a certification program for design-build professionals and custom training programs that "have influenced virtually every federal agency that builds anything, and many state agencies, especially [in] California," according to Kunnath. All of these efforts were "remarkably successful" in helping design-build become accepted, if not preferred, by owners as a project delivery system. Indeed, the cumulative impact of these activities on the industry has been "so profound," Kunnath thinks, that "had we, when we got together for those first couple meetings, ever realized the impact that we would have and the momentum that [the design-build] movement would gain, nobody would have believed it."[32]

The success of the movement may be measured by membership in the DBIA, the growth of which exceeded the wildest expectations of its founders. Integrated design-builders, contractors like CPBL that used outside designers, and A&Es were invited to join. Initially, membership was to be limited to integrated design-builders—firms that engaged in both design and construction. Largely at the behest of Pankow, Stephan, and Kunnath, membership was broadened to include contractors and A&Es. Later, at the urging of Beard, the DBIA expanded its membership categories. When the association held its first annual meeting in October 1993, there were 20 firms paying $10,000 in annual membership dues. By the end of 1999, the list had grown to 667, including regular, associate, and "special individual" members. That year, the DBIA began chartering local chapters. By 2010, the DBIA had more than 1,000 members and 16 regional offices that coordinated the activities of local chapters.[33]

The DBIA provided the Pankow firm with an institutional vehicle through which its professionals could promote design-build. Kim Lum, for instance, notes, "We had a couple of events [in Hawaii, after he returned to Pankow in 1995] to try to get people to come and educate them about design-build." The creation of regional organizations helped to increase the effectiveness of such efforts. When Lum transferred to Northern California in 2001 to become regional manager, he also joined the Board of Directors of the Western Pacific Region, which included five chapters in Arizona, California, Hawaii, and Nevada. Committees at the regional level support DBIA staff in implementing initiatives approved by the national board in five areas: education, programs, membership, legislative, and public relations. In addition to serving on its board of directors, Lum, who earned his DBIA designation, chaired the region's education committee. In 2010, he joined the national board of directors.[34]

These institutional efforts helped design-build converge on design-bid-build as the delivery system of choice for nonresidential construction in America. Of course, as Kim Lum notes, "The best way to [promote design-build] was to go out and do design-build work," which the Pankow firm, he believes, continued to do "quite successfully."[35] At the same time, the increasing interest of the public sector in using design-build to deliver projects constituted a development that the founders of the DBIA did not expect and provided CPBL with new opportunities to demonstrate the value of the methodology.

Public Sector Design-Build Procurement and Delivery

Public sector interest in design-build grew in the decade leading up to the formation of the DBIA. Progress resulted largely from experimentation on a case-by-case basis, however, as laws at the federal, state, and local levels restricted its use. Notwithstanding the growing support for design-build from architects and engineers, competitive bidding regimes remained largely intact into the 1990s. The so-called Brooks Act of 1972, which amended the 1949 Federal Property and Administrative Services Act that had allowed US agencies to negotiate architectural and engineering services for nonmilitary projects, required federal agencies to procure architectural and engineering services based only on qualifications and demonstrated performance. The law made it impossible to combine design and construction services. Almost every state had followed suit by passing "mini–Brooks" acts. Nevertheless, on the eve of the formation of the DBIA, design-build was "starting to change the way governments, designers, and contractors work in the public-sector market," announced *Civil Engineering*, the ASCE's flagship publication.[36]

The General Services Administration and the US Postal Service were important "first movers" at the federal level. Like private sector owners, the GSA looked to CM in the 1970s to address problems associated with fragmented project management and adversarial relations among members of the building team. In the early 1980s, the agency turned to design-build, primarily out of an interest in reducing procurement and delivery cycles. As James Stewart, director of the GSA's Office of Design and Construction, later put it, "The main reason we went into design-build was to buy some time." The GSA was "basically pleased" with the results of two early demonstration projects, involving border stations in Arizona and California. The GSA used the results of these projects to develop a guide and a model contract for use on future projects. At the same time, the USPS let three experimental design-build contracts, with mixed results. The projects did not allay its concern that the guaranteed maximum price formula used in the contracts created incentives for the builder to cheapen design. Thus USPS officials devoted more attention to reviewing working drawings than would be the case normally. Nevertheless, the USPS increased its use of design-build on a case-by-case basis for large projects of $10 million or more. Said John Wiernicki, Stewart's counterpart in the USPS's Office of Design and Construction: "We plan to use it when it makes sense and when there are indications that it will be in our best interest to do it." After several years of using the process satisfactorily, Wiernicki found that project administration "is far easier with design-build. It's easier to reach an agreement with one entity."[37]

Yet design-build overall made little headway at the federal, state, and local levels for much of the 1980s because of the laws that mandated competitive bidding. Public entities that incorporated design-build into their projects invariably bid the construction work competitively, which limited the effectiveness of the approach. At the local level, officials continued to abide by tra-ditional practices even if their city council or board of supervisors "passed ordinances [giving] them the right to negotiate construction management services," observed Peter Winchell, operations manager in the Tampa, Flor-ida, office of Ellerbe Builders.[38]

Design-build began to make inroads into the public sector during the re-cession of the early 1990s, when tax revenues declined, even as demand for projects remained steady or increased. For example, Volusia County, Florida, selected Ellerbe Builders to evaluate sites for a new civic center. Ellerbe's de-sign arm designed the building; Ellerbe Builders constructed it. Similarly, the City of Chesapeake, Virginia, hired Centerra Corporation to design and con-struct its city hall after conducting a selection process that included 25 firms. In both of these cases, local government had to comply with laws on competi-tive bidding, but ordinances allowed them to negotiate professional services. At the state level, the State of Massachusetts used design-build to produce the Suffolk County Jail after procurement regulations were revised to encour-age "alternative methods of construction." Likewise, the State of Washington procured its Natural Resources Building in Olympia through a design-build competition. At the federal level, Congress directed the Defense Medical Fa-cility Office to experiment with design-build for new medical clinics. It also authorized the Department of Defense and the US Corp of Engineers to let three design-build contracts annually. After five years of positive results, Con-gress lifted these restrictions. Design-build was attractive to public agencies, not only because it promised to shorten procurement and delivery schedules, but also because it relieved them of the costs of oversight. For under the ap-proach, execution risk shifted to the contractor.[39]

By no means did the cumulative effect of this activity constitute a univer-sal mandate of design-build in the public sector. Indeed, in some instances, governments acted specifically to reverse the trend. In 1989, for instance, the attorney general for the State of Texas ruled that design-build could not be used on public contracts. In August 1991, the New York Department of Edu-cation, which had responsibility for professional licensing, banned the use of design-build indirectly by restricting the ability of architects and engineers to delegate design work to non-licensed firms. The bright line separating design and construction at the federal and state levels, drawn by the Brooks Act and various mini-Brooks laws, had hardly dimmed. At the time that the DBIA was formed, the impact of using design-build in the public sector on an ad hoc or experimental basis was measurable. But the legal regime ultimately would have to change at the federal, state, and local levels if the public sector were to use design-build as the private sector was increasingly doing. Most of these changes would occur only after the passage of the Federal Acquisition Reform Act of 1996, which authorized design-build on a limited basis.[40]

Implementing design-build across the public sector involved changing the procurement process, from one that separated design and construction and awarded construction contracts strictly on price, to one that unified design and construction and considered the quality of the design, the qualifications of the firm, and other factors in awarding a contract. Drawing on private sector examples and the resources of professional societies like the AIA, public officials embraced design-build competitions in the 1990s. Over time, these competitions became increasingly refined and more elaborate in terms of the criteria used to evaluate proposals.

The GSA led the way. Its model design-build approach awarded contracts on the best combination of qualifications and price. So-called design criteria professionals under contract with the agency prepared a preliminary design program and technical specifications (based on performance standards rather than the detailed descriptions traditionally found in an RFP [request for proposal]). The agency then sent the documents to potential design-build bidders, who made two sealed submissions. Under this "two-envelope system," one package described the firm's technical qualifications and a project plan, and enclosed vitae of the people who would be involved in the project. The other detailed a fee schedule. Officials evaluated the qualitative material before opening the envelopes containing the prices. After choosing the winner, the GSA held a predesign conference to review preliminary drawings. "After that, he's on his own," said GSA's Hal Buehler: free to proceed on a fast-track basis or wait until design was complete before commencing construction.[41]

State-level design-build competitions typically followed the GSA model. As might be expected, the specific criteria used in design-build competitions varied from state to state and, indeed, from project to project within states. As they developed in the public sector, competitions typically awarded points in three categories: qualifications, design, and price. The most important qualifications included local participation, diversity, lack of litigation, reputation, special expertise, and time to completion. Countless project-specific factors were included as well. The public agency or government typically established point ranges for the selection jury to use to rank proposals. States, if not municipalities, invariably used the GSA's "two-envelope" system to award the contract. The AIA, AGC, and DBIA all considered this approach to be a design-build hybrid. The DBIA labeled it design/design-build and debated it for several years before issuing a policy statement. Still, the approach moved the public sector some way along the road from design-bid-build to design-build.[42]

Public Sector Design-Build Competitions: CPBL's Early Experiences

In the early 1990s, CPBL secured two public projects awarded through design-build competitions: the Special Events Arena at the University of Hawaii, Manoa, and the expansion of Boalt School of Law on the Berkeley campus of the University of California. The former constituted the State of Hawaii's first use of design-build; the latter, "a unique departure" from the way in which the University of California designed and constructed its projects.[43] Both projects provided the company with a way of showcasing design-build. As Kim Lum explained, "We thought if we could get the State [of Hawaii] to do a design-build project, maybe people would recognize it as another way to do construction."[44] Competing for public sector work helped the company fill its order book during the recession of the early 1990s. As such, it was no coincidence that these projects were pursued out of Hawaii and Northern California—the two regions most in need of work.

Like many of the firm's private sector engagements, the UH (University of Hawaii) Arena was a salvage job. The State of Hawaii first used the traditional approach to find a contractor to replace the cramped and not-air-conditioned Klum Gym, which had been built in 1956. After the bids that the state received "significantly exceeded" the project's budget, CPBL approached the office of Governor John D. Waihee III. Recalls Al Fink, who was regional manager at the time: "We [recommended that the state] try a different way, a way that has been successful. [We] explained to them what [design-build] was." As part of this process, Dean Browning, the project's sponsor, spent "a great deal of time" explaining design-build to the Department of Accounting and General Services. (An official from DAGS represented the state on the project, assuring compliance by attending weekly meetings and otherwise staying in contact with the building team.) CPBL's presentation persuaded Governor Waihee, who needed no reminding that his state's building projects invariably busted budgets and schedules. He, in turn, recommended to DAGS that the state hold a design-build competition to procure the arena. It is likely that the governor approached Roy Y. Takeyama, the chair of the university's board of regents and "the perennial chair of [its] Committee on Physical Facilities and Planning," as historian David Yount writes, and convinced him of the merits of proceeding with the project on a design-build basis.[45]

The RFP, Fink explains, "provided all the basic parameters. . . . They had an idea of what the budget was, also when they wanted it delivered, and a fair amount of detail on how this would work." The state set performance standards in lieu of providing detailed instructions. For example, for the HVAC systems, the RFP stated something to this effect: "On the hottest day in the middle of the day we want to be able to have a packed arena and have it cooled down to 75 degrees." The points system approved by the regents weighted heavily the design of the arena. Reflecting the dire financial situation in which the State of Hawaii found itself in 1992, however, the RFP also stated that if no proposal met the target price, then the jury would decide the competition strictly on price. This provision would prove decisive in awarding the contract to the Pankow team when all of the bids exceeded the ceiling set in the RFP.[46]

CPBL joined Heery International, an integrated design-build firm founded in 1952 by George T. Heery, and Kauahikaua & Chun Architects in developing a seating and structural system that facilitated tilt-up and on-site precast construction. As Browning explains, the 10,031-seat arena was contemplated as a "modified tilt-up project. . . . The structural skeleton for the project [would be] built on the ground [slab-on-grade] and stood up." Thirty-two vertical bents, or structures that supported the seats, each comprised of three columns, would be cast in horizontal 16-inch forms in stacks of three (fig. 84). Sixteen ring beams that would connect the bents—each one 64 feet in length—also would be precast on-site. The contractor would form and pour the concourse deck while subcontractor HD&C Precast, a subsidiary of Hawaiian Dredging & Construction Company, manufactured the beams and planks for the seating system in their yard. A white, aluminum, geodesic dome with a free span of 320 feet would top the structure. Temcor, the Gardena, California–based pioneer designer and manufacturer of such structures, would fabricate the dome on the ground and lift it into place (fig. 85). The approach would be less expensive and safer to construct ("because you don't have guys working up in the air," notes Al Fink) than the poured-in-place method that was typically used for this type of structure. As Fink explains, the latter required lifting "all of the concrete in the building in some fashion, either by a bucket lifting it up, or doing it all in one big piece on the ground and lifting it up." Creating

Figure 84. Workers prepare 64-foot bent to be lifted into place, Special Events Arena, now the Stan Sheriff Center, University of Hawaii, Manoa. (Reproduced by permission of Ed. Gross–The Image Group © 2013.)

the formwork to cast-in-place the 55-foot-high and 50-foot-wide bents would have been "expensive and very time-consuming," Browning noted. Under the proposed system, two cranes would pick the 95,000-pound bents from their molds and set them into place. The proposal promised to deliver the arena for $32.2 million, well below the cost of the original design.[47]

The approach enabled the project team to save time as well. Construction began in December 1992. The on-site precast operation took about seven weeks to complete, including 10 days to install the bents and ring beams. As the H-1 Freeway was adjacent to the job site, erecting the arena's structural system quickly became a local spectacle. Recalls Bill Bramschreiber, one of the project's two superintendents: "When we started doing this, everybody was really slowing down to watch. They drive by this every day for years and there's nothing there, and then within two weeks there's this whole superstructure skeleton up there, which was the arena, out of nothing. . . . It was in all the newspapers and the TV, because it was kind of amazing." CPBL finished the project two months ahead of schedule, in August 1994. In his remarks at the dedication of the arena, Governor Waihee commented that no state building projects had been completed on time and within budget in his 12 years as

Figure 85. Elevated view of Special Events Arena, University of Hawaii, Manoa, building site shows seating and structural system under construction. (Photo taken 1 September 1993. Reproduced by permission of Ed. Gross–The Image Group © 2013.)

lieutenant governor and governor. "We can take a great deal of pride in what we have accomplished here," he said. Exactly one month later, the women's volleyball team marked "the dawn of a new era" by dispatching San José State University in four sets before a sold-out crowd in the arena's inaugural athletic event: "a special night," remarked athletic director Hugh Yoshida.[48]

For their part, University of California officials decided to procure the Boalt School of Law expansion project through a design-build competition after they failed to attract a bid that fell within their budget. In doing so, they abandoned the approach they had traditionally used on campus projects. The school needed more library and departmental office space. Some 40 percent of the project would involve expanding the 1950s building that lay at the southeast corner of the campus by 50,000 square feet. Overall, the effort would entail a five-level expansion to accommodate basement parking, a library for rare books, two office floors, two elevators, and stair towers, plus the conversion of adjacent Simon Hall, a dormitory constructed in the 1960s, into office space. They adopted a design-build approach to control costs.[49]

University officials put out an RFP for purposes of prequalifying teams. Noted Rik Kunnath, who was executive vice president and Northern California regional manager at the time: "They said this is what we need; this is how much money we have. Whoever solves our problem best is the winner." Unlike the UH Arena competition, price was not included in the selection criteria. "In this process, you knew only through added quality or added value would you get selected," Kunnath said. [50]

CPBL joined The Ratcliff Architects (now simply Ratcliff) of Emeryville and structural engineers KPFF Consultants in responding to the RFP. Since 1989, Ratcliff had been renovating the massive Life Sciences Building on campus—the largest such effort undertaken by the university. The project, which would produce "a totally new facility within the shell of the old," writes architectural historian Woodruff Minor, was nearing completion. Ratcliff's Crodd Chin, who headed the design team on the renovation project and who would reprise the role for the Boalt Hall expansion, believed that CPBL would make a suitable partner because of its reputation for lift-slab concrete construction. The submission of the Ratcliff-Pankow-KPFF team was one of many received by the selection committee. [51]

The committee conducted interviews to narrow the list to six finalists, three of which it subsequently gave one month to submit "full blown proposals complete with renderings and drawings" that addressed its wish list and met its predetermined cost of the project. As Chin explains, he and project sponsor Dick Walterhouse proposed constructing a North Addition behind the library's stacks as the team's solution. In the middle of Boalt Hall lay an 8-story stack whose shelving supported the structural slab. The RFP stipulated that the project could not alter this part of the building. Stair towers included in the addition would permit the removal of the stairs and elevators in the existing structure, which constituted a fire hazard of unprotected steel openings from floor to floor. Their elimination, Chin notes, was the key factor in enabling the team to meet the university's budget, as it would create space for six years of library expansion (fig. 86). The North Addition would provide an additional four years' worth of space, providing the law library with room to increase its holdings for a decade. [52]

As Kunnath noted, "each of the three teams had very innovative solutions, and they were all very different solutions." But the Ratcliff-Pankow-KPFF proposal was the only one that satisfied all of the programmatic requirements within the university's budget. Chin delivered a final presentation to the selection committee, which soon thereafter declared that the team had won the competition. Construction of the $11 million addition, whose "overall look

Figure 86. Library stacks in the expanded Boalt Hall, University of California, Berkeley. Ratcliff, architect; KPFF Consultants, structural engineer. (Photo Malcolm Lubliner. Courtesy of Charles Pankow Builders, Ltd., Pasadena, California.)

. . . complements and is consistent with the existing building's architecture," enthused Kunnath, began in July 1995 and was completed in October 1996 (fig. 87).[53]

Controversy in Public Sector Design-Build Competitions

According to their critics, design-build competitions lacked transparency. Even as their sponsors became enamored of them, competitions generated criticism from the firms that spent countless hours and tens of thousands of dollars to submit proposals, often, as it turned out, with little chance of success. Poorly defined and vaguely worded selection criteria, critics contended, rendered the selection process highly subjective. As Dean Stephan put it, "If you have four bids, it's not too hard to figure out who's got the low number. But if you've got four designs and you have four different prices, you have then the whole concept of most for your money, what's the most pleasing aesthetically, what gives you the most facility, what gives you the best bells and whistles for the money, and that then becomes subjective." And public agencies incurred no responsibility for outcomes. Without well-defined input from the sponsor, one designer put it, "the architect is left whistling in the

Figure 87. Boalt Hall under construction, January 1996. Elevated view, looking southeast. The limited access of the job site necessitated the use of the tower crane. Simon Hall, also renovated as part of the project, sits behind Boalt Hall in the immediate background. Behind Simon Hall is International House, at the corner of Bancroft Way and Piedmont Avenue. (Photo Malcolm Lubliner. Courtesy of Charles Pankow Builders, Ltd., Pasadena, California.)

wind." Legal action provided one avenue of possible redress for disgruntled teams, albeit an expensive one. Yet when public agencies were beginning to embrace design-build competitions, the commercial real estate market was such that designers and builders jumped at the chance to compete for contracts. Therein lay a major weakness in the way that many competitions were configured: too many entrants.[54]

As there was consensus within the design and contractor communities that design-build competitions, however imperfect, were "here to stay," professional groups that were unwilling to wait for the mechanism to mature through trial and error recommended ways of improving it. An ASCE task committee, for instance, called for more consistency among federal agencies in letting design-build contracts. The AIA, AGC, and DBIA all endorsed the two-phase selection process, similar to the one held for the expansion of Boalt Hall. The AIA and AGC jointly issued guidelines for public sector design-build procurement. The two associations called on sponsors to pay stipends, both to attract high-quality participants and to compensate unsuccessful teams in the second phase of competitions. They also asked sponsors to make clear the weights assigned to each item in the RFP's selection criteria, include agency staff and outside advisers on selection juries, and identify all of the members of the selection jury in the RFP. The DBIA endorsed these recommendations and issued its own guides for public sector project procurement: *Design-Build RFQ/RFP Guide for Major Public Sector Projects* and *Design-Build RFQ/RFP Guide for Small-to-Medium Projects*. At the end of the twentieth century, however, there was no consensus among sponsors and participants on the "one best way" to structure design-build competitions.[55]

Still, the public sector's conversion to design-build drove its diffusion across the construction industry. It constituted perhaps the most surprising trend in project procurement and delivery since the establishment of the DBIA, according to Kunnath.[56] As of 2010, design-build accounted for 40 percent of nonresidential construction in America. Public sector projects accounted for much of the increase in market share since 1990. Given Charlie Pankow's exclusive focus on private sector clients, it is likely that his company made its largest contribution to the proliferation of design-build overall through institutional channels like the DBIA.[57]

"A Remarkable Breakthrough": The Precast Hybrid Moment-Resistant Frame

The chapter shifts its attention to the public-private partnership that produced a breakthrough in structural frame technology. It highlights the unusual role of the Pankow firm as a contractor investing in basic research. As was the case with Charlie Pankow's involvement in engineering education and research at Purdue, the company was both self-interested and professionally interested in the research effort. Given its investment relative to its size in terms of employees and revenues, it can be argued that the broader interest in advancing structural performance outweighed any direct return on investment that the company may have hoped to realize, and that the decision to invest in the research was anchored in the Pankow Way.

The built environment represents a substantial investment in labor, materials, and capital. In a seismic event, it can vanish in seconds, as developers, owners, tenants, lenders, and investors learned when a 6.8 magnitude earthquake struck Northridge, California, on 17 January 1994. The temblor damaged some 112,000 structures, with reinforced concrete buildings erected before the state building code was revised in 1981 suffering the worst effects. The California Seismic Safety Commission pinned much of the damage on shoddy design and construction, lax reviews of plans by inspectors, and noncompliance with building codes. Yet the quake severely damaged steel-framed high-rises that met the requirements of the building codes then in effect and were generally considered to be invulnerable to seismic events of this magnitude.[58]

From the early 1970s, the typical tall commercial structure in a seismic zone utilized a so-called steel special-moment-resistant frame, held together by welds and bolted joints. (Moment is a measure of bending, typically calculated by force times distance.) In dissipating the energy of the Northridge earthquake, however, frames sacrificed their integrity as they saved lives. Some 150 high-rise steel buildings suffered catastrophic brittle failure, their frames tearing like paper rather than bending like paper clips. With the average cost of repairing a joint pegged at $10,000, it was estimated that investors took up to a $20 billion hit from the moderate quake. Northridge "gave us a wake-up call," Dean Stephan wrote in its wake. "The harsh reality is that unless changes are made in how we build buildings, these impacts can be expected again and again."[59]

At the time of the Northridge earthquake, the Pankow firm already had expended several years of effort on a new approach to constructing high-rise concrete structures in seismic zones. The company's motivation, explains Stephan, lay in the building code, which restricted the height of the concrete structures that Charlie Pankow liked to build: "I wanted to be able to extend our techniques to taller buildings, because I thought we'd blow steel out of the water." Stephan was also struck by the naïveté of the financial industry: "In the lender's mind, the building—big earthquake—but it's going to be there and it's going to be fine. In the engineer's mind, the building's not going to collapse but it may have zero economic value after the event."[60] As he noted at the time, "The sole criterion used by designers was that there be no loss of life, and that's the way the building codes were written." The Northridge earthquake, then, exposed "an absolutely ridiculous engineering principle at work."[61] With the support of Charlie Pankow, who was always looking for "the new new thing," to borrow Michael Lewis's phrase, in concrete construction, Stephan began to search for "a better way to build" high-rise structures.[62] The result of the company's collaborative efforts was the precast hybrid moment-resistant frame.

The PHMRF uses mild (ductile) steel reinforcing bar and post-tensioned high-strength steel cable to enable precast concrete beams to yield under the lateral loads produced by a seismic event and then return to their original alignment. A column-to-beam connection that utilizes steel reinforcing bar dissipates and absorbs the seismic forces independently of the integrity of its structural members. The post-tensioned strands provide shear and moment resistance (fig. 88). Explained David Seagren, CPBL's director of research and development: "The elasticity of the joint allows it to open and close, much like a spring-loaded door, [to] accommodate the seismic ground motion." Added structural engineer Suzanne Dow Nakaki, who developed guidelines for its use in the design community and helped researchers devise the methodology used to test it: "The [PHMRF] uses traditional materials in a unique way that vastly improves both the reliability and performance of a building." Designed to meet the requirements of the Uniform Building Code for Seismic Zone 4, the technology represented "a remarkable breakthrough," according to University of Washington Professor of Civil Engineering John Stanton, who worked with CPBL to develop the system. Rather than sustain costly repairs or suffer demolition, structures utilizing the PHMRF would require only inspection and minimal or no repairs following an earthquake. Buildings constructed with the PHMRF would save both the structure and its occupants.[63]

The PHMRF was also "as inexpensive and fast to build as any other [commercial] structural system," noted Seagren. Its design was transparent to the architect. It accommodated both poured-in-place and precast construction. Using the PHMRF with precast concrete, however, promised to save as much as ten dollars per square foot in an office building compared to the cost of either steel or cast-in-place structures (including the interest saved on the construction loan). Observed H. S. Lew, structural division chief at the US Department of Commerce's National Institute of Standards and Technology (NIST), "Virtually any type of building [could] take advantage of this connection system."[64]

The PHMRF was the outcome of a cooperative effort of NIST, the University of Washington, CPBL, and Los Angeles–based structural engineers John A. Martin & Associates, which had worked with Pankow teams on several projects, including Chase Plaza, Shoreline Square, and Waikiki Landmark. In 1987 NIST initiated the development of a precast moment-resistant frame with

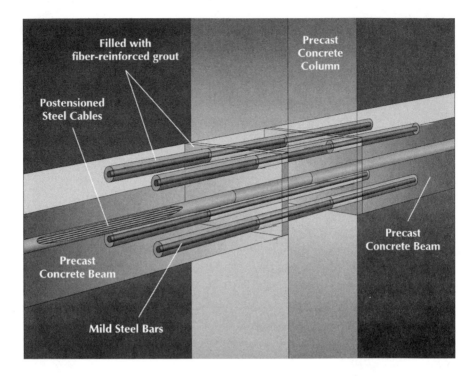

Figure 88. The precast hybrid moment-resistant frame. The precast concrete beams yield under lateral loads and return to their original alignment. The column-to-beam connection dissipates and absorbs seismic forces independently of the integrity of its structural members. The post-tensioned steel cables provide shear and moment resistance. (Courtesy of Charles Pankow Builders, Ltd., Pasadena, California.)

tests of one-third scale models of column-to-beam connections under cyclic inelastic loads at its Gaithersburg, Maryland, laboratory. The experiments examined precast, prestressed joints fabricated from bonded prestressing steel, alone and in combination with reinforcing steel bar. In 1990 Stephan learned about the research at a workshop sponsored by the National Science Foundation under the joint US-Japan Seismic Structural Systems Program (PRESSS). As he puts it, "Immediately the light went off." Soon thereafter, Stephan traveled to Japan with University of Washington Civil Engineering Professor Neil M. Hawkins as part of a PRESSS team. Hawkins, who had recently completed a term as chair of the department and was now serving as Associate Dean of Research, was recognized as an expert in reinforced and prestressed concrete structures subjected to static and dynamic loads. He was a two-time recipient of the ACI's Structural Research Award and the ASCE's State-of-the-Art, Structural Research, and Lin awards. During the trip, Stephan discussed his ideas for a precast hybrid moment-resistant frame with Hawkins, who agreed that they made sense structurally and advised Stephan on how CPBL might proceed. With Hawkins's encouragement, Stephan and Seagren met with University of Washington civil engineering researchers in San Diego. Together they generated a proposal for a hybrid framing system—one that incorporated steel into the design—and agreed to collaborate in its development. In 1991 Stephan proposed to NIST that it fund the development of the PHMRF for commercial application. NIST joined the consortium, agreeing to match resources in-kind to test the concept as an extension of its initial program.[65]

The consortium was not yet in place, however. Owing to CPBL's lack of experience, both in developing design procedures and researching mechanical properties, Stephan brought the ACI's Concrete Research and Education Foundation (ConREF) into the consortium, and assigned its Concrete Research Council with the responsibility of technically reviewing the program. Irvine-based structural engineers Robert E. Englekirk and Suzanne Dow Nakaki were retained to develop design guidelines. Nakaki had worked with NIST since 1987 and with CPBL on several projects, including Catalina Landing, Shoreline Square, and Tyler Mall.[66]

With the consortium in place, each member performed its assigned role. To carry out its work, CPBL created the position of director of research and development to coordinate the firm's relationships in the consortium and assigned Seagren, a Purdue engineering graduate, to the job. Seagren also headed efforts to develop office floor plans to assist in the development of a prototype for NIST to use in one-third-scale tests. CPBL also conducted economic feasibility studies and investigated the possibility of constructing the PHMRF at a precast plant. For their part, John A. Martin & Associates calculated maximum design level moment and shear forces and specified frame and column component sizes, while both NIST and University of Washington researchers developed column-to-beam connection concepts, detailed the reinforcement of columns and beams in the frame, and carried out performance assessments.[67]

NIST subjected the PHMRF to a first round of testing during 1992 and 1993. At the same time, NIST tested cast-in-place concrete frames built to meet the requirements of contemporary building codes. As a result of these one-third-scale model tests, the consortium refined and simplified the PHMRF.

During 1993 and 1994, NIST conducted a second round of tests. Stanton and his colleagues reported that "the hybrid system performed very well." In the fall of 1994, Dean Stephan, in his capacity as president of both CPBL and the ACI, announced that the PHMRF might be introduced early in 1995. He reported that NIST's tests showed that the PHMRF was "100 percent more effective" than existing steel and concrete systems. In the context of the Northridge earthquake, he explained: "We have focused our efforts on developing a system which will not only meet current building codes, it will far exceed the tougher standards we expect to be introduced in the months ahead."[68]

Readying a new technology for nonproprietary commercial use meant little, however, unless building officials, engineers, and insurance underwriters accepted it. The Uniform Building Code did not approve the use of precast, prestressed concrete building systems in seismic zones. As Stanton noted, the "closest possibility [was] a reinforced concrete special-moment-resistant frame, but the hybrid system violate[d] several requirements for that category." The PHMRF also lay outside the ACI's building code requirements for reinforced concrete (ACI 318). Getting the new technology written into the codes would take time. As Suzanne Dow Nakaki explains, the culture of groups such as the ACI and the ICBO (International Conference of Building Officials) puts a premium on consensus: "What that means is everyone has to agree 100 percent, not 90 percent, not 99 percent, but 100 percent." Dean Stephan noted that elitism could trump pragmatism in these professional deliberations. He also appreciated that "reluctance and fear of admitting we did something wrong" in the wake of the Northridge earthquake might have an impact on industry acceptance of the PHMRF. As Ron Hamburger, then senior vice president at EQE International, a San Francisco–based structural engineering firm, noted, "Engineers rely heavily on proven success of similar structures for past earthquakes." A similar culture apparently prevailed within the insurance community. Nakaki recalls the response of one group of underwriters to Stephan's PHMRF presentation: "They looked at us blankly, and they said, 'But what box do we check?' They didn't know how to categorize it on their forms, and they said, 'Whatever you do, don't check that box called "Other," because that sends up red flags and blinking lights and the whole thing.'"[69]

Surmounting these cultural barriers would not guarantee that owners and other contractors would use it. To illustrate the initial lack of interest that owners showed in the PHMRF, Nakaki recalled talking to the owner of a project in San Diego about changing the frame of his building from a cast-in-place to the hybrid system: "I'll never forget. . . . He looked at me, he said, 'Well, that's all fine, but I don't really care about better performance. That's why I have insurance.'" In 1996 the ICBO would approve the PHMRF, considering it to be equal to a special-moment-resistant frame, as the Uniform Building Code defined it. Three years later, CPBL's marketing vice president, Todd Whitlock, would lament the "tremendous amount of resistance in our industry to new technology and technology transfer."[70]

To promote the acceptance of the PHMRF, Stephan and others within the firm acted on several fronts. They persuaded the ACI to establish a task group to evaluate the PHMRF and draft the appropriate revision to ACI 318. They also submitted the materials generated by the research program to the ICBO's Evaluation Service. Meanwhile, the company deployed the technology commercially for the first time.[71]

CPBL demonstrated that the PHMRF could work in a non-seismic zone when Thornton-Tomasetti Engineers (now Thornton Tomasetti) specified its use in the design of three parking structures included in the renovation and expansion of Roosevelt Field Mall (see appendix A). Charles H. Thornton, a recognized expert in structural analysis, who held a doctorate in structural and engineering mechanics from New York University and who, in 1977, co-founded the structural engineering group with Richard Tomasetti, was keen to test the new system. His firm chose to utilize the system primarily because it met the desire of the developer "for an open plan, free of shear walls." Such a plan "significantly enhance[d] the security of the parking garage," said Thornton. The hybrid framing system also offered "lateral load resistance under wind or seismic loads." Construction of the first parking garage began early in 1995. Pankow completed construction of all three garages by November 1996, by which time the ICBO had issued its report approving the use of the PHMRF in seismic zones.[72]

CPBL now applied the hybrid framing system to the construction of parking structures in seismic zones. In December 1996, it completed construction of the Pearl Street garage in Eugene, Oregon, for the city's Urban Renewal Agency. The public design-build project marked the first use of the PHMRF in Seismic Zone 3. Three years later, Pankow finished a parking structure at Stanford Shopping Center—the first use of the PHMRF in Seismic Zone 4. As was the case with the parking garages at Roosevelt Field Mall, the owners of both of these projects valued the open plan enabled by the hybrid frame. In August 1999, while crews worked at the Stanford site, Pankow began construction of the most complicated structure to utilize the PHMRF to date: a 39-story high-rise building at the corner of Third and Mission Streets in San Francisco.[73]

Built to Last: The Paramount in San Francisco's SOMA District

As a residential project, the Paramount was not the high-rise office structure that Dean Stephan envisioned building when he embarked on his investigation of the PHMRF. The San Francisco Redevelopment Agency (SFRA) originally conceived the project in early 1990s as a 28-story steel-framed office tower, but recession ended office construction in San Francisco until the end of the century. The Related Companies, a New York developer, had originally responded to the agency's RFP for an office building. Changes in market conditions associated with the dot-com boom prompted the SFRA, with the backing of the mayor's office, to rezone the site to residential use. The owner approached CPBL to study the site's potential as a housing project. The contractor worked with design architect Elkus Manfredi Associates and Robert Englekirk Consulting Structural Engineers to design a project that met Related's budget. Plans called for 486 apartments, two levels of retail space, one floor of offices, and parking for 350 cars. At 420 feet, the Paramount would surpass the 32-story 3900 West Alameda Tower in Burbank, a cast-in-place structure completed in 1989, as the tallest concrete building in Seismic Zone 4. It would stand as the tallest residential building in the city until 2008. The construction contract called for project delivery in 25 months.[74]

The dot-com-ignited market revival at the end of the century featured a shift in construction from the financial district north of Market Street to the South of Market (SOMA) district. The geography of the boom reflected land use policies set in motion by the Downtown Plan, which the city council passed in October 1985 in the wake of a decade of frenzied high-rise construction that had fueled protests against the "Manhattanization" of the central business district. The Plan restricted future growth and set standards for new projects. It lowered the height limit on buildings, from 700 to 550 feet; reduced floor area ratios (the total amount of floor space in a project relative to the size of the site); and allowed a maximum of 950,000 square feet of office space in buildings of 50,000 square feet or more in size. To preserve the look and feel of the area at street level, the Plan designated 250 structures as historically or architecturally significant and established six conservation districts, each one with its own design criteria aimed at retaining the character of the built environment. The Plan also required developers to contribute to open space and public art, integrate their projects into transportation management plans, and minimize the wind currents that their projects generated at ground level. Marathon Plaza, the 900,000-square-foot complex that CPBL completed early in 1988 for Vancouver-based Marathon U.S. Realties, was the first major expression of the intentions embedded in the Plan. But then the market collapsed.[75]

The stock market–fueled boom of 1999–2001 coincided with Mayor Willie Brown's administration (1996–2004). Changes in the city charter invested greater power in the mayor's office, and Brown took advantage of them to promote large-scale development of SOMA. Brown, the longtime speaker of the California State Assembly who was forced to change careers by term limits, was well connected with developers and the city's powerbrokers. He vigorously backed mass transit, mixed-use development, and civic projects to establish SOMA as the city's cultural center and an extension of the financial services, hotel, and shopping clusters north of Market Street. Of course, the transformation of SOMA was now decades in the making (see chapter 2). But Brown leveraged his connections to push through high-rise projects that ignored the guidance provided by city planners in the 1972 Urban Design Plan and constituted an "end run" around the Downtown Plan. The construction of the Paramount was just one indication that private investors and redevelopment officials were realizing their three-decade-old dreams. The immediate area around the site, which included an expanded Moscone Center and Yerba Buena Gardens, the Contemporary Jewish Museum, the San Francisco Museum of Modern Art, the Museum of the African Diaspora, and the Yerba Buena Center for the Arts, was a becoming "a 24-hour neighborhood," enthused William Witte, a principal partner of the Related Companies.[76]

The Paramount was also "a noteworthy project" for its mixed-income housing aspect, according to Elizabeth Seifel: 20 percent of its units were affordable to people earning less than 50 percent of the area's median income. These units rented initially for as little as $500 per month. At the same time, the Related Companies offered one- and two-bedroom flats at market rates of $1,995 and $3,195, respectively, per month. The incorporation of the below-market units reflected the intention on the part of planners to promote inclusionary housing policies.[77]

CPBL project sponsor Joseph Sanders described the structural solution for the multi-use structure as a tube system, with a frame that had no beginning or end. To execute it, the design incorporated both the PHMRF and the other precast moment-resistant frame system then approved for use, a proprietary system produced by Dywidag Systems International. One side of the Paramount's façade was slightly curved and the typical floor plan featured several "cutout" corners not found in rectangular buildings (fig. 89). Since the post-tensioned strands associated with the PHMRF were more effective and cost efficient over longer distances, CPBL used them to tie together multiple bays. It deployed the Dywidag system for the single bays associated with the reentrant corners. Executing the full-frame perimeter system was facilitated by the inaugural application of an elbow-sleeve system that allowed post-tensioned strands "to turn a corner and continue along the adjoining side." This technology, which the University of Washington's structures laboratory tested during 1999, made possible the implementation of the PHMRF in a high-rise context.[78]

The Dywidag system used precast elements, too, but bolted together the column and beam connections rather than bound them with post-tensioned steel strands. Its developers, which received National Science Foundation support, included Robert Englekirk, who held the patent on its connectors. Englekirk liked both systems, but noted that they differed in their labor and materials costs. The PHMRF was more labor intensive, owing to the effort required to apply tension to its steel strands. Dywidag's ductile connectors required more steel. Workers could implement the Dywidag system more easily, but they had less room for error in lining up the connectors. The PHMRF did not have issues related to exposed hardware, as did the Dywidag frame, but workers confronted the problem of grouting the column-to-beam interfaces without obstructing the blockouts created for the post-tensioned cables. Overall, CPBL's Whitlock calculated, the PHMRF cost less to build than Dywidag's frame and, in the case of an earthquake, would cost less to repair. In any case, the building team estimated that using both systems in tandem saved $4 million and as much as four months in time over a steel-framed system.[79]

The City of San Francisco required a special peer review of the plans for the Paramount before allowing the project to proceed. Said Tom Hanson, manager of the division responsible for reviewing building plans, "You cannot find it in the building code at this time. [The project is] a one of a kind for a structure of this height." The review panel, chaired by Jack P. Moehle, professor of civil engineering at the University of California, Berkeley, ren-

Figure 89. The Paramount under construction at Third and Mission streets, San Francisco, 3 August 2001. The curvilinear façade and "cutout" corners prompted the building team to incorporate both the PHMRF and Dywidag Systems International's precast moment-resistant frame into the structural system. Construction began in August 1999. The building topped out on 5 June 2001. Kwan Henmi Architecture/Planning, executive architect; Elkus/Manfredi Architects, design architect; Robert Englekirk Consulting Structural Engineers, structural engineer. (Courtesy of Charles Pankow Builders, Ltd., Pasadena, California.)

dered a favorable verdict, citing the PHMRF's "economics, seismic response, and the tendency of its post-tensioning cables to keep the structure plumb." The advocacy of structural engineer Englekirk was crucial to securing the panel's approval. Officials, who agreed beforehand to defer to the findings of the panel, issued a foundation permit in December 1999.[80]

Using the PHMRF saved the developer money primarily because its precast components doubled as the building's exterior skin. In fact, it was the capacity to deploy the PHMRF's components in this dual role that convinced the Related Companies to use the frame. Explained Jim Fong, an architect with Kwan Henmi Architecture/Planning, the executive architect for the project: "The precast component [on the perimeter] is the finish component. It doesn't require the separate finish that is applied over steel." Producing the precast elements for the Paramount, however, presented Pankow with a formidable challenge.[81]

Quite simply, the company could not identify a precast manufacturer capable of satisfying both the exacting structural specifications of the PHMRF frame and the high quality architectural finish demanded by the developer and the designer. Fabricators of structural beams and columns did not have quality controls in place to enable them to meet the finishing standards required for the exterior of a signature downtown tower such as the Paramount. And the precasters who manufactured architectural panels lacked the experience and expertise in post-tensioning needed to fabricate the PHMRF.[82]

Out of necessity, the Pankow firm manufactured the precast elements for the project—more than 2,200 pieces in all. Of course, the company did not lack for experience with precast operations, but producing the elements for the PHMRF posed an unprecedented challenge, not least because 22 of the 29 beams associated with the typical floor plan had different dimensions. With no room to set up casting operations on or near the building site, CPBL's executives scrambled to find a suitable location. After evaluating their options, and with an eye to the future, they acquired some two dozen acres of land in Corcoran, a dusty Central Valley town that lay halfway between Los Angeles and San Francisco, and erected a casting yard.[83]

Owing to the mixed uses associated with the structure, not all of the Paramount's structure was built using the new framing systems. The residential tower sat atop eight floors that included an underground parking level, a basement, and six floors of retail, office, and parking space (fig. 90). To comply with the building code's prescriptions for occupancy separations and to ac-

Figure 90. Varying slab elevations and six floors of retail, office, and parking space complicated the construction of The Paramount. These floors were cast in place with concrete moment frames and shear walls. (Photo taken on 3 October 2000. Courtesy of Charles Pankow Builders, Ltd., Pasadena, California.)

commodate varying slab elevations, structural engineers prescribed casting in place these floors with concrete moment frames and shear walls, which complicated construction.[84]

Primarily because of labor shortages associated with the dot-com-fueled construction boom and the failure to plan for them, CPBL had to take extraordinary measures to complete the Paramount on schedule. Implementing the PHMRF on a curvilinear high-rise in Seismic Zone 4 would have been difficult enough under normal circumstances. Doing so with inexperienced crews working multiple shifts helped to ensure that the cost of constructing the Paramount exceeded the price that CPBL negotiated with the owner. Nevertheless, the project, completed in October 2001, demonstrated to Charlie Pankow that the PHMRF could fulfill its promise as "a revolutionized method [of building] where the ability to withstand earthquake forces is substantially increased," as he wrote with pride in the company newsletter.[85]

The development of the PHMRF illustrated Charlie Pankow's passion for improving building technology and the leadership role that he continued to play in the firm. While Dean Stephan may have headed the company's participation in the collaborative research effort, it was Pankow's decision to invest more than $1 million in the development of the hybrid frame, not including the salaries of the individuals who devoted long hours to the project.[86]

Construction of the Paramount occurred during a technology-fueled economic boom whose epicenter was located in Northern California. The dot-com collapse left both the commercial construction industry and Charlie Pankow's business model in tatters, as half a dozen or more planned major office and residential projects evaporated into thin air. The company would adapt to the abrupt downturn by bringing to fruition the diversification strategy that Rik Kunnath and others had been pursuing surreptitiously and in ad hoc fashion since the early 1990s, when market changes produced by the deep recession indicated that CPBL would have difficulty surviving on a strict diet of so-called shell-and-core concrete construction negotiated in lump-sum contracts with private sector clients.

CHAPTER 7

Adapting to Market Change, 1991–2004

W ell before the construction industry began to recover from the deep recession of the early 1990s, it was unclear whether the business model that had sustained Charlie Pankow's company for almost three decades would enable it to thrive when the economy rebounded, much less survive the next downturn. The founder had no doubt that it could. Recall that, when a reporter asked him, in May 1988, to predict what work his firm would be doing in the future, Pankow replied: "Design-build in heavy commercial buildings for the private sector. That's it."[1] Convinced that a recession "had no bearing whatsoever on our business," Pankow explained the decline in contract volumes during recessions in terms of "not selling our approach adequately." Never would he acknowledge that his business model might have lost its relevance.[2]

The success of George Hutton and Russ Osterman as developers had insulated Pankow's business model since the mid-1970s. As Rik Kunnath explains, "They sold a lot of work [to] a very small number of highly prolific clients. . . . That work produced revenue and income at a level that, for the relatively small number of owners in the company, was more than adequate for the owners and more than adequate to sustain the organization." Pankow had no reason to think that he might have to change his business to sustain its profitability. Kunnath elaborates: "Over time there was so much success that he [became] more and more convinced of the wisdom of [his] approach. And it was working, it had always worked, it was continuing to work, and so why tinker with something that is obviously successful?" As the primary beneficiary of the financial and professional rewards of an approach that he had promoted for decades, Charlie Pankow was the last person who would have seen the need for change.[3]

Charlie Pankow's way of doing business benefited commercial real estate developers, who placed a premium on realizing their projects as soon as possible. The Pankow firm accepted execution risk and handled project administration. Charlie Pankow's design-build variant offered certainty of delivery, cost, schedule, and quality. The business model had remained constant, even as the market for commercial real estate changed. Its reliance on repeat business from a small number of customers made it even more susceptible to changes in market conditions. For, as Kunnath observes, "Clients retire and change [contractors] for [many] other reasons."[4]

The commercial real estate market was becoming service oriented, requiring contractors to adapt to the needs of owners. For instance, hospitals typically employed professional staff to administer their capital building projects, so their executives were not interested in ceding project administration to the contractor. Speed to market was less crucial for hospital managers than for owners who needed to let office space or sell condominiums to repay their construction loans, since hospitals generally held ample cash reserves. Ironically, as design-build became more popular as a project delivery methodology, the company needed new ways of differentiating itself among contractors, too, but Charlie Pankow had no interest in modifying his "one-size-fits-all" solution. Kunnath and other second generation managers in the firm recognized what Charlie Pankow, as a first generation founder of the company, did not, namely, that the firm needed to adapt to the market if it were to survive, much less thrive.[5]

It was more important for Charlie Pankow to hold onto the philosophy and beliefs on which he had founded his company than contemplate the need for change, however. In fact, suggests Kunnath, he might have been willing to invest his capital in the company to keep it viable, were the business to falter, much like one might underwrite a hobby. Since Pankow exercised "unilateral, complete, iron-clad control" of the firm, those who were interested in change had to fly cautiously under his radar, so to speak, to avoid retribution.[6]

Those who advocated new directions in the company's business positioned them as experiments, temporary ways of retaining people during recession, or adjuncts to the large building projects that historically filled the firm's order book. Charlie Pankow might tolerate the new initiatives or look the other way, but he never would acknowledge their value, even when they became profitable. And while he might come to respect the success of these ventures, he would take no personal pride or interest in anything other than so-called shell-and-core construction for private developers. Accordingly, he would not recognize the managers and engineers who engaged in what would become known within the firm as "special projects" in the same way that he honored those who built large, concrete commercial structures. Nevertheless, the new initiatives pursued by Kunnath and other executives would make a significant contribution to the financial success of the company. In little more than a decade of diversification, Charles Pankow Builders, Ltd. (CPBL), would reverse course and become a service-oriented contractor, leaving it better prepared to weather future collapses of the commercial building sector.[7]

Charlie Pankow's health was a significant factor in explaining how those who sought to adapt the company's business to changing market conditions were able to implement their initiatives. The aforementioned teenage bout with scarlet fever eventually led to renal failure. Pankow received a kidney transplant in January 1994. Son Steve was the donor. Almost a decade later, his heart condition deteriorated with atherosclerosis, and he was very ill in the last year of his life. Even so, whenever he felt better, as he did in the wake of his transplant, Pankow asserted his executive veto power. Had Charlie Pankow remained in good health throughout the 1990s, the initiatives discussed below would not have seen the light of day, and "there wouldn't be a company," Kunnath believes. "As it turned out [they] frankly saved the company." Given the collapse in construction associated with the bursting of the dot-com bubble, the company may well have died with its founder, who suffered heart failure on 12 January 2004 at his Altadena home, several weeks after undergoing surgery at the University of California, San Francisco.[8]

Product Diversification by Necessity: Pankow Special Projects

Given the absence of tower cranes in downtown San Francisco and Oakland in 1991, it should come as no surprise that the initial product diversification initiative, which focused on tenant improvement (TI) projects, was launched from Northern California. At the same time, the replacement of Brad Inman with Rik Kunnath as regional manager was no less important to the development of this initiative, which needed a champion within the organization. For Charlie Pankow was not alone in his skepticism of TI and related work, as one might expect from engineers steeped in a culture forged in the construction of concrete high-rise buildings. Says Dick Walterhouse, whom Dean Stephan would put in charge of a newly formed Special Projects division within CPBL: "There [was] a general attitude across the company that we were wasting our time doing small project work, that [it] wasn't what we were all about."[9]

There were few prospects for work for CPBL in the San Francisco Bay Area after the completion, in 1988, of Marathon Plaza, as developers would not think again about adding a commercial structure to the skylines of either San Francisco or Oakland for another decade. Meanwhile, owner Marathon U.S. Realties was looking for a contractor to build out the tenant spaces in both Marathon Plaza and 595 Market, a hexagonal office tower designed by Skidmore, Owings & Merrill. Marathon had tried another contractor, but had become disenchanted with its performance. It then approached CPBL. Kunnath jumped at the chance.[10]

In the depressed market, major local contractors, such as Dinwiddie and Swinerton & Walberg, also moved into this segment. In doing so, they acted defensively. At the very least, TI work helped to cover overheads. Yet it made sense strategically, too. For TI work was typically a countercyclical business. Owners competed more vigorously for tenants during periods of recession. For instance, they absorbed the cost of elaborate and expensive interior work to sell vacant space. In good times, space was at a premium. Owners were more inclined to shift the cost of improvements to tenants, who typically devoted relatively less capital to the design and construction of their work spaces. This aspect of the business, however, was not top of Kunnath's mind when he approached Charlie Pankow about adding TI work to the company's product portfolio.[11]

Kunnath did not sell his boss on the value of the work itself. In fact, he conceded that TI projects did not fit with the company's culture. Rather, he argued that doing such work would maintain client relationships, generate revenue, and give engineers experience that would prepare them for the responsibilities involved in supervising large building projects. Kunnath recalled telling Pankow: "If we let another contractor in there to do this work [it's] going to be years before we get another opportunity, and by the time that occurs, the relationships won't be [there]. I can't maintain that relationship [with Marathon] as intensely as anybody working with these guys directly day after day." He added, "Do you realize that the tenant work in the building is going to probably double in value the cost of our construction?"[12]

Charlie Pankow gave his conditional assent. As long as the venture was profitable, he would give Special Projects, as the business would become known, latitude to operate. But Pankow would never develop any interest in it and so for all intents and purposes would ignore it, no matter how beneficial to his company it might become. As Kunnath explains, "[He] got no personal satisfaction from it at all [and so] really didn't even want to know what we were doing." Charlie Pankow was always keen to visit the site of a large concrete building project. Yet he never would ask to see any of the hundreds of jobs that the Special Projects group would eventually complete. "It wasn't his world," explains Wally Naylor, whom Kunnath hired soon after securing Pankow's permission to launch the initiative.[13]

Since the company did not have anyone who specialized in TI projects, Kunnath looked outside the company for someone to manage the work. Kunnath recruited Naylor, who had developed his expertise over nine years with Swinerton & Walberg. Naylor was preparing to leave Swinerton, having failed in his bid to become the head of the company's TI group, when he learned of Kunnath's interest in launching a TI group from the company's marketing director in Northern California, who had been a superintendent at Swinerton.[14]

While he had awaited word on who would run Swinerton's TI group, Naylor had prepared a business plan in anticipation of starting his own firm. In meeting with Naylor, Kunnath vetted its components. Once he became satisfied that it was viable, Kunnath used the plan as a blueprint. From Naylor's perspective, he was being given the opportunity to do what he was planning to do anyway, but with the backing of a firm that enjoyed a first-class reputation among local owners: "I was excited about not having to worry about cash flow running my own company." Ultimately, Naylor joined CPBL because he felt that there was more opportunity to grow the business within the organization than on his own. For one, he was impressed with the quality of the people: "Everybody I ran into was really kind of higher caliber intellectually than I'd seen in the business. And I thought that at the time when I left Swinerton that, by and large, there were some very good people [there], but I thought that Pankow was definitely a step up." From Kunnath's perspective, Naylor was the right person to teach the company how the TI market functioned and what clients expected: "It was like going to school in a fairly low-risk way about [how] the rest of the [commercial] world that we never were really exposed to [worked]." Naylor joined Pankow in November 1991.[15]

As president, Dean Stephan issued Naylor two directives that were consistent with the Pankow Way of recruiting capable professionals and letting them manage their work as they saw fit. Acknowledging that the company had no TI expertise, Stephan deferred decision making to Naylor. The second directive established the financial goal: "We're going to give you a pot of money in the beginning of the year, and at the end of the year come back with a bigger pot of money." Stephan was as skeptical as Charlie Pankow about this new direction, but saw little risk in dabbling in it as long as it did not divert resources away from the company's core business. For the moment, that concern was not an issue for the San Francisco office.[16]

Special Projects would eventually expand to include health care facility upgrades, seismic renovations, adaptive reuse of historically and architecturally significant structures, and other specialized products, but TI work accounted for 80 percent or more of the business in its early years. (Thus "special projects" became synonymous with "tenant improvements" among employees outside the group, even as it expanded into other areas.) Within six months of launching the business with the Marathon U.S. Realties account, Naylor brought aboard a number of his previous clients. In short order, he signed master agreements to perform all of the TI work that these owners needed, covering seven buildings in downtown San Francisco and Oakland. In the first year, Naylor completed about 100 special projects, which brought in some $6 million in revenue—enough to suggest the wisdom of Kunnath's decision to initiate the venture, but not enough to raise eyebrows elsewhere in the firm.[17]

Kunnath soon expanded Special Projects into the health care sector. Again he looked outside the firm for a senior manager and found Larry Malone. Like Naylor, Malone enjoyed an outstanding reputation: Whereas Naylor knew all of the building owners in town, Malone seemed to know every doctor in the Bay Area. And he was looking to make a career move. Once in place, he "opened up the doors for us in a sector that we just were not in at all," Naylor explains. Working from his office in Oakland, Malone proved to be as successful as Naylor in launching a new line of business within the Pankow organization.[18]

Through the efforts of Naylor and Malone, Special Projects grew at an annual rate of more than 30 percent. The success of the group led Kunnath to argue that the Honolulu and Altadena offices should market special projects work as well. The Hawaii office, which had yet to recover from the departure of George Hutton, was receptive. After landing the office's first TI contract—a Warner Brothers Studio store in the Ala Moana Shopping Center—Naylor left in place Arne LePrade, who had spent much of his career at Hawaiian Dredging, to head the group. In his capacities as both president of the company and as regional manager for Southern California, however, Dean Stephan resisted the idea of incorporating special projects into the business model, not least because he could spare no personnel. "We were stretched," he explains. "Everybody was just busting their fanny."[19]

Stephan's concerns were threefold. First, Special Projects was distracting Kunnath from his primary responsibility, namely identifying and developing opportunities to do the type of projects that Pankow had been doing since 1963. From Stephan's perspective, time devoted to special projects work was money lost, given the size of so-called base building projects. Second, Stephan felt that special projects work would corrode company's culture and reputation as it grew, since all of the people engaged in the work had been recruited from other companies and had not been inculcated in the Pankow Way. Third, and most importantly, at least from a short-term perspective, Stephan had no time or inclination to manage an incipient Special Projects group in the Altadena office. "I fought it because I did not want to have to manage a tenant group when I had all this other stuff going on, and we [the Southern California region] were the golden goose of the company," Stephan explains. "We were the only people making money, and so I fought it very hard." Stephan approached Charlie Pankow, arguing that the founder had to either create a separate Special Projects entity or shut it down.[20]

From Kunnath's perspective, Stephan's point of view made sense only if it were impossible to expand Charlie Pankow's business model without fatally compromising it. But, he argues, it was a false choice, especially for Northern California, where Naylor and Malone were making money and no developer was planning a large commercial project. Yet Stephan's views reflected the general belief within the company that the kind of work that Naylor and Malone were doing was residual. "There was never any recognition that it's a strategy in itself," Kunnath notes. "Nobody saw it for anything other than, 'Oh, God, if I'm running out of work, maybe I'll be able to do some of this piddly stuff, but I don't want to.'" Wholly integrating special projects work into the business model would require another deep recession that would drive home to all employees the financial value of the work and the necessity of diversifying the business, if the company hoped to survive.[21]

Stephan and Kunnath compromised. Special Projects became a separate division within CPBL. The move, implemented in August 1995, freed Kunnath to focus his attention on developing leads on the kind of projects that the company had always built and provided structure that the group needed to succeed. Kunnath thought, "We don't need marketing. We don't need anybody to expand the business. We need somebody to basically put a circle around our operation methodologies and help these guys [Naylor and Malone] with some risk management and some other things." Still, Stephan felt that he had lost the argument. "Fortunately," he reflects. "I'm glad Rik had the foresight to get us into that area of business, because without it, I don't know if the company would have made it through the [late] nineties, because there was a long period of time where we just didn't have any work."[22]

Kunnath and Stephan tapped Dick Walterhouse to run the division (with Charlie Pankow's approval, of course). Walterhouse was the "perfect person" to do the job, according to Kunnath, given his ability to organize work and adhere to processes and procedures. But Walterhouse was stunned. At the time, he was a project sponsor without a project, spending his time trying to sell work in a down market. Even so, he felt as if he were being demoted, since he considered himself to be one of the firm's top sponsors and had no experience with either TI work or upgrades of medical facilities. When Stephan approached him about the position, he wondered: "Why do you want to take me away from doing big project work and put me in charge of doing this small stuff?" When Stephan explained to him what the company expected him do accomplish, however, Walterhouse understood why he was the perfect person for the job and "jumped at the chance" to run, in essence, his own construction company.[23]

Confident of Walterhouse's abilities, Stephan gave him carte blanche to run the division. "It's your group," Stephan told him. "You can close offices, you can open offices. You can go to North Dakota. You can do whatever you want. If you make it a success, it's your company, your baby. If it's not a success, we're going to try to sell it." Stephan explained that the "tenant group," as he called it, was burdened by accounting and other internal systems that were designed for large projects. It was also saddled with subcontracting procedures and bonding requirements that were inappropriate for projects that cost tens of thousands, rather than millions, of dollars and took weeks, instead of years, to complete. In sum, Stephan charged Walterhouse with establishing a separate business within the company with rationalized systems that would allow it to operate out of the existing offices and any others that Walterhouse decided to open. Stephan insisted only that Walterhouse proceed in a manner that was consistent with the Pankow Way.[24]

Inculcating the Pankow Way in individuals from disparate backgrounds posed the biggest challenge for Walterhouse. He soon learned, however, that not every aspect of the company's culture was applicable to the work that Naylor and Malone were doing. Ultimately, Walterhouse ensured that Special Projects operated in a manner consistent with Charlie Pankow's business philosophy, even as it charted a new direction in strategy and developed proprietary job accounting, cost estimation, and scheduling systems.[25]

Many people within the company continued to see the move into special projects as temporary—something that would pay the bills until economic recovery revived the market for type of work that had sustained the company for three decades. The creation of Pankow Special Projects, Ltd. (PSPL), as a subsidiary limited partnership, in 1996, however, established an organizational home for work that combined "small-project efficiency with traditional Pankow technology and innovative problem solving," as the company newsletter explained, and facilitated its diffusion throughout the firm.[26]

Under Walterhouse's direction, PSPL became a $100 million operation in six years. Like the original CPI, the business grew largely on the basis of repeat clients, which comprised some 80 percent of its order book. Walterhouse did not open an office in North Dakota, but established branches in Newport Beach, California, and Seattle. He added staff. In time, he began to recruit university graduates into the business, rather than hire people from outside the firm or transfer them from CPBL. The success of PSPL was rooted in the professionalism embedded in the Pankow Way. It built its reputation in the market by completing work on time and adhering to detail. Owners accustomed to dealing with "pickup truck guys," as Walterhouse puts it, also appreciated the fact that the larger Pankow organization stood behind PSPL.[27]

The definition of a "special project" evolved. When Stephan installed Walterhouse as general manager, a special project typically involved renovating an occupied building. The project generally cost $5 million or less. Over time, projects grew in size, to $15 million or more, but then fell back to $10 million. Projects still involved existing buildings, but often they were unoccupied. At the same time, the operating model remained constant. A special project was a two-person operation, with a manager in the office and a superintendent in the field. This contrasted with the configuration of project management on projects built by CPBL, which remained essentially unchanged from the day Charlie Pankow that started the company (even if the titles had changed): a project sponsor in the office and five or six people in the field, including a project superintendent, a field superintendent, a project engineer, and two or three field engineers.[28]

Under this division of labor, PSPL and CPBL might perform similar work. In the adaptive reuse of two historic buildings, for instance, PSPL was responsible for restoring the Charles Dickey–designed Rotunda Building in downtown Oakland, while CPBL converted the Eastern Columbia Building in downtown Los Angeles into luxury condominium lofts. Both buildings had opened as department stores, in 1914 and 1930, respectively. The latter project, a stunning example of zigzag moderne architecture, was one of dozens of examples of the adaptive reuse of commercial and financial space in downtown Los Angeles left vacant by the recession of the early 1990s. Organizationally, it fell to CPBL because the transformation of the Claud Beelman–designed landmark structure required as many as 120 workers and a full complement of Pankow field staff. The project involved gutting the building, except for the passenger elevators and terrazzo main lobby floor; seismically retrofitting the foundations; widening and adding footing; adding a concrete shear wall that extended from the basement to the roof; and installing a swimming pool, exercise room, and sun deck on the roof. The size and complexity of projects determined which Pankow entity would be responsible for the work.[29]

The Summer 2001 issue of the company newsletter celebrated the decade-long growth of Special Projects. The main article noted that PSPL expected to complete 80 projects during the year and highlighted a number of projects then underway, including renovations and seismic upgrades of 180 Sutter Street, a historic office building constructed in 1908 in San Francisco's financial district, and Barker Hall on the University of California, Berkeley, campus; upgrades at all three California Pacific Medical Center campuses; and construction of an auto dealership in Honolulu and an Old Navy store in Santa Monica. The newsletter also featured stories on high-rise residential projects in progress, including the Paramount, Yerba Buena Lofts, Hawthorn Place at 77 Dow, and the Aurora in San Francisco. But never before had it lavished such attention on Special Projects.[30]

On the face of it, the company's order book appeared to be bulging with both special and traditional projects. In short order, however, the market for the latter collapsed, particularly in the Bay Area, where dot-com-driven commercial real estate boom was concentrated. Dot-com stocks had peaked in March 2000 and were plunging in the summer of 2001. In the wake of the terrorist attacks of 9/11, the commercial property market seized up. For CPBL, projects under construction were completed. But those in the planning stages evaporated. In Northern California, notes Kim Lum, who was regional manager at the time, "We had six office buildings on the drawing boards in 2001, looking like we're just going to be great for the next 10 years. . . . Every one of those projects went away, and we're sitting there with no work, and it's all because we were so focused on a single market."[31] Renovation work and the construction of public buildings, such as the San Mateo Public Library, offered lifelines until the commercial market revived, validating the product diversification strategy that Rik Kunnath had initiated more or less by stealth during the previous recession.[32]

Building Headquarters for Public Agencies: Gateway Center

Southern California was the first Pankow region to rebound from recession. By the mid-1990s, the Altadena office was "chockablock full with shell-and-core work," as Dean Stephan puts it.[33] Two of its most significant projects rose at Union Station, on the northeast edge of the central business district. The headquarters of the Los Angeles County Metropolitan Transportation Authority, completed in 1995, was the first high-rise structure added to the downtown skyline in the 1990s. It was the largest project that the company undertook in Charlie Pankow's lifetime. Three years later, CPBL completed a headquarters building for the Metropolitan Water District of Southern California—the second and last downtown high-rise built during the decade. Just as iconic skyscrapers like the Chrysler, Woolworth, and Transamerica buildings had done to project the identity of their corporate sponsors, these projects made bold statements on behalf of two beleaguered, yet powerful, public agencies. They also constituted distinctive achievements in design and construction. As case studies, they demonstrate the ability of Pankow as contractor to continue to innovate on increasingly complex projects within the building team. Their timely and economical completion stood in stark contrast to the contemporary experiences of other major municipal agencies. Two Broadway, the skyscraper that MTA's counterpart in New York City built in lower Manhattan, for instance, was a "high-profile fiasco," notes Barry LePatner, busting its budget by some $300 million and its schedule by several years.[34]

The MTA and MWD projects altered the geographic trajectory of development aimed at reviving a downtown that had decades earlier become irrelevant in the minds of most Angelenos.[35] The MTA project was the first in a planned set of buildings aimed at enhancing the value of the largest intermodal transport facility in America.[36] Together, the structures were to form an archipelago of commercial real estate that stood apart from, and to the northeast of, downtown at historic Union Station, "the last, and most pointless, magnificent train station ever built in the United States," as William Fulton, then editor of the California Planning and Development Report, put it.[37] Ultimately, the commercial cluster did not materialize as planned, but, separately, MWD officials located a new headquarters for their agency on the 52-acre site.

CPBL participated in both projects in a unique (for the company) joint venture with San Francisco–based Catellus Development Corporation. Catellus had begun its corporate life as the real estate subsidiary of the merged Santa Fe and Southern Pacific Railroads, and thus owned Union Station and the land around it. In an environment characterized by heated criticism of the projects and their owners, CPBL delivered both projects ahead of schedule and for costs that compared favorably with similar corporate and public efforts. CPBL also acted as merchant builder, mobilizing its financial resources to salvage a bankrupt subcontractor to ensure the successful completion of the MTA project. It is telling, too, that the clients were public agencies, and that both of them procured their buildings through public competition, not negotiation. The projects foreshadowed the type of work on which the company increasingly would rely during the recessions of the first decade of the twenty-first century.

On one level, the transformation of industrial "wasteland" into office space was functional. Both agencies sought to consolidate employees in either an existing or a new building. Officials at the MTA, which had been created in 1993 by the merger of the Southern California Rapid Transit District (RTD) and the Los Angeles County Transportation Commission (LACTC), hoped that centralizing operations under one roof would help to repair relations among staff of the two agencies, which had been quarreling over control of public transit and highway policy ever since voters approved Proposition A in 1980. Passage of the ballot measure triggered the feud by giving the LACTC control over public transit investment and policy. The interagency rivalry was also grounded in culture. The RTD was a large organization with a unionized staff committed above all to the existing bus system. The LACTC was much smaller, with a white-collar, university-educated staff that backed a countywide light-rail system. Before they merged, the LACTC and the RTD had sought new and separate accommodations for their respective staffs, both of which were mostly housed downtown, but in very different environments. Befitting its blue-collar identity, the RTD operated out of a windowless Skid Row building. The LACTC sprinkled its staff among several shimmering glass towers. For their part, MWD officials began searching for a new location for its headquarters in 1990, when the water agency was occupying cramped quarters on Sunset Boulevard that inspectors had deemed to be unsafe. It vacated that space in 1993, relocating to offices in the newly completed skyscraper, Two California Plaza. All the while, agency officials searched for a permanent headquarters that consolidated staff. They followed their MTA counterparts in constructing a home near Union Station after the transportation agency's search for office space materialized as a portion of a $300 million project "dedicated to the proposition that somehow or other riding on trains can save Los Angeles," as Fulton put it.[38]

And save the MTA, he might have added. On another level, the transformation of industrial "wasteland" into office space was institutional. It would reshape the identity and restore the image of a battered and embattled agency that was struggling to erect a commuter rail system in Los Angeles County. Proposed by transportation experts R. F. Kelker and Charles DeLeuw in 1925, but shelved by planners and municipal leaders in favor of a five-decade-long frenzy of freeway construction, the quest to build such a system revived almost immediately after the last Red Car trolley line shut down in 1961. Downtown and Wilshire Corridor business leaders, who had sought in vain to fund the central subway portion of the Kelker-DeLeuw plan with private capital—a new Red Line—lobbied successfully in 1964 for the creation of the RTD. The state legislature officially charged the RTD with operating Los Angeles's bus lines, the city's sole mode of public transportation, but also empowered the agency to submit propositions to voters in support of light-rail plans. No political consensus for a rail system aimed at reviving the downtown and Wilshire business districts materialized, however, before exasperated state legislators created the LACTC in 1976. Four years later, voters approved Proposition A, authorizing a half-a-cent sales tax to fund both a Red Line subway and a countywide light-rail system (which would eventually comprise Blue, Green, and Gold lines). After years of false starts and innumerable, often embarrassing, tribulations, the system was beginning to materialize. The Gateway Center Intermodal Transit Center would be its linchpin.[39]

The transit center was the product of a private-public partnership between Catellus and the MTA. The former's predecessor, the Santa Fe Pacific
Realty Corporation, had been formed in 1984 as a subsidiary of the Santa Fe
Pacific Corporation, the parent of the Santa Fe and Southern Pacific Railroads, which had agreed to merge a year earlier. The company handled the
Santa Fe Pacific's non–railroad-related real estate activities until 1990, when
the financially burdened parent spun off the business as Catellus Development Corporation. The new firm inherited 1.5 million acres of land, mostly
in California, but also $800 million of the $2.5 billion in debt carried by Santa
Fe Pacific. Property development immediately became a strategic imperative
for Catellus, though its managers had far more experience and expertise in
selling properties. Yet the latter could easily appreciate the potential of the
long-neglected area around Union Station when transportation officials made
known their interest in reshaping Angelenos' perception of, and mobilizing
their interest in, public transit with a project that would link existing systems
with new ones that were coming into operation.[40]

With the backing of RTD (and, later, MTA) board member Nick Pats-
aouras and East Los Angeles councilman Richard Alatorre, both of whom
touted its democratic benefits, Gateway Center evolved from "a rather nar-
rowly defined transportation project," linking El Monte bus line and Amtrak
intercity riders to new Metrolink commuter trains and the Red Line subway,
"into a dramatic public environment" that provided "a 'first class' environment
for transit users," explained Steven M. Nakada, a principal with Ehrenkrantz,
Eckstut & Kuhn Architects (EE&K), designer of the project's master plan and
public architecture. The latter featured the East Portal, a 56,000-square-foot
bus and rail terminal, and Metro Plaza, a two-level space surrounding the
East Portal that EE&K designed as a retreat for commuters and office workers. Featuring a dramatic steel and etched glass skylight that refracted light
onto "an intricately patterned sunburst design" on the main lobby floor and
a "sweeping" limestone archway along Metro Plaza, the East Portal provided
commuters access to an underground parking garage, an elevated bus plaza,
and, via an underground concourse, Union Station itself. The expansive terminal offered Angelenos "an uplifting example of civic architecture in the tradition of the world's great train stations," according to Nakada. Indeed, MTA
officials would promote Gateway Center as the "Grand Central Station of the
West." Yet it was the MTA's headquarters building, towering above the portal and plaza, that attracted the most attention to the project and its owner.[41]

The public competitions held by both agencies to procure their headquarters were unlike any of the design-build competitions described in the previous chapter. Rather than solicit proposals from teams of architects, engineers, and contractors for a building to be constructed at a predetermined location, agency officials sought owners with sites to develop or buildings to either sell or lease on a long-term basis. They specified the amount and type of space that they needed, and myriad other criteria, such as proximity to public transit, accommodation of future expansion, and the extent to which the building represented the identity that the agencies wanted to project. Bidders who met the criteria effectively competed on the basis of rent. CPBL joined Catellus in a joint venture known as Union Station Partners at the invitation of Elizabeth Harrison and Ted Tanner, vice presidents of development for Catellus. Tanner had been introduced firsthand to the Pankow firm's work by project sponsor Joseph Sanders, who walked him through La Tour, a luxury condominium completed in 1991 in West Los Angeles. Under their arrangement, Catellus handled entitlements and other site-related issues and set the price of the land sale. CPBL took responsibility for the delivery of a turnkey project and supplied the construction figure. Together, the price of the land and the building, arrived at independently, equaled the bid. In December 1990, Union Station Partners responded to the RTD's RFP (request for proposal). Four months later, the RTD held a design competition for the headquarters building. In June 1991, the winning building team held its first planning meeting.[42]

CPBL assembled a building team that included Los Angeles architects McLarand, Vasquez & Partners and structural engineers John A. Martin & Associates, its partner in the development of the precast hybrid moment-resistant frame (PHMRF). In their design of the tower, the architects sought to establish "a distinctive architectural vocabulary" derived in part from Union Station, "with numerous contextual influences of the imagery" embedded in that structure and the adjacent US Post Office annex. In all, designers expressed their "distinctive architectural vocabulary" by integrating art deco, art moderne, and so-called Moorish-Spanish elements, featuring 2-story arches and strong vertical lines. They executed the design in buff-colored Minnesota limestone, Italian granite, and English brick. The structure was essentially rectangular in plan, with a slight radius on the short faces. A third-story concourse linked the 28-story tower to the East Portal. Reflecting the intentions of the developers to construct a cluster of office buildings, the building team incorporated a four-level subterranean structure to serve as a common base for future buildings. Hence, the basement, constructed with precast, reinforced concrete beams and columns and a cast-in-place concrete slab, extended beyond the tower's 118-by-165-foot footprint. To resist both lateral and shear seismic and wind forces, the structural engineers designed a lateral system fabricated of widely spaced beams and columns tied together with spandrel beams. The flooring system was comprised of composite beams that spanned the 41-foot distance from the core to the exterior and a 3-inch metal deck topped with 3¼-inch coating of lightweight concrete.[43]

Critics of the project raised their objections in step with the construction of the tower, which rose as a solitary structure 1 mile to the northeast of downtown core, beginning in August 1993 (fig. 91). Architects worried that the scale of the transit center and its accompanying tower would overwhelm Union Station as a historic place. Planners wondered why the MTA needed a new building, given a glut of downtown office space. They also argued that the construction of a new commercial cluster so distant from the central business district was self-defeating from a policy point of view, given the billions of dollars that public and private investors devoted to the revitalization of downtown. Key state politicians were appalled by MTA's apparent ostentation and audacity. Quentin Kopp, chairman of the Senate Transportation Committee, was "flabbergasted" when he saw the MTA tower as it neared completion. "I was in awe of such conspicuous consumption," he said. Richard Katz, Kopp's counterpart in the Assembly, called the building a "monument to wasted transit dollars." Citing the materials used to produce it—the Italian granite in particular—and MTA's "misplaced priorities," one critic labeled the tower a "Taj Mahal."[44]

William Fulton has observed that it is "always dangerous from a public relations point of view for a large public agency to build a monument." It was all the more so in the case of the MTA. The agency began its organizational existence bankrupt, riven by staff divisions, and scandalized by an inability to control the quality of work of its Red Line subcontractors, who were tunneling underground along Hollywood Boulevard. Over the next two years, the MTA's budget woes persisted, forcing the agency to scale back its light-rail plans. And problems with the construction of the Red Line continued unabated. In the fall of 1994, each of its three tunnels was found to be misaligned. As the MTA tower neared completion, the Green Line to Long Beach, along the Century Freeway, opened; agency officials conceded that few commuters would use it. In August 1995, one month before CPBL completed the

Figure 91. The MTA tower nears completion at historic Union Station, north of the downtown Los Angeles core. Union Station is in the near background, across the railroad tracks from the East Portal, to the left of the tower. Completed in September 1995. McLarand, Vasquez & Partners, architect; John A. Martin & Associates, structural engineer. (Photo Warren Aerial Photography. Courtesy of Charles Pankow Builders, Ltd., Pasadena, California.)

MTA tower and two months before Gateway Center opened, the Los Angeles County Board of Supervisors dealt the MTA a severe financial blow by convincing state legislators to amend state law to authorize the transfer of $150 million from MTA coffers to the County to close the latter's yawning deficit. All the while, legislators considered reorganizing the agency. At a minimum, the sight of a "Taj Mahal" looming over Union Station exacerbated the woes that the agency faced, given the impression that it made on politicians like Kopp and Katz, on whose goodwill it depended.[45]

In this context, the delivery of the MTA tower on time and on budget without any change orders constituted the first bright spot in the agency's short history. Certainly, it stands out as an unprecedented achievement among major public efforts in Los Angeles. Under the scrutiny of a six-member public review board that included future mayor Antonio R. Villaraigosa, the project demonstrated the benefits of design-build to local municipal leaders in much the same way that the University of Hawaii Arena project impressed officials in Hawaii. Argues Tom Verti, who was vice president and Southern California regional manager at the time, the scrutiny that the MTA project received allowed CPBL to "highlight a process that was second nature to us but unique in the [public sector]." At $100 per square foot, the cost of constructing the tower's shell and core was not extravagant. The $500,000 spent on Italian granite was a small part of the budget (and, in fact, was cheaper than any granite that could have been supplied from domestic sources); Minnesota limestone constituted 87 percent of the stone used on the building. Designers selected English brick for its durability.[46]

Nevertheless, for their part, MWD officials were determined to avoid criticism that their headquarters was a "Taj Mahal," and so worked "closely and conscientiously" with the building team, as MWD General Manager John R. Wodraska explained, to procure a permanent home that was "economical yet attractive" and projected a strong civic presence that reflected, but did not overplay, the agency's role as one of the most powerful institutions in the Southern California growth machine. In an age of water scarcity, subtly projecting confidence while shedding the arrogance and insularity for which it was well known presumably would help the agency cope with external challenges to its control over regional water supplies and internal organizational dislocation that had left it "mired in inertia [and] ineptitude," as one observer put it.[47]

From its founding in 1928, the Metropolitan Water District promoted limitless growth and promised adequate water supplies to support it. Its pledge to produce water—formalized in 1952 by the so-called Laguna Declaration—was grounded in an engineering mentality that dispensed with economics. Beginning in the 1960s, decades of water abundance gave way to an age of scarcity, owing to a series of supply shocks, as well as to US Supreme Court rulings on allocations of Colorado River water, the demise of the Peripheral Canal (at the hands of voters in 1982), and soaring demand linked to unchecked population growth. Drought from 1987 to 1991 exposed the agency as an institution bereft of ideas for managing demand or controlling supply; it also evaporated its longstanding excess supplies. Still, the agency did not waver from its ideological commitment to growth. As representatives of the de facto monopoly wholesaler for the region, MWD officials attempted to repair strains in relations among member agencies with a series of ad hoc solutions. None of them addressed the contradictions embedded in MWD's noneconomic approach to water delivery. Unsurprisingly, the agency's inability to deliver supplies to member agencies that were equally, and incompatibly, bent on growth sparked revolt. In 1995, as construction on the MWD headquarters building was about to begin, the San Diego County Water Authority, the MWD's biggest customer, signed an agreement to buy water from the Imperial Irrigation District. This shot across the bow rocked the MWD, already "fly[ing] apart" organizationally, owing to high turnover among top management and its board of directors. Little wonder, then, that Wodraska—hired in 1993—saw in the materialization of the new headquarters an opportunity to stabilize the MWD as an organization and restore confidence among staff in its mission, however misplaced the commitment to growth may have been.[48]

For their part, Catellus and CPBL reprised their Gateway Center partnership to deliver the 1.1 million-square-foot MWD complex at Union Station two months ahead of schedule and $2 million under its $135 million budget. At $95 per square foot, the cost of construction was well below that of similar private and public sector projects. At the same time, the structure met the MWD's requirement that the agency's home "would perform better than any other office building in a seismic event," as Barry Schindler of structural engineers Martin Huang International explained. Moreover, the design of architect Gensler & Associates satisfied the MWD's expectations, "expressing permanence and pride while being sympathetic to the scale and characteristics of the adjacent Union Station," according to President Ed Friedrichs (fig. 92). The result was "an elegant, but not ostentatious, durable, high-quality building that [did not] command extremely high costs," just as MWD officials desired, explained Verti. The water agency preempted critics by promising that it could save $6 million dollars annually by consolidating and reducing (through attrition) staff scattered in downtown Los Angeles, City of Industry, and Pasadena, and by owning, rather than renting, its office space.[49]

After considering "dozens" of possible buildings and sites, the MWD concluded its search for a headquarters solution in April 1995 when it acquired a four-and-a-half acre site next to Union Station, with the rights to develop the property, from Catellus for $14 million. At the same time, the agency signed a development agreement with Union Station Partners, the Catellus-Pankow joint venture, for an office building to be delivered by 1 December 1998. Preliminary plans called for as much as 550,000 square feet of office space and up to 356,000 square feet of garage space for parking and vehicle maintenance. The agreement contemplated $90 million in construction costs. Under the direction of project sponsor Joe Sanders, the contractor brought architects Gensler and structural engineers Martin Huang International into the building team.[50]

Figure 92. Metropolitan Water District headquarters at Union Station, Los Angeles. Aerial view, 1 September 1998. The building topped out on 6 March 1998. Underground parking accommodated the historic railway station. Gensler & Associates, architect; Martin Huang International, Levin Seegel Associates, and Tsuchiyama & Kaino, structural engineers; HNA Pacific, parking structure engineer; Melendrez Associates, landscape architect. (Photo Warren Aerial Photography. Courtesy of Charles Pankow Builders, Ltd., Pasadena, California.)

Taking their cue from their MTA counterparts, MWD officials initiated a design-build approach, according to Tom Verti. Their interest in closely monitoring the building team to ensure full value for their investment resulted in a more formal and extensive structural selection process than CPBL typically encountered on projects for private owners. Whereas contractor and engineer might consider two or three structural systems for a private client, MWD officials insisted on analyzing six: two variations of three moment frames (structural steel, cast-in-place concrete, and precast concrete), one with, and another without, a concrete shear wall core. To assure MWD officials that CPBL would select the most appropriate framing system for the project, rather than the one that would yield the highest margins for the contractor, Sanders had recommended that the agency submit its initial options to structural design experts for peer review. MWD officials liked the idea so much that they persuaded the contractor to develop half a dozen options for consideration. The peer review panel included three structural design firms, Robert Englekirk, EQE, and Nabih Youssef. The water agency, which assigned as many as seven lawyers to scrutinize the process, placed the highest priority on seismic performance and cost control. Given the reluctance of local building officials to approve structural steel frames in the wake of the Northridge earthquake (see chapter 6), the panel discarded both steel options. From the remaining options, it fashioned a hybrid solution, combining cast-in-place moment frame elements for lateral resistance with precast beams and columns to carry gravity loads. All slabs would be cast in place. At Sanders's suggestion, precast panels would serve as the formwork for columns and beams—a technique that CPBL had deployed on many of its early projects, but had not used for almost 15 years.[51]

The architects designed the headquarters building to complement Union Station, arranging the 12-story tower and 5-story wing at a 90-degree angle to form two sides of a plaza that the building shared with the railway terminal (fig. 93). For the portion of the tower that lay above the 5-story base, the architects used columns that extended to the roof, separated by narrow strips of glass and concrete. Dominated by strong vertical lines, the façade recalled the most celebrated examples of art deco architecture in downtown Los Angeles, such as the aforementioned Eastern Columbia Building and the Sun Realty Building (1930)—also designed by Claud Beelman—and captured the ethos of modernity and progress that animated the MWD at its founding, which the agency was grasping to hold onto. Elsewhere, the designers strove to complement Union Station. The L-shaped base of the building that wrapped around the courtyard and plaza emphasized horizontal lines with wider panels and windows than were found in the tower. Shorn of the decorative details that adorn the Eastern Columbia and Sun Realty buildings, the art moderne exterior of the base was designed to deflect attention to allow a colonnade at ground level to frame the courtyard and plaza in a manner that complemented the Spanish Colonial Revivalism of the railway station. Cladding the exterior in precast concrete panels created the appearance of limestone more in keeping with the muted, buff-colored Title Guaranty Building (1929–1931), designed by John and Donald Parkinson, than Beelman's vibrant and dramatic façades (fig. 94). With expansive spaces inside the building, including a 2-story rotunda with an aquamarine skylight, the design made it clear to employees that no longer would they have to endure cramped quarters. Satisfied by its "practical yet pleasing appearance,"

Figure 93. Metropolitan Water District headquarters at Union Station, Los Angeles. Aerial view, 1 September 1998. The 12-story office tower and 5-story wing form two sides of a plaza that the building shares with historic Union Station. (Photo Warren Aerial Photography. Courtesy of Charles Pankow Builders, Ltd., Pasadena, California.)

MWD's Wodraska insisted, with self-conscious reference to the MTA tower that literally and figuratively cast a long shadow over his project, "This building can't be called a 'Taj Mahal.'"[52]

The MWD was a demanding owner, reflecting the agency's determination to complete the project on time and within budget, ensure the high quality of the product, and avoid the criticism that the MTA had endured. Superintendent Bill Tornrose and field superintendent Red Ward—the same team that had completed the MTA project—managed the building site. Kevin Smith (fig. 95) managed precast operations at a yard that lay along the railroad tracks. MWD officials required the contractor to submit construction process documents for each task. CPDs, as they were called, detailed every step that a crew would take, for example, to erect a column. Once MWD approved it, crews had to adhere to the CPD, or file a new one to accommodate proposed changes. The agency's oversight of precast operations was equally meticulous. Workers produced

Figure 94. Precast concrete panels create a limestone appearance on the exterior of Metropolitan Water District headquarters. The verticality of the façade recalls the most celebrated examples of art deco architecture in downtown Los Angeles to capture the ethos of modernity and progress that animated the municipal water agency at its founding. (Photo Warren Aerial Photography. Courtesy of Charles Pankow Builders, Ltd., Pasadena, California.)

2,150 beams, 325 columns, and 2,200 architectural panels. The pieces were cast in a sequence that enabled workers to stack them only once. Manufacturing the panels posed another challenge. The panel configuration "was very unique and difficult to cast," conceded project engineer Brian Liske. Crews used 10 steel beds that accommodated the various sizes needed to cast the panels. A special color mix was used to emulate limestone: The project team dedicated 20 trucks to the delivery of concrete to ensure that the color of panels would not deviate between pours. (Both the architect and the agency inspected each panel for uniformity of color and texture.) MWD Executive Officer and Project Executive Gilbert Ivey, for one, was "extremely pleased" with the outcome. In fact, the agency's approach to quality control prompted CPBL's engineers and superintendents to incorporate its core elements into the firm's methodology.[53]

Architect Steve Nakada had never heard of design-build before his employer assigned him to work with Tom Verti, Joe Sanders, and Bill Tornrose on Gateway Center. In the course of completing the project, Nakada observed closely as the Pankow team held subcontractors to the letter of the design doc-

Figure 95. Superintendent Kevin Smith. Hired in 1978, the Purdue civil engineering graduate was eventually promoted to superintendent. Reflecting his preference for directing work in the field, Smith chose to pursue his career with CPBL in this position. (Photo taken in 2012. Courtesy of Charles Pankow Builders, Ltd., Pasadena, California.)

uments and took preemptive measures against possible water intrusion and other matters that might have resulted in litigation—tasks that the architect traditionally performed. He also watched as the Pankow firm "stepped up" financially when O'Keeffe's, the manufacturer of the East Portal skylight, filed for chapter 11 bankruptcy before Nakada and his EE&K colleagues completed shop drawings. It became clear to Nakada that, for everyone in the firm, project sponsorship meant delivering both a high quality product and a successful project for the owner. Impressed by the mentality of the people with whom he worked and the performance of the organization, Nakada became a convert to the Pankow Way. One rarely sees a firm "in any industry with that kind of integrity," he notes. "It's just amazing." Nakada would soon establish his own practice and deploy design-build to deliver his projects. In the meantime, as a principal architect with EE&K, he would work with CPBL on restoring Pasadena's historic civic core with Paseo Colorado, an open-air, Mediterranean-style "urban village," mixing retail, entertainment, and multi-unit residential uses: a final detailed case example that illustrates how the Pankow firm used design-build to deliver a variety of large commercial projects over its four decades under the leadership of Charlie Pankow.[54]

Restoring the City Beautiful with a Next Generation Shopping Center

"In the new world of 1990s' retailing," observed urban planner Fulton, "shopping centers either had to be cheap or they had to be 'experience-based,' providing shoppers with restaurants, entertainment, and lots of other experiences that would want to make them return." Plaza Pasadena, a 2-story, box-shaped, enclosed structure, was neither. Paseo Colorado, the project that replaced it, was a pioneering example of revitalizing space "deadened" by shopping centers constructed from the 1960s through the 1980s. Completed in the fall of 2001, the project represented the Pankow firm's most significant contribution to the "de-malling" of America, involving the transformation of regional shopping centers into mixed-use, often open-air centers: the most significant trend in fin de siècle retail development. It also was a fitting capstone to a record of shopping center construction that began with MacArthur Broadway Center. One of the largest projects of its kind, Paseo Colorado mixed retail, entertainment, and residential uses in an effort by the developer and municipal leaders to revive the commercial fortunes of a city center burdened by Plaza Pasadena, by then a dying mall. It also mitigated some of the damage that Plaza Pasadena inflicted on Pasadena's historic civic district.[55]

Designed by celebrated retail architects Jon Jerde and Ron Altoon, Plaza Pasadena garnered several architectural and design awards. From the day that it opened, however, residents viewed Plaza Pasadena as an affront to their city's sense of place. The mall "was born in controversy," explains Bill Bogaard, who has been mayor of Pasadena since 1999. Much of the local criticism centered on Plaza Pasadena's obstruction of the so-called Bennett Plan for Pasadena's civic center, which the city had adopted in 1925. Reestablishing the integrity of the master plan developed by Edward Bennett, a protégé of Daniel Burnham, America's most-acclaimed urban planner of the early twentieth century, was a major goal of city leaders and planners. The impetus for what became Paseo Colorado, however, was Plaza Pasadena's failure to fulfill the economic expectations of its sponsors. The solution mixed 560,000 square feet of open-air retail and entertainment space with 387 upmarket residential units. Constructing Paseo Colorado required the contractor to apply its experience in building previous generations of shopping centers to problems associated with mixed-use construction.[56]

Plaza Pasadena was the culmination of a public-private effort to arrest "major decline" in downtown Pasadena, as evidenced by "substantial blight and physical decay," declining assessed values, and vacant stores, explained Gerald M. Trimble, executive director of the Pasadena Redevelopment Agency (PRA), which supported the project. The Pasadena City Council approved Plaza Pasadena as the linchpin of an effort to recoup sales tax revenues that were being siphoned by shopping centers in nearby communities. The PRA contracted for the $100 million mall, which would be anchored by J. C. Penney, the Broadway, and May Company department stores. Deploying the idiom of urban renewal, which seems to have remained constant across both time and space, Trimble explained that 600,000 square feet of "new, viable, competitive space" would replace some 350,000 square feet of "worn out space." Reducing retail "fragmentation" would revitalize Pasadena's downtown and make it regionally competitive. PRA deployed eminent domain to acquire a three-block area in the civic center district. Clearing the site resulted in the relocation or bankruptcy of some 160 businesses, and involved the demolition of the historic store and office buildings that they occupied. Only the Pacific Southwest Trust & Savings Bank Building, designed by Curlett & Beelman and completed in 1924, was spared the wrecking ball (fig. 96). The PRA issued $62 million in tax allocation bonds to help fund the project.[57]

Intended to provide consumers with a reason to return to downtown and retailers with a more regionally competitive commercial district, Plaza Pasadena eventually proved to be a bust. It opened with all of its retail area leased. Its incipient success had multiplier effects, as businesses moved into nearby historic buildings. By the late-1990s, when city and business leaders convened a task force to rethink the project, however, half of its leasable space lay vacant. Penney's and the Broadway had led the flight of retailers from the mall. Plaza Pasadena was losing business to other regional malls, such as the Glendale Galleria, and had not kept pace with the changing tastes of the increasingly affluent local community.[58]

For residents, having to endure a structure that "stood fortress-like on Colorado Boulevard, a monument to good architectural intentions gone awry," as John Woolard wrote in the *Los Angeles Business Journal*, was bad enough. Worse was the mall's intrusion on the civic center district. Plaza Pasadena sat astride the north-south axis of the Bennett Plan, blocking the view along Garfield Avenue. Ernest W. Hahn, its developer, in conjunction with Carter Hawley Hale Stores, had promised city officials that the project would maintain visibility between historic civic buildings, but the mall blocked the sight line between the civic auditorium to the south and city hall to the north of the mall. As Morris Newman complained, Plaza Pasadena's "banal design, including multilevel stretches of windowless walls, made it a pocket of mediocrity amid the civic center's 1920s-era Beaux Arts-style [buildings]." Ironically, many residents now viewed the shopping center built as a solution to blight as a blight on their city.[59]

Figure 96. Paseo Colorado, Pasadena, California, aerial view looking south, December 2001. The project restored the sight line along the Garfield Avenue axis of Edward Bennett's master plan for the civic center district. Plaza Pasadena had blocked the view of the beaux arts civic auditorium, in the upper left of the photograph, from the public library and city hall. The Pacific Southwest Building, spared the wrecking ball in the 1970s by the Pasadena Redevelopment Agency, sits to the right of Paseo Colorado, at the corner of Colorado Boulevard and Marengo Avenue. (Photo Warren Aerial Photography. Courtesy of Charles Pankow Builders, Ltd., Pasadena, California.)

Meanwhile, the Old Town district to the west of the mall, along Colorado Boulevard, Pasadena's "Main Street," was demonstrating that adaptive reuse of a cluster of historic buildings, encouraged by the Federal Historic Preservation Tax Incentives Program of 1976 and amendments to the National Historic Preservation Act of 1966, could be remarkably successful as a revitalization strategy. As Plaza Pasadena emptied, national retailers opened stores in Old Town. The district's emergence as a vibrant entertainment and retail area in the 1990s exposed Plaza Pasadena's competitive disadvantages, which were readily apparent to city officials and TrizecHahn Development Corporation, which had become the mall's owners through a series of buyouts. (Trizec Corporation, a Calgary-based developer of office buildings, acquired Ernest W. Hahn and its 32 malls across the Southwest soon after Plaza Pasadena opened in 1980. In 1991 Horsham Corporation, a gold mining and oil refining firm, bought Trizec and formed TrizecHahn. In 1998 TrizecHahn Development Corporation, a retail and entertainment property division, became the sole owner of Plaza Pasadena.)[60]

By then, some two dozen people had convened a task force on Plaza Pasadena. Encouraged by Mayor Chris Holden (1997–1999), the group began its deliberations without expecting to call for a major project as a remedy. Rather, it merely hoped that something might be done to provide greater visibility between the civic buildings. Yet the group's thinking evolved toward consideration of an open-air retail and entertainment complex that reestablished the sight lines of the Bennett Plan and provided retail stores that fronted Colorado Boulevard.[61]

At the same time, TrizecHahn prepared to sell the traditional malls that it had acquired from Ernest W. Hahn as part of a strategy to "break the mall paradigm," as senior vice president of development Richard Froese put it, with projects that mixed retail with other uses. As someone who grew up with Plaza Pasadena and shared residents' views on the mall's impact, Steve Nakada persuaded TrizecHahn's David Malmuth and Lee H. Wagman to study the repositioning of Plaza Pasadena as such a project. (The architect knew Malmuth from the latter's days at Disney, when EE&K performed master planning for two projects, Hollywood Boulevard and Ocean Sea in Long Beach. When Disney decided not to pursue the projects, Malmuth left the company.) The TrizecHahn executives agreed. The developer also began participating in the discussions of the Civic Center Task Force. Influenced in part by the success of the Old Town district, which suggested "that a correctly conceived project could be successful," as Froese put it, the developer hired EE&K, whose Los Angeles office included almost two dozen people who specialized in mixed-use development, to analyze the project financially and then design it.[62]

Known for its ability to "cut through red tape," EE&K also assisted TrizecHahn in working through a lengthy entitlement process with the mayor's office, the planning and design commissions, and the PRA. The biggest challenge lay in convincing planners to allow the architects to vary the heights of buildings. The site had been zoned for Plaza Pasadena with a height limit of 60 feet. The city eventually agreed to set 60 feet as the *average* height for all of the buildings. In return, EE&K and TrizecHahn agreed to use "civic" materials, such as stone, and architectural styles appropriate to the civic center district in the design. As Mayor Bogaard explained, the city "handled [Paseo Colorado] differently from most other projects that preceded it because TrizecHahn accepted [the city's] preference[s]."[63]

EE&K's design embraced the idea of a Mediterranean town square, consistent with the romantic, picturesque, and idyllic version of the region's Spanish past that been embedded in much of its built environment for nearly a century (fig. 97). Designers used fountains, courtyards, vine-covered paseos, balconies, and patios to evoke scenes of a charming rural village, in much the same manner that, for instance, the Community Arts Association's Plans and Planting Committee promoted the rebuilding of Santa Barbara following a devastating earthquake in 1925 and architect Richard S. Requa and his associate, Lillian Rice, realized Rancho Santa Fe, in San Diego County, during the 1920s. (While post-earthquake Santa Barbara and Rancho Santa Fe drew exclusively on Spanish Colonial Revivalism, EE&K's design was more regionally generic. Its town square might have been located just as easily in Tuscany as in Andalusia.) To create its open-air spaces, EE&K broke the project into seven separate buildings. Designers used walkways to incorporate city streets into the design. To integrate the project's storefronts with the historic structures that lined Colorado Boulevard, EE&K developed 15 types of façades and arranged them to complement the eclectic architecture found in these buildings. The design also conformed to the Bennett Plan, with a 70-foot-wide paseo that opened sight lines along Garfield Avenue (see fig. 96). Designers used stone cladding, a grand staircase, and a clock tower to integrate the façades of Paseo Colorado's buildings along Garfield Avenue with the beaux arts classicism of the civic center district. In "creating contemporary architecture," EE&K tapped regional cultural memory and recovered as much of Pasadena's architectural history as structural and commercial parameters allowed, explained EE&K studio director Elaine Nesbit.[64]

Multi-unit housing was not part of the task force's initial discussions. It became part of the mix after TrizecHahn advised city officials that the project was not financially viable solely as a redesigned retail and entertainment center—on the eve of the latter's expected release of the environmental impact statement for such a project. "We were thinking in a narrow box," explained Froese. "We couldn't expand beyond the street. Then we realized we could go up [vertically] and include retail space." Residents would pay rent, of course, but also could be counted on to patronize Paso Colorado's stores. "Residential is a good anchor," Froese argued. "It brings customers to [mer-

Figure 97. Architect's rendering of Paseo Colorado as Mediterranean town square. (Courtesy of EE&K, a Perkins Eastman Company.)

chants'] doorsteps." EE&K prepared schematic designs for a revised configuration of four levels of market-rate residential space atop the two-level retail and entertainment complex. TrizecHahn lined up three candidates, including Atlanta-based Post Properties, Trammel Crow, and the Related Companies, developer of the Paramount, to compete for the housing portion of the project and sent each one the bid package prepared by EE&K. Post Properties won the competition. TrizecHahn then sold Post the air rights over Paseo Colorado. Post brought in their architects, Dallas-based RTKL, to complete EE&K's schematic design. RTKL consulted the EE&K's architects as it implemented the project that the latter had entitled.[65]

Several writers suggested that the mixing of residential and commercial uses at projects such as Paseo Colorado was a recent phenomenon in Southern California. Eric Lassiter, for one, noted that a pattern so fundamental to the urban histories of London, New York, and San Francisco was now taking hold in Los Angeles. Morris Newman wrote that mixed-use was "unusual" for Los Angeles and "untested" in Hollywood. But Paseo Colorado was actually part of an established, if interrupted, tradition of dense urban living in the city and its older suburbs.[66]

In the 1920s, when oilman Ralph Lloyd was developing his holdings in Hollywood and on the edges of downtown, greater Los Angeles was booming in population and real estate development. It may have been bustling with thousands of automobiles, but it was also organized around a busy network of streetcar and interurban railway lines that linked the region. A healthy demand existed for affordable rental units near convenient commercial services and located along public transportation lines—and developers happily supplied them. Apartment buildings from small to large could be found in many areas, and multi-story structures of flats over ground-floor shops lined certain stretches of the city's thoroughfares.[67]

City planners cooperated as well: Los Angeles's first comprehensive zoning ordinance, passed in 1921, encouraged multi-family housing and did not prohibit the commingling of housing and business along thoroughfares. Though creating a suburban homeowner's paradise was a major planning objective at the urban periphery, planners also made room for conventional, bustling urban life normally associated with cities "back East."[68]

Construction of new mixed-use buildings virtually halted in the more than half century between the 1930s and 1990s in Southern California, for several reasons: the contraction of public transit systems, the rise in popularity of low-rise garden apartment living, more restrictive zoning, and resistance by lenders to finance "outdated" development types. Reflecting a kind of collective historical amnesia, the truly varied urban development of a century ago was now remembered only for its residential subdivisions divided by tree-lined streets and dotted with detached bungalows. Hence, when developers began to respond to intense demand for housing in and around Los Angeles with high-density housing at "infill" locations, often set above ground-floor retail, observers could tout a project such as Paseo Colorado as novel and innovative.[69]

Nevertheless, the inclusion of residential units did not make Paseo Colorado financially feasible. Making it so required the reconfiguration of parking to generate revenues that could underwrite additional public investment. The City of Pasadena owned the below-grade parking structure associated with Plaza Pasadena. It had leased two levels to Ernest W. Hahn. Now city officials agreed to buy back the lease. The developer would reconfigure a large portion of the parking area—which had been free to Plaza Pasadena patrons—as municipal garage space while reserving the balance for residents. Rearranging parking in this manner allowed the PRA to issue $26 million in redevelopment bonds in support of the project.[70]

To construct the $200 million project on time and within budget, Steve Nakada, the principal architect, persuaded TrizecHahn to adopt a design-build approach and include CPBL in schematic design. Based on EE&K's two dozen design concepts for the project, the contractor produced cost estimates for 18 mixed-use configurations.[71]

Building Paseo Colorado was a complex affair, for several reasons. The design retained the concrete columns and slabs of Plaza Pasadena's first level (fig. 98). These structural elements were incorporated into the new structural steel seismic moment frame. They had to be strengthened to support the Post residences. This was accomplished by wrapping the columns in Tyfo, a glass fiber reinforced membrane manufactured by the Fyfe Company. As project engineer Rick Schutter explained, the Fibrwrap system was easier and less costly to install than conventional steel jackets. To support the residential tower, the project team used a frame composed of panelized studs and metal-backed shear panels (fig. 99). At the same time, the footprint of the old mall had to be extended to the property line along Colorado Boulevard so that stores would front the sidewalk. Reconfiguring parking for revenue and segregating and securing an area for residents meant moving all of the elevators. The demolition of the mall, undertaken in the summer of 2000, also spared the Macy's department store, which remained open during construction. But it was the concurrent construction of the retail-entertainment and residential portions of the project that posed the greatest challenge to the contractor.[72]

The problem had nothing to do with installing systems associated with vertically related residential and retail functions, however. "You would think that would be the problem," said superintendent Kevin Smith. "That was not the problem. The problem was getting the two owners to decide what they wanted to do." Daily disputes over scheduling and other matters "caused us more frustration than the physical challenge of coordinating [for instance] mechanical piping and venting through each [owner's space]. [It] was not an easy job."[73]

As a mixed-use project, Paseo Colorado involved a tremendous amount of research and effort, Nakada notes, but it paid off. Of the three projects that TrizecHahn developed, namely Desert Passage, a mall that opened with the adjacent Aladdin Resort and Casino in Las Vegas in August 2000; Hollywood and Highland, a 1.2 million-square-foot retail-restaurant-hotel-theater complex named for the streets bordering it; and Paseo Colorado, only the latter met its schedule and budget (however untimely its opening just after the 9/11 attacks may have been). Hollywood and Highland, for instance, started out as a $280 million project but escalated to $615 million. It, too, opened in the

Figure 98. Paseo Colorado under construction, August 2000. Aerial view, looking north along Garfield Avenue. The design retained the concrete columns and slabs of Plaza Pasadena's first level. The civic auditorium is in the foreground. (Photo Warren Aerial Photography. Courtesy of Charles Pankow Builders, Ltd., Pasadena, California.)

wake of 9/11. In 2002 TrizecHahn wrote down the value of the property twice for a total of $404 million. In early 2004, it sold the property for $201 million.[74]

According to Tom Verti, the Post residences were the most significant part of the project. Recognizing the "great need for more housing," he predicted that "retail with housing above it will be an important prototype for future residential development." Verti argued that CPBL was well positioned to construct more mixed-use projects, noting, "We've been pioneering economically competitive methods to build five- and six-level residential projects like Paseo Colorado." It was not long before the firm validated Verti's argument.[75]

Soon after completing Paseo Colorado, Pankow began work on Sunset + Vine, an adaptive reuse and mixed-use project located at the corner of one of Hollywood's most famous intersections. The project had been languishing for years. Originally planned as Hollywood Marketplace, a three-block retail and

Figure 99. Post Properties residences under construction atop retail and entertainment spaces, July 2001. The building team chose a frame that is composed of panelized studs and metal-backed shear panels. (Photo Warren Aerial Photography. Courtesy of Charles Pankow Builders, Ltd., Pasadena, California.)

entertainment complex, that project fell through when a major tenant, The Good Guys, pulled out and Mann, the theater operator, declared bankruptcy. In 1999 Santa Monica–based Bond Capital bought the site with the intention of developing a less ambitious project that mixed retail and residential uses. The architects and retail consultants that it hired produced a project that was $12 million over budget. Saddled with some $260,000 per month in carrying costs, the owner engaged CPBL to rethink the project.[76]

Joe Sanders and project manager Brad Whitaker assembled a team that included Nakada & Associates; HNA/Pacific, a designer of parking structures; and Suzanne Dow Nakaki, now a principal with The Nakaki Bashaw Group, as structural engineer. Nakada & Associates was Steve Nakada's new design firm, which he left EE&K to start before the completion of Paseo Colorado. Sunset + Vine was its first job. As Nakada recalls, CPBL came to him and basically said, "You just left your firm, you don't have any money, you don't have a job. You work this on spec for us a couple of months and if you can save $12 million, the job is yours. We'll support you on that."

Most recently, Nakaki had worked with Pankow's engineers on the Paramount. Nakada, Nakaki, and HNA/Pacific's Scott Herman "worked in the shadows," according to Nakada, to simplify a design that contemplated residential units on the ground floor and floors three through six, retail on the second floor, and underground parking. They eliminated the ground-floor residential units without reducing the square footage devoted to residential use and reconfigured the parking as a stand-alone structure. The residential space wrapped around the garage, whose levels lined up with the units, accommodating residents with vehicles. Removing the below-grade parking was critical to making the project pencil, for it reduced the cost per stall by half. Signage associated with the building also contributed to the pro forma budget: Nakada estimates that advertising generates quarterly revenues that exceed the annual rent from all of the residential units. With CPBL's encouragement, the design incorporated the streamline moderne façade of the abandoned structure on the site—once the home of ABC Radio and later, from the 1970s, Merv Griffin's Celebrity Theater—as the main entrance to the residential units. In Nakada's estimation, CPBL took a chance on getting the design done behind closed doors before they could reveal to the owner how it could save it $12 million. It took "a lot of guts" to tell developer Larry Bond that he had to start over. As it had done time and again over four decades, Pankow salvaged the project, realizing Sunset + Vine as a 5-story retail and residential complex on a single block.[77]

Sunset + Vine employed a structural metal stud framing system, demonstrating again that design-build could be deployed independently of the concrete construction techniques with which the Pankow firm was still closely associated. The building team chose the patented system, composed of 3,000 prefabricated panels, to save time and improve the quality of the finished product. Bellevue, Washington–based Inter-Steel Structures, Inc. (ISSI), the patent holder, subcontracted the production of the panels to a shop in Portland, Oregon. Manufacturing the panels in the four months allotted by the schedule, however, overwhelmed the fabricator. CPBL assumed responsibility for producing the panels on time. It identified a local shop that used building information modeling, a nascent information technology, to generate panel designs and contracted with the company to manufacture them.[78]

Construction began in June 2002. Before it finished, in May 2004, the company confronted the issue of leadership succession in the wake of Charlie Pankow's death.

Answering the Leadership Succession Question

With his estate planning and reorganization of the company, Charlie Pankow addressed the crucial issue affecting the welfare of family firms generally—leadership succession—by eliminating any doubt that company executives, rather than his children, would run the company after he died. The reorganization also addressed the issue of entrepreneurial declension after the death or retirement of the founder—another problem with which family firms have grappled generationally, often unsuccessfully. Even as he acted to remove all vestiges of a family business from it, however, Charlie Pankow continued to run his company in the manner of a first-generation founder, wielding the absolute control that came with majority ownership. Like most of his counterparts in family businesses, at least in America and Britain, he engaged in no formal succession planning, which propensity is linked in the literature to an inability to delegate authority. At the time of the reorganization, it seemed to Pankow that there was no immediate need of doing so, as he had no intention of retiring. And why would he? After all, as son Rick Pankow observes, "He loved his work and he loved his company."[79] Yet the conflation of self and firm, on the one hand, and the idea that the company would carry on his legacy as a master builder, on the other, created a dilemma. Either Pankow the company would die with its founder or the founder would have to decide how the company might run without him. From the time of the reorganization, Charlie Pankow continued to work even as he withdrew from the day-to-day operation of the company, allowing him to postpone resolving the matter indefinitely.[80]

The reorganization may have resolved issues related to employee ownership, but it did not influence corporate governance. For Charlie Pankow, the company was his firm and he would continue to control it as he saw fit, as long as he was mentally and physically able to do so. As noted, for age- and health-related reasons, he did not always exercise his authority as he might have done, allowing others to take steps to diversify the business. Nevertheless, to the extent that he faced his mortality and considered how his company might be run after his death, Pankow envisioned a handpicked successor running the company just as he had always done.

The succession question was exacerbated by the lack of managerial hierarchies. To be sure, the company no longer lacked for managerial titles. There were regional managers, general managers, vice presidents, senior vice presidents, an executive vice president, and a president. Yet they remained meaningless, at least as far as corporate governance and leadership succession were concerned. Managers continued to report directly to Charlie Pankow, even as the founder distanced himself from operations. And the founder was accountable to no one. The lack of organizational structure may have suited Charlie Pankow, but it obstructed succession planning—almost fatally, as far as the company was concerned.

In this context, Charlie Pankow named Rik Kunnath as his successor (fig. 100). Kunnath had risen to the top of the generation of engineers hired before the reorganization. He had demonstrated his leadership abilities as Northern California regional manager and national chairman of the Design-Build Institute of America (DBIA), and as the driving force behind PSPL. In February 1996, Pankow invited Kunnath and Dean Stephan to dinner at Le Petit Trianon, the 30-odd-room mansion on Washington Street in San Francisco that he had acquired in 1982 to house his burgeoning art collection—more about that in the epilogue—and provide him with living quarters during the work week. Stephan joined Pankow in telling Kunnath that he had been promoted to executive vice president (of which there was one in the company). An option agreement was signed, wherein Pankow provided for the transfer of his ownership in the stock of Pankow Management, Inc., the corporate general partner (see fig. 71). The agreement identified Stephan as the "initial grantee." If Stephan were no longer employed by CPBL, then ownership would transfer to Kunnath as the "substitute grantee." As Stephan had been discussing his early retirement with Charlie Pankow, because it was clear that the latter was not interested in relinquishing control in the near future, the option agreement effectively designated Kunnath as Pankow's successor. He would have the power to control the company as the founder had done. The decision became widely known and accepted within the company. The issue of succession would remain settled until the last year of Charlie Pankow's life—with one wrinkle.[81]

Charlie Pankow failed to name a replacement for Stephan in a timely manner when the latter resigned as CPBL's president in May 1997, giving people within the firm the impression that the founder might have not settled the question of leadership succession after all. In particular, Tom Verti campaigned for the open position. At one point, he gathered signatures on his behalf and presented them to Pankow. The latter gave no weight to the petition, remarking to Kunnath that he was not holding a popularity contest. Nevertheless, in 1999, Pankow appointed Verti president (fig. 101). At the same time, the

Figure 100. Richard M. Kunnath, CEO of the Pankow companies since 1999. (Photo taken in 2010. Courtesy of Charles Pankow Builders, Ltd., Pasadena, California.)

founder stepped aside as CEO and tapped Kunnath to lead the company. Pankow remained chairman of the board. In short order, however, both Pankow and Verti came to a mutual realization that the latter was far better suited to business development, his former role, than to operations and strategy. Pankow demoted him to executive vice president and named Kunnath both CEO and president. There matters rested until "the eleventh hour and fifty-ninth minute," as Kunnath puts it, when several employees persuaded Pankow that his company might well disintegrate were Kunnath to take the reins.[82]

The group of dissatisfied employees who approached Pankow was led by Christopher J. Turner. A 1980 graduate of the civil engineering and construction management program at Ohio State University, Turner was hired as a field engineer in 1984. He worked on projects in Southern California, including Chase Plaza and Shoreline Square, before he was promoted to project sponsor in 1992. Seven years later, he was named assistant regional manager. In that capacity, he supervised preconstruction and construction activities for the Paramount, the project around which the balance of the group coalesced.[83]

Notwithstanding its significance as the first high-rise structure in Seismic Zone 4 to utilize the PHMRF, the Paramount was "a very, very bad job" financially, according to Kunnath. "Scary bad." It lost money and reflected negatively on the project team associated with it. Though the job site lay within a few hundred yards of the San Francisco office, it was run out of Altadena.

Figure 101. Thomas D. Verti. Vice president and Southern California regional manager, Charlie Pankow named him president of the company in 1997. At Pankow's behest, he relinquished the position in 1999. From 1999 to 2004, he served as senior vice president. In 2009 he retired as president of Pankow Operating, Inc., the general partner of Charles Pankow Builders, Ltd., ending a career with the company that spanned four decades. (Courtesy of Charles Pankow Builders, Ltd., Pasadena, California.)

As Kunnath delicately explains, there was "some inappropriate competition" from that office by individuals who were determined to show others within the firm how well they could run the project. As a result, the individuals associated with the Paramount did not share information in a way that might have helped the company manage the execution risk that the contractor assumed in a design-build relationship. This mattered, because construction occurred during an unprecedented period of stress within the organization. Charlie Pankow and those who worked for him had always prided themselves on their ability to assume the risk of completing projects on time and within the owner's budget (on which the construction contract was based). Now, for the first time in its history, the company accepted more work than it could handle with existing staff. With half a dozen high-rise projects to deliver on schedule in the heated Bay Area commercial real estate market of 1999–2001, CPBL found it impossible to hire engineers and superintendents who met its expectations and needs. Furthermore, there was a dearth of trades workers and laborers available to perform the work. The Paramount suffered more than other local projects, all of which were run from Northern California. The productivity assumptions used to estimate the complex job proved to be wildly inaccurate. There "was a complete misunderstanding of how oversold this market was," notes Kunnath, who would have said so, had anyone associated with the project bothered to ask him. As noted, CPBL was forced to produce the structural elements for the Paramount's complex frame itself. The plant managers hired to run the new casting yard in Corcoran were "amateurs," according to Kunnath, with no experience producing precast concrete for a project on this scale. Costs soared. The Paramount may have been cheaper to build with the PHMRF than with a steel frame, but it still cost far more to construct than CPBL had budgeted. (Ironically, reducing labor costs associated with using crews in high-wage areas, such as San Francisco, was a strategic reason for establishing Mid-State Precast in a soybean field in Corcoran.) The project's managers took extraordinary measures, including using workers on multiple shifts, to complete the project on time. A worried Charlie Pankow wondered what the hell was going on. Kunnath concludes: "You can't let competition between individuals get in the way of running a company in the best possible way, and it got to that point."[84]

The individuals who questioned Charlie Pankow's choice of Kunnath as his successor solidified their bonds as they struggled to complete the Paramount. In addition to Turner, they included Bill Hughes, Jeff Lucas, a well-regarded project sponsor, and Roger Stevenson, an estimator in the Altadena office. Hughes was one of Pankow's most revered superintendents, noted for his expertise in meeting demanding schedules and solving complex construction problems. Since 1981, he had supervised many of the projects that have been discussed in these pages. In 2000 Charlie Pankow unexpectedly promoted Red Ward over Hughes as operations manager. Hughes had served as Ward's mentor, but the apprentice apparently had demonstrated better management skills and a better aptitude as a conduit between field and office. The decision rankled Hughes (even if the exact role it played in motivating him to challenge Charlie Pankow's succession plan remains unclear).[85]

The grievances of the group related to the changes that Kunnath had implemented to diversify the business away from a model in which they were emotionally and professionally invested. It is not clear whether any of them gave much thought to special projects or public sector contracting during the 1990s. The increasing reliance of the company on such work after the dot-com-driven commercial real estate boom collapsed, however, played a significant role in motivating them to act. As they later stated, each of them "began to feel that [the] company had begun to drift away from the type of negotiated, repeat customer work and project deliveries that drove their passion for the industry." More to the point, they had little interest in TI work, medical facility upgrades, renovations of historic buildings, or public sector projects subject to bid. The impact of the Paramount project on their reputations aside, they concluded that their post-Paramount careers might suffer under a Kunnath regime. As evidence of this, they could point to a realignment of financial interests that Kunnath had undertaken to make possible the recruitment of staff to replace individuals hired in haste during the boom (who either had left or were now being shown the door). The move required existing staff to give back a small part of their ownership stake in the company. It was well received by the great majority of unit holders, but poorly received by a few of them, including Turner and Hughes, who seemed—from Kunnath's perspective—to see the program in personal, rather than strategic, terms.[86]

Turner and Hughes told Pankow of their dissatisfaction with the changes that Kunnath had been implementing. In fact, they painted a picture of general dissatisfaction within the company and blamed Kunnath. They suggested to the ailing founder that people were about to leave en masse. Charlie Pankow took their grievances to heart and wavered on his choice of Kunnath as his successor. At the same time, he had no intention of replacing him with Turner. When this became clear to them, Turner and Hughes decided to form their own company.[87]

Were he in robust health, Kunnath suggests, Charlie Pankow would have paid Turner and Hughes no heed. Now, in 2003, the founder's health was rapidly deteriorating. In these circumstances, the narrative woven by Turner and the others gave him pause. A stubborn man who remained wedded to his business model, Pankow readily agreed that there was no reason to change the business. Fearing that his company was unraveling, Pankow wavered and reconsidered his succession plan. He did so even though his failing health prevented him from assessing the situation accurately, according to Kunnath. Confined to his home, he lacked the information he needed to act wisely. He knew only what the people who visited him told him. Nevertheless, he now struggled to assert control over the issue he considered most important to his legacy. Under the best of circumstances, it would have been difficult for him to act in the best interests of the firm. For he had been disengaged from the daily operation of the business for almost two decades. Ailing, it was impossible for him to do so.[88]

Charlie Pankow held fast to the idea that a single individual would run his company as he had done. With the assistance of CFO Kim Petersen (fig. 102), Pankow considered other candidates to succeed him, including people who had left the firm years earlier, most importantly Bill Tornrose. The

Figure 102. Kim Petersen, CFO, Charles Pankow Builders. Ltd. Hired as an assistant controller in 1983, he became one of two division controllers in Pankow Construction Company, a subsidiary, the following year. In the early 1990s, he was promoted to corporate controller. In 2000 he replaced Timothy P. Murphy as CFO. (Photo taken in 2009. Courtesy of Charles Pankow Builders, Ltd., Pasadena, California.)

former superintendent was now selling insurance. Yet, at the eleventh hour, Pankow reached him by phone and offered him the CEO position. When other executives learned of this decision, they told Pankow one by one that they would leave the firm, were he to fire Kunnath. Ultimately, identifying a suitable replacement proved fruitless. As Petersen explains, "Charlie would support singular individuals as being a successor and, over a period of time, would become disenchanted with that person and transfer it to the next person. I went through several of our senior people that way, such that there was no real heir apparent." Simply put, there was no other Charlie Pankow to be found.[89]

Petersen then worked though leadership alternatives that involved multiple individuals. These efforts were ultimately stillborn. For Charlie Pankow never envisioned leadership succession in any other but authoritarian terms. For every scenario that he presented, Petersen explains, "There would be a lot of analysis and a lot of what about this, what about that, can we change this, can we change that, and [Charlie] just could not in the end accept the fact that [the alternative] was too much of a democracy. [He felt] that the company should be run, in an ideal environment, by a single individual." But when the discussion turned to consideration of a specific individual, the process stalled, "because there was nobody" to replace him. Pankow may have "intended that the company would not skip a beat" after his death, as Tom

Verti later stated, but the leadership succession question remained unanswered at the time of his death.[90]

Meanwhile, Turner, Hughes, Lucas, Stevenson, and former corporate director of marketing and new business development Todd Whitlock, who had left Pankow a couple of years earlier to join Southland Mechanical, one of the largest companies of its kind in America, formed WEST Builders. Several Pankow engineers joined them, but the mass exodus about which Turner warned Charlie Pankow did not materialize. The group left with work in hand, but the loss of the project and the owner relationship attached to it was not as devastating as the departure of Webcor Builders founders David Boyd and Rosser Edwards three decades earlier.[91]

The project was the Montana, a luxury condominium across from Paseo Colorado along Colorado Boulevard. The developer was Berkshire Hathaway chief counsel and vice chairman Charles T. Munger. Steve Nakada & Associates designed it. Grateful that the contractor had invited him to participate in Sunset + Vine, Nakada brought the project to CPBL. Contractor and developer signed a construction contract valued at more than $60 million. During this time, Chris Turner apparently developed a "very good" relationship with Charlie Munger, enabling him to secure Munger's agreement to engage his company as contractor, should he and his colleagues leave the Pankow firm.[92]

The loss of the people associated with WEST Builders, the Montana project, and developer Munger ultimately had little impact on the Pankow organization. To be sure, losing capable, long-term employees that, most everyone in the company believed, had been well treated, was disappointing. Yet, as Kunnath notes, the people who formed WEST Builders were not going to be happy if they remained with the firm. Had they not left, "there would have been a significant amount of energy that would have to go into realigning them and getting them comfortable with the new management team [that took control of the company after Charlie Pankow died]." Because CPBL thought that it had a strong relationship with Munger, there was "some feeling of betrayal" toward him when he built the Montana with WEST. But Munger did not provide Turner and company with the kind of support that Bill Wilson provided Webcor Builders. In fact, Munger's other project on the drawing board at the time that WEST Builders formed, Stanford Law School, was not built. In Kunnath's view, the departure of Turner, Hughes, and the others who formed WEST Builders ultimately proved to be a "healthy cleansing." The company hired qualified replacements. The people who stayed with the company were committed to it. New management was able to get off to a "fresh start."[93]

The answer to the leadership succession question was a controlling general partnership. Just before he died, Pankow thought he had come up with a way of dividing control of the company between six executives and tapped Dick Walterhouse to assume the role of CEO. (Kunnath suggests that Pankow might have made this choice because Walterhouse was the only one of the six men who had not worked closely with him and therefore had not had the chance to do anything to alienate him.) But then Pankow died. In the wake of his death, Kim Petersen briefed senior managers on the modeling exercises that he had performed with the founder. They decided on a governance structure that contemplated six co-equal general partners who would run the company and a group of limited partners who enjoyed no control but shared in the profits. The six—CEO Rik Kunnath, Northern California Regional Manager Kim Lum, CFO Kim Petersen, Southern California Regional Manager Joe Sanders, Executive Vice President Tom Verti, and PSPL President Dick Walterhouse—agreed on the need for a CEO whom the others had the power to remove. They voted unanimously that Kunnath should hold that position. Since the remaining officers held more than 51 percent of the voting stock, the CEO had to work by consensus. For the first time in the history of the company, the CEO was accountable to others. Having answered the leadership succession question, the Pankow firm was poised to carry on the founder's legacy. Announced Tom Verti, in the context of a recovering commercial real estate market, "We're ramping up for a significant, but controlled expansion of the company."[94]

EPILOGUE

Free of the contract provisions, minimum fees, and other constraints embedded in the founder's business model, Charles Pankow Builders was able to compete on equal footing in the commercial segment of the industry during the frenzied boom that peaked in the spring of 2006. Indeed, the favorable market helped the company adjust to life without its founder. "Things took off," remarks CEO Richard M. Kunnath. Contract volumes topped $500 million and the company enjoyed record profits. Perhaps no project was more symbolic of the excess of the market than the Montage, a five-star luxury hotel in Beverly Hills, which surpassed the MTA tower as the company's largest project to date. But while the Montage was under construction, the US economy crashed. Given the central role that real estate played in fueling the boom, the construction industry was hit especially hard. The annual rate in the value of construction put in place, which had leapt 41.5 percent from the beginning of 2002 to March 2006, plummeted more than 37 percent over the next four years. In this environment, CPBL relied on public sector work, whose volume actually increased by more than 16 percent during this period, to fill its order book. Had Charlie Pankow lived a few more years and had insisted on adhering to his business model, Kunnath suggests, the company likely would not have survived the Great Recession.[1]

Under Charlie Pankow's leadership, the company never grew in the manner of the (largely manufacturing) organizations that Greiner observed in his 1972 *Harvard Business Review* article.[2] In contrast, under the direction of Kunnath and the other general partners, the company became as sophisticated as the largest contractors in the industry in terms of managerial structure and systems. Almost immediately, the six executives engaged in strategic planning and developed a business plan—exercises that had never taken place. They implemented state-of-the-art information systems to support these endeavors. A personnel management system incorporated job descriptions, annual reviews, and rationalized compensation schedules—all lacking during the Charlie Pankow years, when salaries and bonuses were subject to the whim of the founder. The goal of these initiatives, according to Kunnath, was to produce strong, entrepreneurial managers in an organization that would be much more adept at selling work across the commercial segment of the industry.[3]

Of course, Kunnath notes, it will take years to overcome the organizational legacy of relying on Charlie Pankow and his business model for more than four decades. Pankow may have been a charismatic, even visionary leader, according to Kunnath, but the atrophy of the organization under his direction left the company at a disadvantage in the marketplace. Narrow in its range, Pankow's business model discouraged legions of otherwise good sales people from closing deals. Moreover, Pankow did not systematically reward, much less identify, sales skills in his people. Disputes over the kind of work that the company would perform dated from the earliest days of Charles Pankow, Inc. (CPI), when co-founder Lloyd Loetterle departed before the first annual meeting. Redirecting the organization was made all the more difficult, according to Kunnath, because Charlie Pankow's insistence on doing projects for private sector clients on a fee basis, delivered through a particular design-build approach that was not readily adaptable to public sector or institutional work, was embedded in the Pankow Way. Grappling with organizational change takes time. The Great Recession, however, forced the general partners to take extraordinary steps to buy more time, if they hoped to realize their goal of changing course strategically.[4]

Construction is a lagging industry. With construction financing secured, large commercial projects typically proceed even as the economy spirals downward. Though construction ultimately is less insulated from recession than other sectors, the industry is buffered at the beginning of the downturn in the business cycle by its backlog of work. The negative effects of recession manifest themselves once work in progress has been completed. And so it was with CPBL. Work on the Montage, for instance, continued throughout 2008. Only in 2010 did the effects of the Great Recession hit home. For the forty-seventh year in a row, the company made a profit—but just barely.[5]

Retrenchment, beginning as early as 2008, included freezing salaries, eliminating the company's match of employees' 401(k) plans, cutting executive compensation by 10 percent, and canceling the holiday party. There were "a few" layoffs. Retrenchment also included a reorganization. Under the "One Pankow Initiative," all of the company's subsidiaries were folded into CPBL, with Kim Lum installed as president. The general partnership expanded by two with the additions of Dave Eichten, regional vice president for Southern California, and Senior Vice President Mike Helton. PSPL (Pankow Special Projects, Ltd.) became a division after failing to perform as expected during the downturn. In 2007 its annual volume reached $130 million, up from $100 million in 2000. Early in 2008, Dick Walterhouse stepped down as head of the group to take the position of chief risk officer for CPBL. Though PSPL had thrived during previous recessions, and, indeed, had come into being to counter the effects of recession on CPBL's business, it did not fare well under Helton during the latest one. "I don't know what happened," Kunnath concedes. At the same time, Mid-State Precast emerged during the Great Recession as the company's most successful business, accounting for 50 percent or more of the firm's profits in 2011.[6]

The Birth and Growth of Mid-State Precast

The highly visible and successful demonstration of the PHMRF on the Paramount, Charlie Pankow was confident, would persuade owners to use the frame in their buildings. But they did not do so to the extent that he expected, leaving the company with an underutilized casting yard in Corcoran. To salvage an investment in land and equipment that had been driven solely by the Paramount project, Kunnath convinced Pankow to convert the property into a subcontracting subsidiary.[7]

Charlie Pankow's hopes for the PHMRF were likely encouraged by two projects that CPBL constructed concurrently with the Paramount: Pacific Plaza, in Daly City, a suburb immediately south of San Francisco, and Westside Media Center, in West Los Angeles. For each project, the owners and designers selected the PHMRF because it allowed them to project the image desired by technology-oriented tenants during the height of the dot-com boom. HKS, architects on Westside Media Center, designed interior office spaces as lofts with 14-foot ceilings and exposed columns, beams, and decks. As such, the building team substituted the hybrid frame in place of shear walls "for the moment connections in the longitudinal direction," noted marketing vice president Todd Whitlock. In the case of Pacific Plaza, Joe Sanders explained, "Most of the high-tech tenants wanted exposed structure."[8]

It soon became clear, however, that demand for the PHMRF would not justify the company's Corcoran investment—an investment made, as was the case with each aborted attempt to open a branch office, without a business plan behind it. Kunnath decided to utilize the yard in a subcontracting capacity. The biggest obstacle to embarking on this course was Charlie Pankow, who, as we have seen throughout this book, showed no enthusiasm for initiatives that lay outside the business model on which he had founded his company. As had been the case with PSPL, Kunnath argued that he was merely deploying otherwise idle resources. He also started with small projects that demonstrated the capacity of the yard to function as a subcontractor without disrupting the company's operations. The first such project was a culvert job for Homer J. Olsen, Inc., a contractor based in Hayward, California. That Homer Olsen had worked at Peter Kiewit Sons' and knew Charlie Pankow well likely helped Kunnath "sell" the latter on the idea of establishing a precast operation that would not be wholly dependent on CPBL for its contracts.[9]

Mid-State Precast came into its own as a stand-alone business with the hiring of David B. Dieter as operations vice president and general manager in 2005. Dieter had more than two decades of experience in precasting with Omaha, Nebraska–based American Concrete Products and Dallas-based Enterprise Concrete Products. He was soon promoted to president. Under his direction, Mid-State Precast became the most profitable business in the company. It produced precast elements for structures that utilized the PHMRF, including another parking structure on the campus of Stanford University, a medical office building for White Memorial Medical Center in Los Angeles, and the 11,000-seat, $130 million Citizens Business Bank Arena in Ontario, California. Mid-State Precast could not have succeeded as a business, however, if it had confined its operations to the use that Charlie Pankow had intended for it.[10]

As much as 90 percent of Mid-State Precast's annual sales have derived from producing precast elements for projects that did not involve CPBL as contractor. Dieter developed strong relationships with local processors, resulting in projects such as the Leprino Foods Manufacturing Facility in Lemoore. Other high-profile projects have included the Galen Events Center at the University of Southern California, the Matthew Knight Arena on the campus of the University of Oregon, the Save Mart Arena at Fresno State University, the Construction Innovation Center on the campus of Cal Poly San Luis Obispo, the El Cajon Safety Center, the Willow International Center, Fresno, the Stockton Event Center, the Annenberg Community Beach House on the Marion Davies Estate in Santa Monica, and the British Telecom Parking Lot Tracking System, El Segundo. The latter two projects won the Precast Concrete Institute's Design Award for Best Custom Solution, in 2009 and 2010, respectively.[11]

Kunnath singles out Dieter's leadership as the key to Mid-State's success. Like Charlie Pankow, "Kelly" Johnson, and other leaders of so-called Great Groups, Dieter assembled an exceptional group of highly capable people around him and motivated them to work as a creative team. That he was able to recruit people to the rather inhospitable San Joaquin Valley makes his achievement all the more remarkable. In Kunnath's view, it shows "what one fine leader can do" to get the people to believe in what they can do. Dieter also understands the business and knows how to execute projects for clients, adds the Pankow CEO. For all intents and purposes, the Great Recession did not hit Mid-State's operations, which, as the aforementioned projects suggest, relied heavily on institutional and public sector work.[12]

In fact, such work was the lifeblood of the Pankow firm during the Great Recession. Public sector projects as a share of the firm's order book grew to 80 percent. Institutional clients, such as hospitals, accounted for much of the remaining 20 percent, meaning that the private sector commercial work that Charlie Pankow relied on almost exclusively almost dried up completely. While 2011 was another difficult year, contract volume recovered and, on the strength of Mid-State Precast's performance, recorded a profit. It shows that "we're making it," Kunnath observes.[13]

An Art Collection Funds a Foundation

As the Great Recession has hammered home, the life of a commercial contractor is precarious. Should the company that Charlie Pankow founded cease to exist, professional awards, endowed chairs, and two concrete laboratories at Purdue University will carry on the founder's name and professional reputation. The Charles Pankow Foundation will do so as well, despite the fact that it was almost an afterthought.

Charlie Pankow began to accumulate art as a serious collector in the mid-1970s. He did not know much about art, but he knew what he liked and appreciated the quality and singularity of craftsmanship. For instance, on a trip to Russia, he developed an instant liking of the icons of the Orthodox Church and so, upon his return, acquired half a dozen pieces. He was also fascinated by the Treasures of Tutankhamun, an extraordinarily popular exhibit that toured America for almost three years, beginning in November 1976. Collecting art neatly dovetailed with his love of travel. He may have enjoyed shopping for art as much as he did displaying it. Over a period of some 15 years, Pankow collected about 1,200 pieces in five areas: Asian art, including Chinese bronzes and ceramics, Cloisonné vases, and Japanese woodblock prints; Egyptian antiquities; Pre-Colombian art; Russian and Greek icons; and Impressionist and Post-Impressionist paintings. He spent upwards of $10 million, but many items in the collection, the icons in particular, appreciated in value. He became the largest private collection of icons in the Western Hemisphere. Still, Pankow collected art more for its psychic, than for its financial, benefits. He exhibited parts of the collection at the University of Hawaii, Purdue University (on three occasions), the Triton Museum of Art, Santa Clara University, and the San Diego Museum of Art. He rarely sold an item once he acquired it. As he assembled his collection, it probably did not cross his mind that he might sell it at auction to generate funds to establish a foundation, even one created for the purpose of advancing innovation in building design and construction. Yet, in the end, this is exactly what he did.[14]

In 1982 Charlie Pankow formed Washington Street Associates, a limited partnership, to acquire a mansion in Presidio Heights, a small but affluent neighborhood squeezed between the Presidio and the Laurel Heights district in San Francisco. The property was Le Petit Trianon, a classical revival replica of a mansion constructed for Louis XV at Versailles in the 1760s. Built in 1904 by a local wool merchant, the structure was "in an accelerating state of decay" when Pankow bought it for $1.5 million. He restored the structure to its original grandeur to showcase his art collection.[15]

By the mid-1990s, Charlie Pankow was contemplating the idea of leaving the mansion and the collection to his alma mater. As an important donor to Purdue engineering, he developed a friendship with Dr. Steven C. Beering, who had become president of the university in 1983. Beering was born in Berlin in 1932—the last year of the Weimar Republic. He was the first of two sons in a merchant family that did business in Berlin and Hamburg. His family tried unsuccessfully to flee Hitler's Germany in 1938, following the *Anschluss*—the annexation of Austria. The Allied bombing of Hamburg split the family, as Dr. Beering's father was in Berlin at the time. Reunited only at war's end, the family made its way to Pittsburgh—where Beering's maternal grandparents lived—by way of London, where the family also had relatives. Steven Beering graduated *summa cum laude* from the University of Pittsburgh and then entered Pitt Medical School. In his last year of his studies, Beering joined the US Medical Corps—and got married. During his 12 years with the Medical Corps, Beering distinguished himself as director of the internal medicine program at the Wilford Hall Medical Center in San Antonio, Texas, and as consultant to the US Surgeon General. In 1969 he left the Medical Corps to begin an appointment as professor of medicine and assistant dean at the Indiana University Medical Center. He later served as the center's dean of medicine and its director. Not long into Beering's presidency, Dean of Engineering Henry T. Yang introduced him to Charlie Pankow.[16]

Charlie and Doris Pankow entertained the Beerings many times at Le Petit Trianon. On one occasion, as Beering tells it, Pankow told him, "I'd like to give you this building and its contents, and after I'm no longer around, I want you to use that for Purdue University in any instructional way that the liberal arts people can use it." An enthusiastic supporter of liberal arts, who had focused much of his efforts as president on raising Purdue University's profile as a comprehensive institution, Beering imagined using the collection as the basis for a master of arts program. Pankow discussed the idea with CFO Tim Murphy, who also handled his estate, and others about how best to realize the goal. Ultimately, Pankow concluded an agreement with Purdue University whereby he agreed to donate a share of the limited partnership's interest in the mansion with the understanding that he would gift the collection upon his death. Judy L. Vawter, whom Charlie Pankow hired in 1989 to curate his collection, would continue in that role and also administer the program as an adjunct professor. The program was launched before Dr. Beering stepped down as university president in 2000. "I had high hopes for [it]," he reflects.[17]

Dr. Martin Jischke, Beering's successor, had other ideas for the collection, however. Emphasizing Purdue's traditional areas of strength in engineering, the new university president saw Charlie Pankow's gift more in monetary than liberal arts educational terms. Jischke made his views known in December 2000. Purdue was scheduled to play in the Rose Bowl—its first appearance in 33 seasons. As might be expected after such a long drought, the office of the president sent a large delegation to California, including both Beering and Jischke. As part of their trip, the ex-president insisted that his successor meet Charlie Pankow. The Pankows hosted a dinner at Le Petit Trianon. Early in the evening, Martin Jischke informed Pankow that he did not see how the nascent arts program fit into Purdue's mission as an engineering institution. He wondered if Pankow would object to the university selling the property and the collection, and devoting the proceeds to engineering programs and research. Pankow was taken aback. He explained that there was value in keeping the property and the entire collection together. He offered to write a check, if money was what the university needed. For his part, Beering was "literally speechless." Disenchanted with Martin Jischke's proposal, Pankow reversed course. He could not recover the share in the partnership that he had given to the university, but he withdrew the gift of his art collection. The arts program terminated. Laments Steven Beering: "In all these years I've been at Purdue, there's not a single event that can match my disappointment with this one."[18]

In the wake of Pankow's decision to retain the art collection, Rik Kunnath developed the idea for a foundation after he read Michael E. Porter and Mark R. Kramer's *Harvard Business Review* article, "The Competitive Advantage of Strategic Philanthropy."[19] Struck by the computer science program that Sun Microsystems had established for minority students in community colleges, Kunnath began to think about how a Pankow foundation might help the company leverage its capabilities and relationships. He passed the article to Charlie Pankow. The ideas of professors Porter and Kramer initiated a conversation that led in short order to the establishment of the Charles Pankow Foundation. The financial burden of developing the PHMRF formed the backdrop to the thinking of Kunnath and Pankow as they discussed strategic direction. A properly funded and organized foundation, they agreed, would be able to support building systems research more effectively than CPBL could hope to do on its own.[20]

Modestly funded at the outset, the sale of Charlie Pankow's art collection at auction by Sotheby's in 2004 put the Foundation on a solid footing. The gift was wholly unexpected. "[We] were surprised to realize that he had left a very significant piece of his estate to funding the Foundation. [That] gave us the resources to execute the ideas that [we were talking about] in a way that I don't think [we] would have ever dreamed about," says Kunnath. The Foundation's initial endowment of $300,000 was insufficient to support even one major research effort. With the proceeds from the collection, the Foundation was able to fund more than $4.6 million in research from 2006, when it began to award grants, through 2009, and can fund research at these levels, without raising additional funds, in perpetuity.[21]

The Foundation was established before thinking about its mission had crystallized. Pankow and the other three board members, Kunnath, Murphy, and Vawter, considered funding research, but the founder also saw his foundation as the conduit through which support for civil engineering and construction management education would flow. He remained interested in ensuring that university programs prepared potential recruits. Keeping in mind the analysis and assessments of Michael Porter and Mark Kramer, Pankow and the board considered other ideas that would satisfy the rules that governed charitable giving, "but simultaneously could be beneficial and helpful to the prospects of the company." With Pankow's passing in January 2004, it fell to Rik Kunnath to author the founding precepts on which the Foundation's grant-making goals would be based. In 2005 the board distilled Charlie Pankow's ideas into a statement that defined the mission of the Foundation as an institution dedicated to "advanc[ing] innovations in building design and construction, so as to provide the public with buildings of improved quality, efficiency, and value." Grant-making purposes coalesced during 2008 around two areas, namely, structures and project team tools and practices. The first aimed to improve the quality of large buildings "by advancing codifiable [sic] innovations in structural components and systems." As we have seen, Charlie Pankow preferred large commercial projects, especially those executed in concrete. The second area sought to meet the Foundation's mission by advancing "integration, collaboration, communication, and efficiency" among members of the building team and "new means and methods for project team practices." Decoded, this strategic statement may be seen as an embodiment of the spirit, if not the practice, of design-build as a project delivery methodology. Together, grant making aims to advance Charlie Pankow's ideas on bringing value to the building process and improving the productivity of an industry that continues to be characterized by waste and inefficiency and a systemic lack of demand for innovation.[22]

The Foundation thus represents the legacy of Charlie Pankow apart from the company that he founded. Kunnath considers the Foundation to be the "perfect" vehicle for advancing innovation in the industry. Pankow "would never accept that we couldn't do better the next time, that there wasn't a better idea." The research that the Foundation underwrites, Kunnath contends, will "move the industry forward with better ideas, more efficient buildings, betters tools and technology, [and] better education." As a private institution promoting research that will produce nonproprietary technologies of benefit to the industry and ultimately to the general public, the Charles Pankow Foundation is animated by the culture that spurred the company to participate in the development of the PHMRF. Kunnath concludes: "That really encapsulates [Charlie Pankow's] legacy and who he was."[23]

CONCLUSION

Charlie Pankow assembled a capable group of self-starters in a building division within the Los Angeles District of Peter Kiewit Sons'. Under his leadership, they became experts in concrete construction, and, as contractors, willing and able to assume sole responsibility for project execution under a building team configuration that became known as design-build. Most of these men followed Pankow when he incorporated his eponymous firm. As a stand-alone organization, Charles Pankow, Inc., played an important role in restoring the master builder to the commercial building site. Animating the organization was the Pankow Way: a culture forged within the Kiewit building division that valued creativity, teamwork, and client service. Diffusion of design-build was slow, but five decades after the founding of CPI, design-build achieved parity with design-bid-build as a project delivery system.

The key to the Pankow company's success lay in keeping its clients satisfied. For happy owners were repeat customers. In "A Message for Potential Clients," which introduced the company's mid-1980s marketing brochure, Charlie Pankow wrote that he started his company in the singular belief "that success can be achieved in the construction industry by applying management innovation, engineering creativity, strict cost and quality controls, and state-of-the-art construction techniques."[1] But these were means to the end. In fact, as Robert Law notes, once any discussion with an owner turned to design-build or techniques that "automated" concrete construction, "you [would] immediately see their eyes glazing over."[2] Owners paid far more attention when Charlie Pankow assured them that his company's professionals would solve their problems as they arose; not generate change orders late in the construction cycle; do whatever was necessary to finish their projects on time and within budget; and not walk away from the building site until they had addressed all of their issues. Execution risk might be quantified in the contract, of course, in terms of bonuses, penalties, and the like. But Charlie Pankow and those he led impressed owners by the way in which they met or exceeded their expectations. Owners, in turn, extolled their experiences with Pankow to their fellow developers within local, regional, and national networks. At the heart of Charlie Pankow's legacy and the reputation of his company lay owner satisfaction.

The Pankow firm was midwife to the dreams of architects that otherwise might have been left on the drawing board. Involving the contractor in pre-construction planning typically made designs easier and less costly to construct. The cost of money and labor aside, the traditional manner in which designers conceived and elaborated projects could add 15 to 20 percent, and sometimes more, to the final cost of a project, rendering it financially infeasible, according to Pankow.[3] Making a project viable at the end of its design under design-bid-build often involved "value engineering," which invariably diminished the value of the design and quality of the project. Architects eager to safeguard their leading role on the building team generally resisted design-build. But those who worked with Pankow project teams on design-build assignments generally appreciated its benefits and embraced the approach, if they had not done so already.[4] For Pankow as contractor enabled designers to express their passions and ambitions and have them materialized in the final product.

In finding ways to make projects economical without compromising either inspiration in design or integrity in construction, the Pankow firm helped to shape urban spaces. The owner, of course, conceives a project and is responsible for financing it. And the architect is responsible for representing the owner's vision in a design. But as a participant in all phases of the project, the design-build contractor advises owner and architect and ensures that the project is built as envisioned. Charlie Pankow may not have devoted much attention to the cultural and social significance of the structures that he built. He may not have shared the concerns of planners who, for instance, sought to restore the Bennett Plan for Pasadena's civic center. He may not have been a champion of infill and transit-oriented development. He certainly did not concern himself with the economic impact of regional shopping centers on nearby central business districts. He played no role in developing urban renewal policy or setting height limits on buildings in downtown San Francisco or Honolulu. And while he might appreciate the architectural (or, more broadly, design) significance of a building that he constructed, he did not pause to construct purely functional buildings that garnered no critical accolades, or, indeed, were subject to negative criticism. As former CFO Tim Murphy notes, Charlie Pankow "was a builder, and he pursue[d] jobs that he thought were profitable and made sense for the company."[5]

Still, in collaboration with owners and leading designers, the Pankow firm adapted the regional shopping center to demographic and cultural change. It pioneered mixed-use residential and office projects in Hawaii. It completed residential over retail projects in Hawaii and California that represented the latest thinking of the so-called New Urbanism. It constructed office towers that reshaped the skylines of Honolulu, Los Angeles, Oakland, and San Francisco. It erected condominiums, hotels, and shopping centers that helped to make Waikiki the travel destination it remains today. It constructed residential towers that were part of the redevelopment of the Western Addition and South of Market (SOMA) districts in San Francisco. All of these projects were complicated affairs, involving a variety of actors who together resolved a complex set of planning, design, and construction problems in realizing them. But for the contractor, many of them would not have been built. Because Pankow project teams sought, and found, better ways to build them, they were completed. A few projects, such as MacArthur Broadway Center, have been demolished. Several others, such as Chase Plaza and Penn-Can Mall, have been converted to other uses. The majority of Pankow projects remain part of the built environment, however, and are devoted to the uses intended by their developers. Each represents an intersection of theoretical and practical ideas on organizing cultural, economic, political, and social life specific to their time and place.

Whenever possible, Pankow as contractor used design-build to deliver these commercial projects. The company guaranteed the cost of a project and eliminated designer- and contractor-generated change orders. For its part, the company did not employ designers and structural engineers. Rather, it assembled building teams to meet the requirements of particular projects. Nevertheless, it promised owners that design-build would eliminate adversarial profit incentives and issues related to performance on the part of the contractor and its subcontractors. Early involvement allowed Pankow to make these promises, for it allowed the company's engineers to advise the other members of the building team on the structural systems and other areas wherein lay the material costs of a project. As contractor, Pankow did not simply adhere to working drawings handed to it, but collaborated with architects and structural engineers, using the owner's aesthetic, functional, and quality requirements and expectations, documented in preliminary drawings and specifications, as points of departure. In this context, the company argued that design-build "tends to reverse the conventional attitude where the cost [of a project] is the result of the plans to a more practical method where the plans are the result of cost discipline."[6] That owners often turned to Pankow only after sinking thousands, hundreds of thousands, or even millions of dollars into the development of designs that could not be built within their budgets supported the argument. In advocating design-build, Charlie Pankow sought to restore the role of the master builder to the commercial building site. The idea was woven into the fabric of the Pankow Way. Yet until the late 1980s, at least, what seemed obvious to Charlie Pankow seemed to be lost on the building community.

No less important to Charlie Pankow was "automating" the commercial job site. The "staggering increases in craft wages that caused building costs to skyrocket at a rate far in excess of the economy as a whole" was never far from his mind, as this quote from the company's earliest marketing brochure suggests.[7] Conceding that the building site could not quite replicate the factory, Pankow nonetheless held out the manufacturing assembly line as the ideal configuration of construction work.

As the completion of myriad shopping center projects showed, successful design-build delivery did not depend on the material used to frame the structure. Executing projects in concrete, however, allowed Charlie Pankow to control construction costs and schedules. Structural steel beams and columns had to be manufactured by other companies, often at great distances from the job site. Orders were subject to long lead times under the best of circumstances. Particularly in the early years of Charlie Pankow's career, the best of circumstances rarely prevailed in the steel industry. The industry was an oligopoly. A contractor could not easily turn to alternative sources if he found himself at the end of a production queue. Strikes were common during Pankow's years with Peter Kiewit Sons'.[8] The possibility of a strike was ever present, too. And so Charlie Pankow felt that he could not rely on timely deliveries of structural steel. With the precast concrete industry in its infancy in the 1950s, Pankow directed his Kiewit project teams to set up casting yards at or near the building site. His Kiewit and Pankow teams used slipforms, fly-forms, jump forms, and techniques, such as using precast concrete wall panels as forms for beams and columns and incorporating vertical airshafts into a building's concrete core, to save additional time and labor. Pankow distinguished itself from other contractors by self-performing this work. It did so to control costs and quality. Of course, the flexibility of concrete as a material accommodated the aesthetic visions of designers and functional needs of structural engineers. But above all, using concrete enabled Pankow to guarantee project costs and schedules more confidently.

Even as it sought ways to reduce the quantity of labor it needed to complete a project, Pankow relied on unionized workers to ensure the quality of the final product. A sufficient number of trades workers might not always be available, as George Hutton found to his frustration when he arrived in Honolulu for the Campbell Building project; as several Pankow teams experienced on Winmar projects around the country; and as Tom Verti, Chris Turner, and Joe Sanders discovered when they tried to recruit crews to build the Paramount. Yet in general, Pankow project teams found that union halls were the best place to secure the labor they needed. Employing unionized carpenters and other trades workers helped Pankow project teams meet their productivity goals.

Charlie Pankow's legacy is inexorably linked to the firm that bears his name. Though he started Charles Pankow, Inc., with several colleagues in the Kiewit building division, there was never any question in his mind that it was *his* firm. This mentality framed his thinking on employee ownership, corporate governance, management structure, organizational decision making, and strategic direction—in short, all of the issues that mattered to the long-term viability of the enterprise. During his career with Peter Kiewit Sons', Pankow developed a business model that he applied successfully within a Kiewit organization that resisted it, and then he applied it on his own. Owing to a great extent to the development efforts of George Hutton and Russ Osterman, two of the men who followed Charlie Pankow out of the Kiewit organization, the model delivered profits year after year and made Pankow a wealthy man. The way in which Pankow ran his company, however, made it vulnerable to changes in the condition of the commercial real estate market. Ironically, adhering to the model that had made Pankow a successful contractor in all likelihood would have doomed the company, if not in Charlie Pankow's lifetime, then as a result of the Great Recession.

Charlie Pankow was a leader, not a manager. The managerial structure of the company, such as it was, was not decisive in the success of the company. How, then, did the company manage growth? Charlie Pankow ran his company much like a holding company, not unlike an oil company with a portfolio of producing or potentially producing properties. He (and Russ Osterman, George Hutton, and, on occasion, others, such as Jon Eicholtz) sold the work and then delegated project execution to capable self-starters. Each project was organized essentially as a discrete entity. The company was an aggregation of projects, each managed by a well-defined team. Pankow determined the growth of his company by the number of projects that he approved. The approach was wholly tactical. It assumed that the business model could be sold to owners, regardless of market conditions. Pankow did not engage in strategic planning (for that matter, neither did George Hutton). Unlike with the typical large oil company then, approving a construction project was not linked to capital budgeting. But bonding capacity and a financial return appropriate to the risk that the contractor was accepting were key considerations. Charlie Pankow's business model, including the financial criteria embedded within it, was conservative. The company remained a niche contractor, doing business for relatively few repeat customers.

Making the business work required the recruitment and retention of talented engineers who could operate in a platoon-like environment to complete projects to the satisfaction of owners. Charlie Pankow acted as a general who, through force of character and personality, inspires those who he commands to find ways to complete missions successfully. He set expectations. Project teams invariably met them. As Tom Verti reflects, "You just wanted to be a part of his winning team."[9]

Loyalty was a many-layered attribute of Charlie Pankow's complex personality. Because he conflated self with firm, his responses to perceived disloyalty on a personal level resulted in fractured relationships. In particular, Pankow was reluctant to allow individuals with ownership stakes to leave the company. He equated the decision to leave the firm with disloyalty to him. The departures of Bob Carlson, David Boyd, Rosser Edwards, George Hutton, and Russ Osterman—all of whom followed Charlie Pankow from the Kiewit organization and believed in his business philosophy—were particularly contentious. Often at issue was the value of their shares in the company.

In Osterman's case, the matter went to arbitration, and Osterman prevailed. Among principals with significant stakes in the company, only Dean Stephan, citing health reasons, was allowed to retire without a fight. Employee loyalty was crucial to the success of the company. How Charlie Pankow handled the departures of colleagues suggests the extent to which he ran his company as the first generation founder of a family firm, even as he acted to exclude his children from his succession plans.

The issue of loyalty may have limited the growth of the firm. Charlie Pankow's efforts to rein in the Hawaii operation after George Hutton succeeded spectacularly in growing it independently of Altadena may be understood as a test of loyalty. At least Hutton thought so. From his point of view, the actions aimed at "emasculating" the Honolulu office proved that Pankow saw the company as solely *his* firm. To Hutton, Pankow was demanding subservience, not loyalty, from him. However sound the business reasons for the transfer of much of the staff from Hawaii to California during the 1980s may have been, it is clear that the reallocation of personnel among offices prompted Hutton to think differently about his role in the firm long before he left it. He would no longer have his team—his loyal team that he had handpicked—in place when the market rebounded. When the market rebounded in the late 1980s, things *were* different for Hutton, as he expected. Charlie Pankow's decision to decline the Aloha Tower construction contract after the project's partners excluded him from the development deal left money on the table. Perceptions of disloyalty may have shaped Pankow's thinking. There is no way of knowing how much work Hutton may have generated had he not retired in 1991. (He had at least a decade left in him.) The extent to which loyalty shaped relations between Pankow and Hutton cannot be pinpointed. At the same time, it is not clear that Charlie Pankow considered only business criteria on matters affecting Hutton and his office. If Pankow, in fact, acted to "kill the goose that laid the golden egg" in Hawaii, as Hutton suggests, because he interpreted Hutton's independence as a kind of disloyalty, then loyalty came at a price. The founder was willing pay to the price, however, in order to stamp his authority on the business.

Charlie Pankow retreated from the daily operation of the firm convinced that his business model was robust and impervious to external market conditions. But for the development efforts of Hutton and Osterman and the repeat business of Winmar, the limitations of the model would have been exposed before he did so. Current CEO Rik Kunnath recognized this, even if Charlie

Pankow did not (or would not). The resistance that Kunnath subsequently faced in diversifying the business and the manner in which he overcame it shows how invested a successful founder can become in his or her way of doing business. It is a testament to Charlie Pankow's business acumen that he decided to leave ownership of his company in the hands of employees, and therefore management of his company in the hands of professionals, given the decline that may be expected in the performance of second generation family firms. (Whether this would have been an exception in the case of his family will never be known. What is known is that none of the children expressed an interest in running the business. But ownership interests with voting rights attached may have complicated the succession.) Even so, the most senior of Charlie Pankow's managers could not convince him that the survival of his company demanded strategic redirection. Had Special Projects not succeeded before Charlie Pankow died, the principal vehicle of his legacy may well have not outlasted him.

APPENDIX A

Major Projects Completed on the Mainland, 1963—2004, and in Hawaii after the Retirement of George Hutton, 1992—2004

Project	Location	Year Completed	Description
MacArthur Broadway Center	Oakland, CA	1965	6-story office building atop shopping center; 400,000 sf; demolished 2009
Central Towers	San Francisco	1964	known within the company as "Turk & Eddy": twin 15-story towers, each containing 181 units and ground-floor retail; 300,000 sf
Montgomery Ward	Stockton, CA	1965	200,000 sf department store; anchored the north end of the 500,000 sf Sherwood Manor shopping center
Las Flores Area	Camp Pendleton, CA	1966	barracks and chapel
1625 The Alameda Building	San José, CA	1965	9-story office building; 100,000 sf
Dean Witter Building	San José, CA	1966	20,000 sf office building
Borel #1	San Mateo, CA	1966	5-story office building; 68,000 sf (net leasable area)
Borel Square	San Mateo, CA	1968	"Town & Country" style shopping center; 80,000 sf (net liable area)
Borel #2	San Mateo, CA	1969	3-story office building; 40,000 sf (net leasable area)
Borel #3	San Mateo, CA	1970	6-story office building; 95,000 sf (net leasable area)
Borel #5	San Mateo, CA	1970	92,000 sf office building
First American Building	San José, CA	1967	130,000 sf office building

Project	Location	Year Completed	Description
IBM Building	Menlo Park, CA	1967	45,000 sf office building
Joe West Hall, San José State College (University)	San José, CA	1967	12-story residence hall for 650 students and 2-story dining facility; 150,000 sf
May D & F	Denver, CO	1967	155,000 sf department store
Oxford University Press	Fairlawn, NJ	1968	40,000 sf office building
Winmar Building	Bellevue, WA	1968	13-story office building; 225,000 sf
McCurdy's	Rochester, NY	1969	180,000 sf department store
Sixth & Harvard Building	Los Angeles	1969	5-story office building with grade parking; 80,620 sf
Clementina Towers	San Francisco	1970	twin 13-story towers; 276 units total; 170,000 sf
Evergreen Building	Renton, WA	1970	98,000 sf office building
Holiday Inn	Toledo, OH	1970	19-story hotel with 13-story parking structure; 200,000 sf
Lafayette Park Place	Los Angeles	1970	123,000 sf office building
Oxmoor Center	Louisville, KY	1971	400,000 sf regional shopping center
San José Plaza I	San José, CA	1971	14-story office building; 224,000 sf
San José Plaza II	San José, CA	1973	13-story office building; 230,000 sf
Santa Ana Building	Santa Ana, CA	1971	184,000 sf office building

Project	Location	Year Completed	Description
Citizens Fidelity Bank Building	Louisville, KY	1972	30-story office building; 700,000 sf
City of Ventura Parking Structure	Ventura, CA	1972	4½-story, 569-stall municipal parking structure
First Security National Bank & Trust Building	Lexington, KY	1973	15-story office building; 352,000 sf; 217,728 sf of floor space
Kaiser Permanente Parking Structure	Oakland, CA	1973	540-stall parking structure
New England Telephone & Telegraph	Braintree, MA	1973	60,000 sf office building
Citizens Bank Building	Eugene, OR	1974	10-story office building; 160,000 sf
Kaiser Permanente Parking Structure	Walnut Creek, CA	1974	
Jefferson Plaza	Spokane, WA	1974	4-story office building; 70,000 sf
Pacheco Village	Novato, CA	1974	180-unit low-rise apartment complex; 185,000 sf
Washington Square	Tigard, OR	1974	constructed J. C. Penney's, Lipman-Wolfe, and Nordstrom's, department store anchors for this regional shopping center
Bristol Town & Country	Santa Ana, CA	1975	160,000 sf shopping center; 11,700 sf medical clinic; 37,000 sf medical building
Las Flores Area	Camp Pendleton, CA	1975	marine barracks complex; 280,000 sf

Project	Location	Year Completed	Description
PT&T	San Francisco	1975	7-story office building; 410,000 sf
Chappo Area	Camp Pendleton, CA	1976	bachelor enlisted quarters; 155,000 sf
Parking Structure "A," University of Southern California	Los Angeles	1976	4-level, 980-stall parking structure
Parking Structure "B," University of Southern California	Los Angeles	1976	4-level, 980-stall parking structure
Parking Structure "A" Addition, University of Southern California	Los Angeles	1978	3-level addition to 4-level parking structure; 762 stalls
Penn-Can Mall	Clay, NY	1976	420,000 sf regional shopping center
AT&T	San Francisco	1977	6-story office building; 210,000 sf
South Shore Plaza	Braintree, MA	1977	enclosure and renovation of regional shopping center opened in 1960 and expanded in 1962 and 1967
South Shore Plaza	Braintree, MA	1979	expansion of regional shopping center with parking structure and Sears store
Capitol Court	Milwaukee, WI	1978	enclosure and expansion of regional shopping center opened in 1956 and expanded in 1960 and 1966; demolished 2001

Project	Location	Year Completed	Description
Grove Street Parking Garage	Boise, ID	1978	5-level, 547-stall municipal parking structure
Lions Manor	Monterey Park, CA	1978	6-story, 126-unit seniors apartment building; 127,410 sf
Portland West Parking Structure	Portland, OR	1978	4-level, 480-stall municipal parking structure
Fairmount Terrace II	Los Angeles	1979	6-story, 100-unit senior citizens residence; 118,000 sf
Senior Citizens Housing of Fontana	Fontana, CA	1979	8-story, 151-unit apartment building; 137,000 sf
J. C. Penney's	Tigard, OR	1980	230,000 sf department store, addition to Washington Square shopping center
Pacific First Federal Center	Portland, OR	1981	16-story office building with 2½ levels of underground parking; 294,000 sf
10560 Wilshire	Los Angeles	1982	22-story, 108-unit condominium with 4 levels of parking; 467,000 sf
Crocker Plaza	Long Beach, CA	1982	14-story office building; 327,000 sf
Kaiser Permanente Parking Structure	Hayward, CA	1982	580-stall parking structure
Walnut Center	Pasadena, CA	1983	7-story office building for Kaiser Permanente; 750,000 sf
411 East Wisconsin	Milwaukee, WI	1984	30-story office building; 2-story bank building; 9-story, 915-stall parking structure
2101 Webster Street	Oakland, CA	1984	20-story office building; 535,000 sf

Project	Location	Year Completed	Description
Catalina Landing	Long Beach, CA	1984	three 5-level and one 4-level office buildings with 3-level, 1,500-stall parking structure; 740,000 sf
Oxmoor Center	Louisville, KY	1984	90,000 sf expansion of regional shopping center
Capitol Court	Milwaukee, WI	1985	renovation of Gimbels department store for Target; BATUS, owners of Gimbels since 1973, had closed the store in 1984
Chase Plaza	Los Angeles	1985	22-story office building with more than 460,000 sf of space with 10-level, 920-stall parking structure
Montebello Town Center	Montebello, CA	1985	488,000 sf regional shopping center
Kaiser Permanente Parking Structure	Walnut Creek, CA	1986	5-level, 540-stall parking structure adjoining structure completed in 1974
Nissan	Wilmington, CA	1986	135,000 sf distribution and auto service center
South Coast Plaza II Parking Structure	Costa Mesa, CA	1986	2,308-stall parking structure built in conjunction with expansion of South Coast Plaza, a regional shopping center
YMCA	Oakland, CA	1986	70,000 sf fitness center and 700-stall elevated parking structure
General Telephone Company	Thousand Oaks, CA	1987	annex building to GTE headquarters; 225,000 sf
Hotel Sofitel	Redwood City, CA	1987	329-room luxury hotel and 350-stall parking structure; 225,000 sf

Project	Location	Year Completed	Description
South Coast Executive Centre	Costa Mesa, CA	1987	twin-tower office complex (320,000 sf) and 500-stall parking structure
Valley Fair Mall	Salt Lake City, UT	1987	renovation (80,000 sf) and expansion of food court (20,000 sf) for regional shopping center opened in 1970
Marathon Plaza	San Francisco	1988	twin-tower office complex; 900,000 sf
Shoreline Square	Long Beach, CA	1988	15-story, 500-room Sheraton Hotel; 21-story office building; 1,400-stall parking structure; 30,000 sf retail space; 1.25 million sf total
Westminster Mall	Westminster, CA	1988	addition of food court to regional shopping center; 23,000 sf
Clarion	Gardena, CA	1989	3-story office building and 72,000 sf warehouse
Brea Mall	Brea, CA	1990	renovation of and food court addition to regional shopping center; 4 multi-level parking structures with more than 4,300-stalls
Resort at Squaw Creek	Lake Tahoe, CA	1990	4-story resort plaza complex (150,000 sf); 9-story, 405-room hotel (300,000 sf); skiing and ice skating facilities, golf course, swimming pools, water slide
La Tour (10380 Wilshire)	Los Angeles	1991	21-story, 73-unit luxury condominium; 392,000 sf
Kaiser Permanente Parking Structure	Martinez, CA	1991	4-level, 714-stall parking structure

Project	Location	Year Completed	Description
Kaiser Permanente Parking Structure	San Francisco	1991	7-level, 533-stall parking structure with storage and retail space
Tyler Mall	Riverside, CA	1991	renovation of and addition of second level to regional shopping center; 930-stall parking structure; rechristened The Galleria at Tyler; 662,805 sf
Roosevelt Field Mall	Garden City, NY	1993	addition of second level to and renovation of regional shopping center opened in 1955 and enclosed in 1968
Santa Rosa Plaza	Santa Rosa, CA	1993	renovation of common area and food court of regional shopping center; 127,000 sf
Special Events Center, University of Hawaii, Manoa	Honolulu	1994	10,300-seat multipurpose arena; named for Stan Sheriff, the athletic director who advocated its construction
Gateway Center	Los Angeles	1995	2.2 million sf transportation complex constructed at Union Station, featuring East Portal ticketing terminal; 28-story office building for the Metropolitan Transportation Authority
Meadows Mall	Las Vegas	1995	renovation of regional shopping center; 890,000 sf
Boalt Hall School of Law	Berkeley, CA	1996	renovation and expansion of Boalt Hall and Simon Hall, an adjacent dormitory, on the University of California campus

Project	Location	Year Completed	Description
Citizens Parking Structure	Eugene, OR	1996	3-level, 183-stall parking structure built next to Citizens Bank Building constructed by CPI in 1974
Clackamas Town Center	Portland, OR	1996	renovation of shopping center; 125,000 sf
Glendale Galleria Parking Structure	Glendale, CA	1996	replacement of 2,906-stall parking structure damaged in the Northridge earthquake with parking structure completed in three phases; 725,000 sf
Pearlridge	Honolulu	1996	renovation of regional shopping center opened in 1972 and expanded in 1976; 116,000 sf
Pearl Street Parking Structure	Eugene, OR	1996	4-level, 264-stall municipal parking structure; utilizes the PHMRF
Roosevelt Field Mall	Garden City, NY	1996	"Phase III" renovation of and addition of 210,000 sf of retail space to regional shopping center; construction of 3 parking structures with more than 5,300 stalls, utilizing the PHMRF
Sheraton Harbor Island Resort	San Diego	1996	renovation of East Tower; 718,000 sf
Sherman Oaks Fashion Square	Los Angeles	1996	renovation of regional shopping center opened in 1962; 972,000 sf
Kapalua Bay Hotel	Kapalua, Maui	1997	renovation of 194 rooms; 204,253 sf
The Orchid at Mauna Lani	Kohala Coast, Island of Hawaii	1997	renovation of luxury resort; 287,000 sf

Project	Location	Year Completed	Description
CarBarn Parking Structure	Los Angeles	1998	1,459-stall parking facility providing off-site parking and shuttle service to LAX airport; 549,000 sf
Methodist Hospital of Southern California	Arcadia, CA	1998	5-story Nor and Fran Berger patient tower; 155,146 sf
Metropolitan Water District	Los Angeles	1998	12-story MWD headquarters building (536,000 sf) built over 768-stall underground parking structure
Montebello Town Center	Montebello, CA	1998	renovation of regional shopping center; 140,000 sf
Walt Whitman Mall	Long Island, NY	1998	renovation and expansion of regional shopping center; 112,000 sf
Outrigger Royal Waikoloan Hotel	Waikoloa Beach, Island of Hawaii	1999	demolition and renovation of hotel completed by CPA in 1981; 43,000 sf
Stanford Shopping Center Parking Structure	Palo Alto, CA	2000	3-level, 1,504-stall parking structure; utilizes the PHMRF
The Aurora	San Francisco	2000	8-story mixed-use building, including ground-floor retail, 2 levels of parking, and 5 levels of residential space; 263,510 sf
Paseo Colorado	Pasadena, CA	2001	2-level retail and entertainment center with 400 upscale residential units
Pacific Plaza	Daly City, CA	2001	10-story office building (400,000 sf) and plaza with two 7-story parking structures; uses the PHMRF

Project	Location	Year Completed	Description
The Paramount	San Francisco	2001	39-story residential building with 485 units; residential tower over 8 stories of parking, retail, business, and recreational space; utilizes the PHMRF; 440,000 sf
Provail	Seattle, WA	2001	seismic and structural upgrade of existing office building; construction of 2-story office building and parking structure
Yerba Buena Lofts	San Francisco	2001	10-story, 200-unit apartment building
555 City Center	Oakland, CA	2002	21-story office building; 487,000 leasable sf
Hawthorne Place at 77 Dow	San Francisco	2002	14-story, 83-unit loft apartment building; 116,000 sf
Stanford University Parking Structure	Palo Alto, CA	2002	4-level, 1,508-stall parking garage; utilizes the PHMRF
Westside Media Center	Los Angeles	2002	4-story office building over 2 levels of parking; utilizes the PHMRF; 160,000 sf
Sunset + Vine	Los Angeles	2004	5-story retail, residential, and parking building with 100,000 sf of ground floor retail space; adaptive reuse of TAV Studios

NOTE: Unless otherwise noted, square feet (sf) measured in gross. Excludes projects completed by Pankow Special Projects, Ltd.

APPENDIX B

Projects Completed in Hawaii under George Hutton

Project	Year Completed	Description
James Campbell Building	1967	6-story office building; 175,000 sf
Hawaii Kai Office Building	1968	15,000 sf office building
100 Ward Plaza	1969	130,000 sf office building
Hilo Mall and J. C. Penney's, Hilo	1970	160,000 sf shopping center; 180,000 sf department store
Long's Drug Store, Hilo	1970	160,000 sf retail store
T. Yokono Office Building	1970/1	30,000 sf office complex; built in 2 phases
Bank of Hawaii Building, Hawaii Kai, Oahu	1970	20,000 sf office building
Kauluwela Co-op	1970	21-story, 126-unit apartment building; 126,054 sf
Koko Marina Trade Center, Hawaii Kai, Oahu	1971	50,000 sf shopping center
Kauluwela Low-Rise	1971	fourteen 3-story apartment buildings; 84 units, 107,500 sf
250 Ohua	1971	16-story, 96-unit condominium; 168,600 sf
Kauluwela Elderly Housing	1971	22-story, 175-unit apartment building; 144,000 sf
Kalani Village, Wahiawa, Oahu	1972	three 3-story, one 8-story apartment buildings; 109,100 sf
The Esplanade, Hawaii Kai, Oahu	1973	9-story, 209-unit condominium; 487,000 sf

Project	Year Completed	Description
Haleloa	1973	seven 2-story condominiums; 52 units, 68,100 sf
Kaimana Lanais	1974	24-story, 114-unit condominium; 161,634 sf
Hawaii Baptist Academy	1975	79,000 sf dormitory
Pearl One	1975	22-story, 301-unit condominium and parking garage; 408,385 sf
Waianae Mall, Waianae, Oahu	1975	34,500 sf shopping center
Moiliili	1975	26-story, 180-unit apartment building, with 3 parking levels and recreation deck; 165,400 sf; converted to condominiums; now Hale Kulanui
Pearl Two	1975	32-story, 317-unit condominium and parking garage; 417,923 sf
Kinalau Tower	1976	23-story, 180-unit condominium; 185,705 sf
Waikiki Lanais	1977	22-story, 160-unit condominium; 211,983 sf
Wilder at Piikoi	1978	30-story, 150-unit condominium; 192,274 sf
Century Center	1978	42-story tower with 375 office and condominium units; 368,788 sf
Kawaiahao Plaza	1978	low-rise office building; houses administration and endowment offices for Kamehameha Schools
Makiki Park Place	1978	16-story, 97-unit condominium; 214,061 sf

Project	Year Completed	Description
Pacific Monarch	1979	34-story, 216-unit condominium; 185,787 sf
Brookside	1979	23-story, 198-unit condominium; 264, 823 sf
Waikiki International Plaza	1979	125,000 sf shopping center
Honolulu Club	1980	78,298 sf retail and fitness facility
The Cliffs at Princeville, Kauai	1981	nine 3-story condominiums (202 units) and hotel; 290,926 sf
Sheraton Royal Waikoloan Hotel, Waikoloa Beach, Island of Hawaii	1981	6-story, 528-unit luxury hotel
Craigside	1982	2-tower condominium; 27-story, 189-unit and 27-story, 54-unit towers; multi-level parking and recreational facilities; 492,751 sf
Honolulu Tower	1982	395-unit condominium with multi-level parking and recreational facilities; 619,200 sf
Hale Kaheka	1982	38-story, 175-unit condominium with 5-level parking structure; 251,344 sf
Windward Mall, Kaneohe, Oahu	1982	319,000 sf regional shopping center
Sears at Windward Mall, Kaneohe, Oahu	1982	131,000 sf department store
The Hobron	1983	43-story, 607-unit hotel with 3 levels of parking; 294640 sf

Project	Year Completed	Description
Executive Centre	1984	mixed-use complex, including: 41-story Bishop Tower with 469 office and condominium units (362,504 sf); 12-story, 33-unit townhouse building with 3 levels of parking (422,713 sf)
Maile Court	1984	43-story, 580-unit hotel with 3 levels of parking, retail space, and recreational facilities
Prince Kuhio Plaza, Hilo	1984	336,876 sf regional shopping center anchored by Liberty House, Sears, Woolworth's
Kapalua Bay Hotel, Kapalua, Maui	1987	renovation of hotel and expansion of pool, lounge, and dining area; 149,102 sf
Ala Moana Shopping Center	1989	addition of 60,000 sf food court
Honolulu Park Place	1990	40-story, 437-unit condominium (516,100 sf) with multi-level, 200,000 sf parking structure
Ala Moana Shopping Center	1991	renovation and expansion of regional shopping center opened in 1960; 264,000 sf
Nuuanu Parkside	1991	23-story, 198-unit condominium; 264,823 sf
Waikiki Landmark	1991	189-unit condominium; 2 towers plus 12-story parking structure

NOTE: Projects located in Honolulu, unless otherwise noted. Square feet (sf) measured in gross.

APPENDIX C

*The Pankow Companies: Innovations,
Adaptations, and Tweaks*

Innovation	Description	First Instances
Precast concrete beams and spannal slabs	Support for the spannal was cast into the beams, enabling the permanent structure to act as part of the temporary formwork	San José Plaza
Precast concrete columns	Concrete columns were cast on their side and raised from the casting bed, saving storage space on small building sites	USC & Grove Street parking structures
Slipforming all vertical structural components	Using a slipform to form the columns and beams of the entire structure enabled CPA project teams to achieve three-day floor-to-floor construction cycles	Kauluwela Elderly; Brookside
Flyforming all horizontal structural components	Flyforms used to form floor slabs and beams of the entire structure	Brookside; Craigside
Precast concrete panels used as formwork	Wall panels used to form exterior columns and beams	Borel Estate; San José Plaza
Portable prestressing bed	Ralph Tice designed, and the company patented, a 260-foot-long portable prestressing bed to cast beams and girders for use on projects where commercial prestressed concrete was either unavailable or cost prohibitive	USC & Grove Street parking structures
Air shaft part of slipformed core	Vertical air shafts were incorporated into the concrete service core, reducing the amount of sheet metal ductwork and increasing the numbers of walls available to frame the structure	San José Plaza; PT&T; AT&T
Retail overbuild	Second story of mall was constructed under roof of existing enclosed mall	Tyler Mall

Innovation	Description	First Instances
Prefabricated form slabs	Prefabricated deck forming tables on precast concrete beams mechanized the forming operation	Roosevelt Field parking structures
Seismic concrete core with steel frame	Use of shear and ductile concrete cores reduced amount of steel needed for structure while improving performance of structure on drift and flexibility	Chase Plaza; Shoreline Square
Precast hybrid moment- resistant frame	Beam-and-column system incorporated the advantages of architectural and structural precast concrete while providing a seismic framing system that performed better than structural steel	The Paramount; Pacific Plaza; Westside Media Center

NOTE: Adapted from Robert Law, "Pankow Innovation History," 2008, CPBL, Pasadena, CA.

LIST OF ARCHIVAL COLLECTIONS

Bancroft Library, University of California, Berkeley
    T. Y. Lin Papers
    Tudor Engineering Company Records

Charles Pankow Builders, Ltd., Records and Promotional Materials
    Annual Meeting, binders, 1974–2004
    Charles Pankow, Inc./Builders, marketing brochure [undated, c. 1985]
    Charles Pankow, Inc., profile placed in *ENR Directory: Contractors, 1974–75*
    Charles Pankow, Inc., company profile, marketing brochure [undated, c. 1977]
    Charles Pankow Builders, Ltd., marketing booklet [undated, c. 1996]
    CPI News/CP News/Single Source, company newsletter, 1983–2001
    Charles Pankow, Inc., "Headquarters Building of the Pacific Telephone & Telegraph Co.," brochure [1976]
    "Making It Work (History and Philosophy of Charles J. Pankow)," *40 Years of Building Innovation, 1963–2003*
    Offsite Managers' Meetings, agendas and minutes, 1975–1997
    Project Data Sheets
    Project Technical Reports

Charles J. Pankow Legacy Project Oral Histories, 2008–2011
    Steven C. Beering
    Bill Bramschreiber
    Dean Browning
    Crodd Chin*
    Conan "Doug" Craker
    Vincent P. Drnevich
    Jon T. Eicholtz
    Albert W. Fink
    Joy Haystead
    Robert Heisler
    George F. Hutton
    Brad D. Inman
    Albert Josselson
    Richard M. "Rik" Kunnath, 1 October 2008, 21 October 2008,
        9 July 2009*, 2 July 2010, 17 October 2011*
    Robert Law, 25 April 2008, 21 July 2009*
    Mike Liddiard
    Arthur Love
    Kim Lum
    John F. McLaughlin
    Norman L. "Red" Metcalf
    Alan D. Murk, 15 December 2008, 8 May 2009*
    Timothy P. Murphy
    Steve Nakada*
    Suzanne Dow Nakaki
    Wally Naylor
    Russell J. Osterman*
    Rick Pankow
    Steve Pankow
    Mark J. Perniconi
    Kim Petersen
    Lee Sandahl
    Joseph Sanders
    Kevin Smith

---

*Interview not recorded.

Dean E. Stephan
Bill Trimble*
Judy L. Vawter
Thomas D. Verti
Russell L. Wahl
Dick Walterhouse
William "Red" Ward
William Wilson III*
Henry T. Yang*

Department of Special Collections, Stanford University Libraries
    California and Western Manuscript Collection

Environmental Design Archives, University of California, Berkeley
    Oakland and Imada Collection

Henry E. Huntington Library, San Marino, California
    Lloyd Corporation Archive

Oral History Series, Earthquake Engineering Research Institute, University
of California, Berkeley
    Clarkson W. Pinkham

Regional Oral History Office, Bancroft Library, University of California,
Berkeley
    T. Y. Lin
    Walter N. Shorenstein

State of California Archives, Sacramento
    Department of Public Works, Director's Office Records, Director's
        Records related to the Division of Architecture
    Department of Public Works, Architecture Division Records, Work
        Orders

University of Notre Dame Archives

NOTES

1. Ralf Meisenzahl and Joel Mokyr, "The Rate and Direction of Invention in the British Industrial Revolution: Incentives and Institutions," National Bureau of Economic Research, Working Paper No. 16993 (April 2011).

2. "Charles J. Pankow, ASCE Member," *Transactions of the American Society of Civil Engineers* 169 (2004): 816–7; James McClintock, "Charles Pankow, Building Industry Icon, Dies at 80," *San Gabriel Valley Tribune*, 15 January 2004; "Making It Work (History and Philosophy of Charles J. Pankow)," *40 Years of Building Innovation, 1963–2003*, brochure, CPBL, Pasadena, CA.

3. *Engineering News-Record (ENR) Construction Facts: The Sourcebook of Statistics, Records, and Resources* (New York: McGraw-Hill, 2003), 65–6.

4. "Project Delivery Method Market Share for Non-Residential Construction," figure, RSMeans Business Solutions, a division of Reed Construction Data, May 2011, www.dbia.org/pubs/research/rsmeans110606.htm; James McClintock, "Charles Pankow's Company Has Had a Hand in Many Key Valley Projects," *San Gabriel Valley Tribune*, 15 September 2003.

5. Walter A. Friedman and Geoffrey Jones, "Business History: Time for Debate," *Business History Review* 85 (Spring 2011): 1–8, 3 (quoted).

6. Warren Bennis and Patricia Ward Biederman, *Organizing Genius: The Secrets of Creative Collaboration* (Reading, MA: Addison-Wesley, 1997).

7. Far more scholarly attention has been paid to construction firms that concentrate on so-called heavy industry, that is, dams, bridges, highways, and the like. Cf. Stephan B. Adams, *Mr. Kaiser Goes to Washington: The Rise of a Government Entrepreneur* (Chapel Hill: University of North Carolina Press, 1997); Mark S. Foster, *Henry J. Kaiser: Builder in the Modern West* (Austin: University

of Texas Press, 1989); Joseph A. Pratt and Christopher J. Castaneda, *Builders: Herman and George R. Brown* (College Station: Texas A & M University Press, 1999); Christopher James Tassava, "Multiples of Six: The Six Companies and West Coast Industrialization, 1930–1945," *Enterprise & Society* 4 (March 2003): 1–27; Donald E. Wolf, *Big Dams and Other Dreams: The Six Companies Story* (Norman: University of Oklahoma Press, 1996).

8. Janet Welles Greene, "Sources for the History of the Building and Construction Industry," *Labor History* 46 (November 2005): 495–511, 496 (quoted).

9. CPI, brochure, 1982, CPBL, Pasadena, CA. To institutionalize the core values of their firm, William Hewlett and David Packard presented in written form what became known as "the HP Way" at a two-day managers' retreat in 1957. Not coincidentally, it was a "watershed year" for HP. The company reorganized along divisional lines, concluded an initial public stock offering, and moved into Stanford Industrial Park (Courtney Purrington, "Hewlett Packard: Creating, Running, and Growing an Enduring Company," Harvard Business School Case 9-698-052, 1997, rev. 2000, referencing David Packard, *The HP Way: How Bill Hewlett and I Built Our Company* [New York: HarperBusiness, 1995], 71, 79–81).

10. The film may be accessed from the websites of the Charles Pankow Foundation and Charles Pankow Builders, Ltd. The interviews are archived at Karnes Archives and Special Collections, Purdue University Libraries, and are accessible through its website, www.lib.purdue.edu/spcol/.

11. Charles Pankow Foundation, memorandum, "Charles J. Pankow Legacy Project Objective, Audiences, Themes, and Products," 25 October 2007.

12. "Oral History: Saving the Past," *Civil Engineering* 55 (March 1985): 50 (quote); Jon Eicholtz interview; Richard M. Kunnath interview, 1 October 2008; Robert Law interview; Kim Petersen interview; Dick Walterhouse interview; Darl Williams, Minutes, Offsite Managers' Meeting, Juarez, Mexico, 12–14 May 1975.

13. Victor W. Geraci, "Documenting Cultural History with Oral Sources," *Liwa* 1 (June 2009): 50.

14. Donald A. Ritchie, *Doing Oral History: A Practical Guide* (New York: Oxford University Press, 2003), esp. 117–27. Budget constraints limited the number of recorded and transcribed interviews to three dozen. Another half dozen or more interviews were either conducted by telephone or were not recorded for transcription purposes. Interviewees were prioritized on the basis of the closeness of their relationship with Charlie Pankow and their knowledge of the business and engineering histories of the firm, but also with the goal of obtaining a representative cross-section of employees and associates by level of their employment (in the case of employees) and across time. Several individuals outside of the company were unavailable, owing to scheduling or other issues.

15. Geraci, "Documenting Cultural History with Oral Sources," 56, 55, 52. On the suitability of oral history to the study of business history, cf. Christopher J. Castaneda, "Writing Contract Business History," *Public Historian* 21 (Winter 1999): 11–29; Carl Ryant, "Oral History and Business History," *Journal of American*

History 75 (September 1988): 560–6; and the various case studies described in the Summer 1981 issue of the *Public Historian* (vol. 3).

Introduction

1. Nathan Rosenberg, "Factors Affecting the Diffusion of Technology," in *Perspectives in Technology* (New York: Cambridge University Press, 1976), 191–5. The essay originally appeared in *Explorations in Economic History* 10 (Fall 1972).

2. To avoid confusion, Construction Management is capitalized when it refers to the specific building team configuration and project delivery method that gained acceptance in the 1960s and 1970s.

3. Jeffrey L. Beard, Michael C. Loulakis, and Edward C. Wundram, *Design-Build: Planning through Development* (New York: McGraw-Hill, 2001), 13–21; Luzerne S. Cowles, "The Engineer and Architect Unite," in *Technology and Industrial Efficiency* (New York: McGraw-Hill, 1911), 480 (quoted); Frank Miles Day, "Some Needed Reforms in Contracting," *The Architect and Engineer* 19 (December 1909): 59–60.

4. Daniel Hovey Calhoun, *The American Civil Engineer: Origins and Conflict* (Cambridge, MA: MIT Technology Press, 1960); Cowles, "The Engineer and Architect Unite," 480 (quoted); Day, "Some Needed Reforms in Contracting," 59 (quoted); "Frank Miles Day Dead," *New York Times*, 18 June 1918; "Dividing Line Between Engineer and Architect," *The Architect and Engineer* 60 (January 1920): 104–5; Otto E. Goldschmidt, "The Owner, the Architect, and the Engineer," *The Architect and Engineer* 47 (November 1916): 88–92; "The Relative Positions of the Engineer and the Architect in Designing Commercial Buildings," *The Architect and Engineer* 39 (November 1914): 108–9.

5. Rebecca Menes, "Limiting the Reach of the Grabbing Hand: Graft and Growth in American Cities," in *Corruption and Reform: Lessons from America's Economic History*, ed. Edward L. Glaeser and Claudia Goldin (Chicago, University of Chicago Press, 2007), 64 (quoted), 74–6. The classic essay on municipal reform efforts of the Progressive Era is Samuel P. Hays, "The Politics of Reform in Municipal Government in the Progressive Era," *Pacific Northwest Quarterly* (October 1964) 157–69. On Los Angeles, where Charlie Pankow began his career, see, Tom Sitton, *Los Angeles Transformed: Fletcher Bowron's Urban Reform Revival, 1938–1953* (Albuquerque: University of New Mexico Press, 2005), intro., chap. 1; idem, "Did the Ruling Class Rule at City Hall in 1920s Los Angeles?" 302–18, in *Metropolis in the Making: Los Angeles in the 1920s*, ed. Tom Sitton and William Deverell (Berkeley: University of California Press, 2001).

6. Beard, Loulakis, and Wundram, *Design-Build*, 21–2; "The Lowest Bidder and the Lowest Responsible Bidder," *The Architect and Engineer* 38 (October 1914): 80–1.

7. George E. Burlingame, "The Trouble with the Builder's Business in San Francisco," *The Architect and Engineer* 26 (August 1911): 88 (quoted); Day, "Some Needed Reforms in Contracting," 59 (quoted); Charles Evan Fowler, "Low Bidders

on Contract Work, *The Architect and Engineer* 46 (September 1916): 118; Sullivan W. Jones, "Present System of Estimating an Injustice to the Owner," *The Architect and Engineer* 33 (July 1913): 106.

8. John C. Austin, "Relation of the Architect to the Contractor," *The Architect and Engineer* 45 (May 1916): 96 (quoted); Burlingame, "The Trouble with the Builder's Business in San Francisco," 89 (quoted); "Compensation of Contractors," *The Architect and Engineer* 25 (June 1911): 93–4; Fowler, "Low Bidders on Contract Work"; Frederick James, "Segregation Versus General Contract," *The Architect and Engineer* 26 (September 1911): 95–6; "The Lowest Bidder and the Lowest Responsible Bidder," 80–1.

9. Day, "Some Needed Reforms in Contracting," 59 (quoted), 60. As a contractor, Brassey negotiated fixed-price contracts. He then allocated funds to project superintendents, who typically worked on the basis of a fee plus a share of the profits; if they completed the work below cost, they kept the difference. Brassey would supply additional funds to his agents in the field if they encountered unforeseen problems (Thomas Stacey, *Thomas Brassey: The Greatest Railway Builder in the World* [London: Stacey International, 2005], 9–17; Charles Walker, *Thomas Brassey, Railway Builder* [London: Frederick Muller, 1969], 23–4).

10. A. J. McKenzie, "Contractors and Engineers Are Not Rivals," *The Architect and Engineer* 78 (August 1924): 121 (quoted), 122.

11. Brad Inman, "The Appeal of Design Build," *Urban Land* 46 (November 1987): 21–5; Richard M. Kunnath and Brad Inman, "Who Does the Design Work on a Design/Build Project?" *Concrete Construction* (November 1985); "One Contract Replaces Two or More in Design/Build Work," *Concrete Construction* (November 1985): 909–10; Todd L. Whitlock, "An Argument for Design/Build," *Real Estate Finance* 9 (Fall 1992): 87–90.

12. McKenzie, "Contractors and Engineers Are Not Rivals," 121.

13. Beard, Loulakis, and Wundram, *Design-Build*, 20–1; "General Contractors Have a Grievance," *The Architect and Engineer* 33 (July 1913): 105 (quoted); "Lowest Builder Entitled to Contract," *The Architect and Engineer* 26 (October 1911): 102 (quoted); "The Lowest Bidder and the Lowest Responsible Bidder"; "The Letting of Public Contracts and the Rights of Lowest Bidder," *The Architect and Engineer* 28 (February 1912): 86–7.

14. "One Contract Replaces Two or More in Design/Build Work"; Charles J. Pankow, "Automation on the Job Site," *Journal of the American Concrete Institute* 64 (June 1967): 281 (quoted); idem, Raymond E. Davis Lecture, ACI, 1986 (quoted). A copy of the Raymond E. Davis Lecture was given to the author by Robert Law.

15. Thomas D. Verti, "To Create Opportunity for Concrete Construction Competitiveness and Innovation, Think Like a Master Builder," *Concrete International* 28 (July 2006), www.concrete.org/About/ab_presmemo_verti03.htm.

16. On Ford, see, esp., Olivier Zunz, *Making America Corporate, 1870–1920* (Chicago: University of Chicago Press, 1990), 79–90.

17. On corporate identity and the erection of signature office towers, see Gail Fenske and Deryck Holdsworth, "Corporate Identity and the New York Office Building, 1895–1915," in *The Landscape of Modernity, New York City, 1900–1940*, ed. David Ward and Olivier Zunz (Baltimore, MD: Johns Hopkins University Press, 1992), 129–59; Roland Marchand, *Creating the Corporate Soul: The Rise of Public Relations and Corporate Imagery in American Big Business* (Berkeley: University of California Press, 1998), 26–41.

18. Amy E. Slaton, *Reinforced Concrete and the Modernization of American Building, 1900–1930* (Baltimore, MD: Johns Hopkins University Press, 2001), esp. chap. 4.

19. "Austin Guarantees," advertisement, Annual Mid-Winter Number, *LAT*, 1 January 1925, I:22 (quoted); "Austin Pushes More Ideas for Construction," *ENR*, 27 November 1952, 48; Martin Grief, *The New Industrial Landscape: The Story of the Austin Company* (Clinton, NJ: Main Street Press, 1978), 15–23, 34–7; Slaton, *Reinforced Concrete and the Modernization of American Building*, 163–6. Charles Pankow Builders would later use the phrase "single source" responsibility to mean much the same thing. Indeed, the company eventually used the term to entitle its internal newsletter.

20. Gannett quoted in "Austin Pushes More Ideas for Construction," 50; Charles A. Shirk, *The Austin Company: A Century of Results*, Newcomen Publication 1085 (New York, 1978), 22; Grief, *The New Industrial Landscape*, 55–65. Rankings of US contractors based on contract volume, as reported annually in the *ENR* Top 400 and also, since 1977, the *BD&C* 300.

21. Grief, *The New Industrial Landscape*, 59–60, 65–6, 84–5; Shirk, *The Austin Company*, 12.

22. On Ralph B. Lloyd's oil-related activities, see Michael R. Adamson, "The Role of the Independent: Ralph B. Lloyd and the Development of California's Coastal Oil Region, 1900–1940," *Business History Review* 84 (Summer 2010): 301–28.

23. C. W. Norton to A. A. Schramm, State Superintendent of Banks, 5 November 1926, Port, drawer 1, box 3 (hereafter, for all Lloyd Corporation Archive records, in the form: drawer number-box number, that is, 1-3); Lloyd to E. C. Sammons, US National Bank, 12 February 1927, LCL 2-6 (quote). On Lloyd's Holladay Park acquisitions, see, "R. B. Lloyd to Buy Sullivan's Gulch," *Oregonian*, 28 September 1926, I:1; E. Kimbark MacColl, *The Growth of a City: Power and Politics in Portland, Oregon, 1915–1950* (Portland, OR: Georgian Press, 1979), chap. 12.

24. Lloyd to Norton, 4 March 1927, Port 1-4 (quote); Lloyd to Norton, 9 March 1927, Port 1-4; Lloyd to Wright and Gentry, 19 May 1927, LCL 2-6; Bart King, *An Architectural Guidebook to Portland*, 2d ed. (Corvallis: Oregon State University Press, 2007), 234–5, 261 (quoted), 262.

25. Norton to Schramm, 5 November 1926, Port 1-3; Lloyd to Norton, 29 March 1927, Port 1-4; Lloyd to Austin Co., 19 May 1927, Port 1-3; Lloyd to Wright and Gentry, 19 May 1927, LCL 2-6; "Addition to New Business Center,"

Oregon Journal, 16 October 1927. The design for Lloyd's Walnut Park building incorporated "Spanish Renaissance" and "Moorish" elements similar to those employed by Morgan, Walls, and Clements in their contemporary McKinley Building on Wilshire Boulevard in Los Angeles (Donald E. Marquis, "The Spanish Stores of Morgan, Walls & Clements," *AF* 50 [June 1929]: 902–9). On contemporary mixed-use buildings, cf. Richard Longstreth, *The Drive-In, The Supermarket, and the Transformation of Commercial Space in Los Angeles, 1914–1941* (Cambridge, MA: MIT Press, 1999).

26. Lloyd to R. K. Wood Lumber Co., 3 February 1923, RBL 3–5; Lloyd to Dombrowski, 19 February 1923, RBL 3–3; Lloyd to Wright and Gentry, 19 May 1927, LCL 2–6; Lloyd to Robert W. Hunt Co., 17 September 1927, Port 1–3 (quote); "Plans Completed for New Automotive Structure," *LAT*, 4 February 1923, V:14. The theater project was not built. Universal Films backed away from the deal in 1928 because of "upheaval" in the theater business. At the time, there was "considerable demand" for office space in the planned structure (Lloyd to Edward H. Sensenich, President, West Coast Bancorporation, 16 January 1929, LCL 4–7 [quote]; Lloyd to Norton, 20 February 1930, LCL 5–5; Norton to Dr. Earl Muck, 18 June 1931, LCL 6–3 [quote]).

27. Cf. Lloyd to L. H. Hoffman, 7 February 1928, Port 1–3; Builder's Contract, 2 March 1928, LCR 3–4; Hoffman to Lloyd, 13 July 1928, LCR 3–4; Lloyd to Hoffman, 20 November 1928, Port 1–3; Lloyd to Hoffman, 19 December 1928, Port 1–3; C. L. Peck to Lloyd, 16 September 1941, BI 3–3; Lloyd to Title Insurance and Trust, 5 October 1944, LCR 4–4; C. L. Peck to Claud Beelman, 21 September 1948, LCL 20–3; Beelman, invoice, 15 November 1948, LCL 21–1; Von Hagen to C. L. Peck, 1 April 1950, LCL 23–3.

For all intents and purposes, Lloyd did use the same firm to design and construct many of his Portland projects. Architect Charles W. Ertz and his partner, Thomas B. Burns, doing business as Ertz & Burns, Architects, and Ertz-Burns & Co., Building Contractors, designed and built auto and truck dealerships, a Coca-Cola bottling plant, the Lloyd golf course clubhouse, a grocery store, and 2-story office buildings for the Bonneville Power Administration. Lloyd often bid the builder's contract, especially after 1935, when Ertz relocated to Beverly Hills to open an architectural office, and Burns stayed behind as contractor. But he restricted the bidding to Ertz-Burns & Co. and L. H. Hoffman, another prominent local builder. And, if he didn't like the bid, Lloyd negotiated the price (cf. Charles W. Ertz, "Estimate of Automotive Building for Ralph B. Lloyd, 8 April 1930, LCR 3-1; Builder's Contract, 5 May 1930, LCL 5-1; Ertz to Lloyd, 19 February 1932, LCL 7-2; Builder's Contract, 18 April 1932, LCL 7-2; Lloyd to Ertz-Burns & Co., 11 December 1937, LCR 8-2; Builder's Contract, 10 January 1938, LCR 8-2; Drinker to Von Hagen, 19 October 1945, LCR 10-4; Von Hagen to Lloyd, 25 March 1946, LCL 19-1; Builder's Contract, 1 July 1946, LCL 19-1). On the careers of Burns and Ertz, see Richard Ellison Ritz, *Architects of Oregon: A Biographical Dictionary of Architects Deceased—19ᵗʰ and 20ᵗʰ Centuries* (Portland, OR: Lair-Hill Publishing, 2002), 59–60, 124–5.

The $30 million Lloyd Center (1956–1960), easily the Lloyd Corporation's largest project, was an exception, adhering strictly to design-bid-build. Construction of the shopping center was much delayed, owing to restrictions on construction materials associated with the Korean War, the need to acquire additional parcels to accommodate the size of the project, and difficulties involved in securing the anchor department store and hotel tenants. It was eventually bid out under seven contracts (David Yule, Lloyd Center timetable, handwritten, untitled, n.d., in folder, "Lloyd Center Financing: Prudential Ins. Co.–Preliminary Papers," LC 10-4; Richard G. Horn and W. Joseph McFarland, "Forum and Seminar, Portland Chapter, American Institute of Banking," n.d. [November 1954], LC 10-3; Richard R. Von Hagen to Proctor H. Barnett, Prudential Insurance Co. of America, 7 December 1956, LC 10-5; Larry Smith to Barnett, 7 December 1956, LC 10-5). Bids may be found in LC 6-3. During construction, the Lloyd Corporation approved 309 change orders (out of 417 requests), which may be found in LC 6-4.

28. "American Cement Building's New Ideas Began with New Bidding," in advertising supplement, "The American Cement Building . . . Concrete Symbol of the Future," *LAT*, 11 July 1961 (quoted).

29. Lindy Biggs, *The Rational Factory: Architecture, Technology, and Work in America's Age of Mass Production* (Baltimore, MD: Johns Hopkins University Press, 1996); Slaton, *Reinforced Concrete and the Modernization of American Building*, 1–14; Zunz, *Making America Corporate*, 61–4, chap. 3.

30. Slaton, *Reinforced Concrete and the Modernization of American Building*, 127–9. On the application of managerial innovation to capital-intensive, mass production industries, cf. Alfred D. Chandler Jr., *The Visible Hand: The Managerial Revolution in American Business* (Cambridge, MA: Belknap Press of Harvard University Press, 1977); idem, *Scale and Scope: The Dynamics of Industrial Capitalism* (Cambridge, MA: Belknap Press of Harvard University Press, 1990); David A. Hounshell, *From the American System to Mass Production, 1800–1932: The Development of Manufacturing Technology in the United States* (Baltimore, MD: Johns Hopkins University Press, 1984); Thomas K. McCraw, ed., *Creating Modern Capitalism: How Entrepreneurs, Companies, and Countries Triumphed in Three Industrial Revolutions* (Cambridge, MA: Harvard University Press, 1997).

31. Sampson quoted in "Change: The Building Team Is Getting Together for a Change," *BD&C* 14 (December 1973): 34; "The Industry Capitalism Forgot," *Fortune* 36 (August 1947): 61–7, 167–70, 65 (quoted), 66 (quoted).

32. Steven G. Allen, "Why Construction Industry Productivity Is Declining," National Bureau of Economic Research, Working Paper No. W1555 (February 1985); J. E. Cremeans, "Productivity in the Construction Industry," *Construction Review* (May/June 1981): 4–6; Steven Rosefielde and Daniel Quinn Mills, "Is Construction Technologically Stagnant?" 83–111, in *The Construction Industry: Balance Wheel of the Economy*, ed. Julian E. Lange and Daniel Quinn Mills (Lexington, MA: D. C. Heath, 1979); Richard Vedder, Review of LePatner, *Broken*

Buildings, Busted Budgets, EH.Net Economic History Services, 3 September 2007, eh.net/bookreviews/library/1257.

33. Greene, "Sources for the History of the Building and Construction Industry," 485 (quoted). On labor-intensive, batch-production, and thus fragmented, manufacturing sectors, see Paul Hirst and Jonathan Zeitlin, "Flexible Specialization versus Post-Fordism: Theory, Evidence, and Policy Implications," *Economy and Society* 20 (1991): 1–56; Philip Scranton, *Endless Novelty: Specialty Production and American Industrialization, 1865–1925* (Princeton, NJ: Princeton University Press, 1997); Jonathan Zeitlin, "Flexibility and Mass Production at War: Aircraft Manufacture in Britain, the United States, and Germany, 1929–1945," *Technology and Culture* 26 (1995): 46–79.

34. Peter J. Cassimati, *Economics of the Construction Industry*, Studies in Business Economics No. 111 (New York: The Conference Board, 1969), 26–9; Gerald Finkel, *The Economics of the Construction Industry* (New York: M. E. Sharpe, 1997), 31–5; US Department of Commerce, Bureau of the Census, *Census of Construction Industries: 1967* (Washington, DC: GPO, 1971), table A2; idem, *Commercial and Institutional Building Construction: 2002* (Washington, DC: GPO, 2004), tables 5, 6.

35. Barbara White Bryson and Canan Yetmen, "Why Owners Make the Difference," ENR.com, 28 July 2010, enr.construction.com/opinions/viewpoint/2010/0728-OwnersMakeTheDiference.asp; Barry B. LePatner, with Timothy Jacobson and Robert E. Wright, *Broken Buildings, Busted Budgets: How to Fix America's Trillion-Dollar Construction Industry* (Chicago: University of Chicago Press, 2007), 7 (quoted).

36. Slaton, *Reinforced Concrete and the Modernization of American Building*, 134.

37. Charles J. Pankow, Raymond E. Davis Lecture, ACI, 1986 (quoted); Daniel C. Brown, "Concrete Technology Pushes to New Heights," *BD&C* 24 (January 1983): 62–5; David M. Kinchen, "A Love Affair with Concrete: Charles J. Pankow Places Emphasis on Cost Control," *LAT*, 25 September 1983, VI:1. On labor relations in the steel industry, cf. John P. Hoerr, *And the Wolf Finally Came: The Decline of the American Steel Industry* (Pittsburgh, PA: University of Pittsburgh Press, 1988), chap. 4; Jack Metzger, *Striking Steel, Solidarity Remembered* (Philadelphia: Temple University Press, 2000); James D. Rose, "The Struggle over Management Rights at U.S. Steel, 1946–1960: A Reassessment of Section 2-B of the Collective Bargaining Contract," *Business History Review* 72 (Autumn 1998): 446–77; Paul A. Tiffany, *The Decline of American Steel: How Management, Labor, and Government Went Wrong* (New York: Oxford University Press, 1988). Enacted by the Republican-dominated 78th Congress over President Truman's veto on 23 June 1947, Taft-Hartley "codif[ied] and crystalize[d] all the legal and administrative procedures that had tended to constrain, contain, and discipline the unions" in the wake of a series of strikes in 1941 against firms engaged in defense production, writes Nelson Lichtenstein, biographer of Walter Reuther, president of the United Auto Workers

from 1946 until his death in 1970 (*Walter Reuther: The Most Dangerous Man in Detroit* [Urbana: University of Illinois Press, 1995], 261). Congress amended the Labor-Management Relations (Wagner) Act of 1935 by adding unfair labor practices on the part of unions to the list of practices prohibited by employers under the original legislation. Taft-Hartley also authorized the president to intervene in actual or potential labor actions that might create a national emergency; required union leaders to sign anti-communist affidavits; banned secondary boycotts; and permitted states to outlaw the union shop.

In his 1890 text, *Principles of Economics*, Marshall coined the term "industrial districts," and explained why they formed. He wrote: "The mysteries of the trade become no mystery, but are as it were in the air."

38. Biggs, *The Rational Factory*, 81–5; David Hounshell, *From the American System to Mass Production, 1800–1932: The Development of Manufacturing Technology in the United States* (Baltimore, MD: Johns Hopkins University Press, 1984), chap. 6.

39. "The Birth and Growth of History's Most Exciting Building Material," *Civil Engineering* 47 (October 1977): 118–20; "The Story of Cement, Concrete, and Reinforced Concrete," *Civil Engineering* 47 (November 1977): 63–5. Post-tensioning is often used to cast elements on projects, such as bridges, where their required size makes their transport, and therefore prefabrication, difficult. Pre-tensioned elements commonly include beams, columns, and floor slabs used in buildings. For extended discussion, cf. Edward G. Nawy, *Prestressed Concrete: A Fundamental Approach*, 5th ed. (Upper Saddle River, NJ: Prentice-Hall, 2005). The use of concrete as an architectural medium became widespread only in the 1930s (cf. "Concrete Improves Its Appearance," *ENR*, 9 December 1937, 947).

40. T. Y. Lin, "Tall Buildings in Prestressed Concrete," *AR* 138 (December 1965): 165–70; "Prestressed Concrete Envoy," *ENR*, 7 June 1962, 53–4; C. D. Wailes Jr., "Prestressed Concrete—Dynamic New Force in So. California Construction," *SB&C*, 25 August 1961, 25–6.

41. Allen H. Brownfield, "Growing Pains in Prestressed Concrete Building," *Civil Engineering* 28 (February 1958): 46–9; W. Burr Bennett Jr., "Future of Prestressed Concrete," *Civil Engineering* 39 (July 1969): 41–3; Frederick S. Merritt, "Prestressed Concrete: The Infant Is Growing Up," *ENR*, 6 October 1960, 34–40; "On-Site Prestressing Plant Cuts Concrete Building Cost," *ENR*, 18 April 1957, 58–60; J. L. Peterson, "History and Development of Precast Concrete in the United States," *Journal of the American Concrete Institute* 25 (February 1954): 494–5; Wailes, "Prestressed Concrete."

42. Brownfield, "Growing Pains in Prestressed Concrete Building"; T. Y. Lin, "Prestressed Concrete—Slabs and Shells: Design and Research in the United States," *Civil Engineering* 28 (October 1958): 74–7; idem, "Revolution in Concrete," *AF* 114 (May 1961): 121–7; idem, "Revolution in Concrete," *AF* 114 (June 1961): 116–22, 117 (quote).

43. Lin, "Tall Buildings in Prestressed Concrete."

44. Brownfield, "Growing Pains in Prestressed Concrete Building"; Wailes, "Prestressed Concrete."

45. Peterson, "History and Development of Precast Concrete in the United States," 487–90; Slaton, *Reinforced Concrete and the Modernization of American Building*, 144–6.

46. Charles J. Pankow, "Slipform Construction of Buildings," in *Concrete Construction Handbook*, ed. Joseph J. Waddell, 2d ed. (New York, 1974), 34–1 (quoted); "New Era in Slipforming," *ENR*, 16 July 1964, 92; "Slip-Form Construction Speeds Erection of Apartment Hotel," *AR* 118 (October 1955): 248.

47. Charles J. Pankow, Raymond E. Davis Lecture (quoted).

48. Slaton, *Reinforced Concrete and the Modernization of American Building*, 1 (quoted). On the influence of industrial vernacular structures on Modern architecture, see Reyner Banham, *A Concrete Atlantis: U.S. Industrial Building and European Modern Architecture, 1900–1925* (Cambridge, MA: MIT Press, 1986).

49. Reyner Banham, *Guide to Modern Architecture* (London: The Architectural Press, 1962), chap. 4.

50. Lloyd to Drinker, 23 November 1935, LCR 7–2 (quoted).

51. Dean E. Stephan interview (quoted); Inman, "The Appeal of Design Build"; Kunnath and Inman, "Who Does the Design Work on a Design/Build Project?"; Christopher Olson, "Design/Builders Rapidly Expand Their Market," *BD&C* 25 (June 1984): 72–5, 74 (Stephan quoted).

52. Allan B. Jacobs, *The Good City: Reflections and Imaginations* (London and New York: Routledge, 2011), 44.

53. Cf. Douglas Frantz, *From the Ground Up: The Business of Building in the Age of Money* (New York: Henry Holt, 1991).

54. Rosenberg, "Economic Development and the Transfer of Technology," in *Perspectives in Technology*, 166 (quoted); idem, "Factors Affecting the Diffusion of Technology," 197–9. "Economic Development and the Transfer of Technology" originally appeared in *Technology and Culture* 11:4 (1970).

55. Albert W. Fink interview, 13 February 2009; Brad Inman interview; Robert Law interview (quote); Michael Liddiard interview; Suzanne Dow Nakaki interview; Thomas D. Verti interview; William "Red" Ward interview.

Chapter 1

1. Alex C. Humpherys, "Business Training for the Engineer," in *Addresses to Engineering Students*, ed. John Alexander Low Waddell and John Lyle Harrington, 2d ed. (Kansas City, MO: Waddell & Harrington, 1912), 113.

2. Towne quoted in Biggs, *The Rational Factory*, 39.

3. J. B. Johnson, "Two Kinds of Education for Engineers," in *Addresses to Engineering Students*, 32–3.

4. Matthew H. Wisnioski, "Engineers and the Intellectual Crisis of Technology, 1957–1973," Ph.D. diss., Department of History, Princeton University, 2005, 99–100, citing Robert Perruci and Joel Emery Gerstl, eds., *The Engineers and the Social*

System (New York: Wiley, 1969), 3. In 1960 female engineering undergraduate students were still rare: women comprised less than 1 percent of the engineering population.

5. Humpherys, "Business Training for the Engineer," 113.

6. Ibid., 133 (quoted), 114; Biggs, *The Rational Factory*, chap. 2; Calhoun, *The American Civil Engineer*; Edwin T. Layton, *The Revolt of the Engineers: Social Responsibility and the American Engineering Profession* (Baltimore, MD: Johns Hopkins University Press, 1986 [1971]); David F. Noble, *America by Design: Science, Technology, and the Rise of Corporate Capitalism* (New York: Oxford University Press, 1979).

7. Sellers quoted in Biggs, *The Rational Factory*, 39.

8. Wisnioski, "Engineers and the Intellectual Crisis of Technology"; idem, "'Liberal Education Has Failed': Reading Like an Engineer in 1960s America," *Technology and Culture* 50 (October 2009): 753–82.

9. Rick Pankow, letter to the author, 7 May 2009; Franklin W. Scott, ed., *The Semi-Centennial Alumni Record of the University of Illinois* (Urbana: the university, 1918), 460, 505. From 1914–1918, "aero squadrons" were organized under the Aviation Section of the US Signal Corps—the aviation service of the US Army. Sister Mary preceded Charlie Pankow in birth on 2 October 1920. Brother James was born on 29 March 1926.

10. Thomas E. Bonsall, *The Studebaker Story* (Stanford, CA: Stanford University Press, 2000), chap. 3; Donald T. Critchlow, *Studebaker: The Life and Death of an American Corporation* (Bloomington: Indiana University Press, 1996), 66–93, 87 (quoted); Patrick J. Furlong, "South Bend, Indiana," in *The American Midwest: An Interpretive Encyclopedia*, ed. Richard Sisson, Christian Zacher, and Andrew Robert Lee Clayton (Bloomington: Indiana University Press, 2007), 1176–7; John Palmer, *South Bend: Crossroads of Commerce* (Charleston, SC: Arcadia Publishing, 2003), chap. 6. Erskine became president in 1915. He was ousted after the company went into receivership in 1933.

11. Bonsall, *The Studebaker Story*, 115.

12. After World War II, engineers increasingly hailed from rural areas and blue-collar families. At the same time, they remained overwhelmingly white and male (Wisnioski, "Engineers and the Intellectual Crisis of Technology," 100–1).

13. Bonsall, *The Studebaker Story*, 118–9; William Henry Perrin, *The History of Edgar County, Illinois* (Chicago: William LeBaron Jr. & Co., 1879), 711; "Kelly Family Tree," RootsWeb World Connect Project, updated 25 August 2001, wc.rootsweb.ancestry.com/cgi-bin/igm.cgi?op=REG&db=:596327&id=1179249; Scott, *The Semi-Centennial Alumni Record of the University of Illinois*, 460, 505.

14. Charles Roll, *Indiana: One Hundred and Fifty Years of American Development*, vol. 3 (Chicago: Lewis Publishing, 1931); finding aid to the Sollitt Construction Company collection, Drawing and Documents Archive, College of Architecture and Planning, Ball State University, Muncie, Ind.; "E. Ross Adair Federal Building and U.S. Courthouse, Fort Wayne, Ind.," US General Services

Administration, Historic Buildings website, www.gsa.gov/portal/ext/html/
site/hb/method/post/actionParameter/searchCriteriaForm/buildingId/922/
category/25431.

15. Frank E. Hering, "Concerning a Stadium at Notre Dame," 16 October
1928, UADR 13/79; Hadden to Rockne, 21 May 1929, UADR 13/02; Rockne to
Hadden, 28 May 1929, UADR 13/02; Faulkner to Evans, 29 May 1929; Rockne
to Ryan, 31 May 1929; Green to Rockne, 31 May 1929; Rockne to Green, 3 June
1929; Healy to Rockne, 2 July 1929; Rockne to Healy, 4 July 1929; Haley to Green,
3 August 1929 (all found in UADR 19/157) (all found in UNDA); Rick Pankow
interview; "Notre Dame Stadium Has Legendary History," University of Notre
Dame Facilities website, www.und.com/facilities/nd-stadium-history.html.

16. "A Salute to Charles J. Pankow," dinner program, Third Annual
Achievement Awards Dinner, sponsored by the Northern California Construction
and Real Estate Industries Alliance for the City of Hope, 16 March 1989 (quoted);
Dennis McLellan, "Charles Pankow, 83; Founded Firm that Built MTA Complex,"
Los Angeles Times, 19 January 2004; Michael Stremfel, "All for One," *Los Angeles
Business Journal*, 3 June 1991, 40. A copy of the program, "A Salute to Charles J.
Pankow," was given to the author by Steve Pankow, his son.

17. Bruce Seely, "Research, Engineering, and Science in American Engineering
Colleges, 1900–1960," *Technology and Culture* 34 (April 1993): 344–67; Slaton,
Reinforced Concrete and the Modernization of American Building, 22–5, 35–48.

18. *Bulletin of Purdue University: Catalogue Number* (Lafayette, IN: Purdue
University, 1941), 280–2; Francis E. Griggs Jr., "Joseph B. Strauss, Charles A.
Ellis, and the Golden Gate Bridge: Justice at Last," *Journal in Professional Issues
in Engineering Education and Practice* 136 (April 2010): 71–83; John van der Zee,
The Gate: The True Story of the Design and Construction of the Golden Gate Bridge
(New York: Simon & Schuster, 1986); John van der Zee and Russ Cone, "The
Case of the Missing Engineer," *San Francisco Examiner*, *Image* magazine, 31 May
1992.

19. Slaton, *Reinforced Concrete and the Modernization of American Building*,
44–8, 47 (Ellis quoted).

20. *Bulletin of Purdue University: Catalogue Number* (1941), 190 (quoted), 191–
5, 278–83; Seely, "Research, Engineering, and Science in American Engineering
Colleges," 379.

21. H. B. Knoll, *The Story of Purdue Engineering* (West Lafayette, IN: Purdue
University Studies, 1963), 68–99, 103–4, 255; Seely, "Research, Engineering, and
Science in American Engineering Colleges," 354–7.

22. Knoll, *The Story of Purdue Engineering*, 254, 255 (quoted). Indiana was
arguably the most Republican of the Great Lakes states. Alf Landon won almost
42 percent of the vote in his landslide defeat to Franklin D. Roosevelt in the
1936 general election; it was his best showing in the region. In 1940 Republican
presidential candidate Wendell Wilkie won the state—where he was born—with
50.5 percent of the vote.

23. Ibid., 240.

24. *Bulletin of Purdue University*, 191; Transcript of Charles John Pankow Jr.; Knoll, *The Story of Purdue Engineering*, 105–6. With the assistance of Steve Pankow, the author was able to receive a copy of the transcript from the Office of the Registrar, Purdue University, with the grades redacted.

25. Transcript of Charles John Pankow Jr.; Knoll, *The Story of Purdue Engineering*, 105 (quoted).

26. Albert Josselson interview; Knoll, *The Story of Purdue Engineering*, 105–6.

27. The seminal work on the US occupation of Japan, 1945–1952, is John W. Dower, *Embracing Defeat: Japan in the Wake of World War II* (New York: Norton, 1999). See, also, Michael S. Molasky, *The American Occupation of Japan and Okinawa: Literature and Memory* (London: Routledge, 1999); Howard B. Schonberger, *Aftermath of War: America and the Remaking of Japan, 1945–1952* (Kent, OH: Kent State University Press, 1989); Michael Schaller, *The American Occupation of Japan: The Origins of the Cold War in Asia* (New York: Oxford University Press, 1985).

28. "A Salute to Charles J. Pankow"; Pankow quoted in Strempfel, "All for One," 40; David M. Kinchen, "A Love Affair with Concrete," *LAT*, 25 September 1983, VI:1.

29. Michael R. Adamson, Memorandum of Conversation with Doris Pankow, Steve Pankow, and Rick Pankow, Altadena, CA, 12 February 2009; Albert Josselson interview; Steve Pankow, letter to the author, 2 November 2009.

30. *Bulletin of Purdue University*, 190–6; *The Catalogue of Purdue University, 1946–1947 and 1947–1948* (Lafayette, IN: Purdue University, 1946), 1269–73; transcript of Charles John Pankow Jr.; Knoll, *The Story of Purdue Engineering*, 112–14.

31. Charles Pankow quoted in Kinchen, "A Love Affair with Concrete"; Michael R. Adamson, Memorandum of Conversation with Doris Pankow, Steve Pankow, and Rick Pankow, Altadena, CA, 12 February 2009; "A Salute to Charles J. Pankow"; Clarkson W. Pinkham, *Connections: The EERI Oral History Series*, no. 13, Stanley Scott, interviewer (Oakland, CA: Earthquake Engineering Research Institute, 2006), 13–15.

32. *Life* quoted in Lou Cannon, *Official Negligence: How Rodney King and the Riots Changed Los Angeles and the LAPD* (New York: Times Books, 1997), 3–4; "City of Angels: It's Still an Age of Miracles," *Newsweek*, 3 August 1953, 64–6; Robert M. Fogelson, *Downtown: Its Rise and Fall, 1889–1950* (New Haven, CT: Yale University Press, 2001), 393–4; William Fulton, *The Reluctant Metropolis: The Politics of Urban Growth in Los Angeles* (Baltimore, MD: Johns Hopkins University Press, 2001); 9–10; Roger W. Lotchin, *Fortress California, 1910–1961: From Warfare to Welfare* (New York: Oxford University Press, 1992); idem, *The Bad City in the Good War: San Francisco, Los Angeles, Oakland, and San Diego* (Bloomington: Indiana University Press, 2003); Kevin Starr, *Golden Dreams: California in an Age of Abundance, 1950–1963* (New York: Oxford University

Press, 2009), chap. 1; Arthur Verge, "Daily Life in Wartime California," in *The Way We Really Were: The Golden State in the Second World War*, ed. Roger W. Lotchin (Urbana: University of Illinois Press, 2000), 13–29. All population figures are US Census figures, compiled in the *Los Angeles Almanac*, www.laalmanac.com/population.

33. Charles C. Cohan, "8 Billion Dollar Decade," *LAT*, 3 January 1955, D30–3; idem, "Southland Building at New Peak," *LAT*, 3 January 1956, D44; "Third Successive Annual Southern California Building Record," *SB&C*, 27 January 1956, 43; Ray Hebert, "Vast Growth Predicted in Southland During 1958," *LAT*, 3 February 1958, B1; "Buildings Spur Area's Economy," *LAT*, 16 July 1961, J14; "Construction Second Highest in History," *SB&C*, 22 January 1965, 28; Robert Moore Fisher, *The Boom in Office Buildings: An Economic Study of the Past Two Decades*," Technical Bulletin No. 58 (Washington, DC, Urban Land Institute, 1967), 6–9; William C. Wheaton and Raymond G. Torto, "Office Construction Booms: History and Prospects," *Urban Land* 44 (July 1985): 32–3.

34. Steve Pankow, letter to the author, 20 October 2009; "A Salute to Charles J. Pankow"; Pankow quoted in Kinchen, "A Love Affair with Concrete." In 1961 the Los Angeles District was renamed the Southwest District.

35. The Skunk Works was established in 1943 under Clarence L. "Kelly" Johnson as an experimental group within Lockheed Corporation to develop a jet fighter for the US Army Air Force. For all intents and purposes, it lay dormant from 1945 until 1954, when it was revived to develop the U-2 spy plane and other weapons of the Cold War. See Bennis and Biederman, *Organizing Genius*, chap. 5; Jay Miller, *Lockheed Martin's Skunk Works*, rev. ed. (Leicester, UK: Midland Publishing, 1995); Ben R. Rich and Leo Janos, *Skunk Works: A Personal Memoir of My Years at Lockheed* (Boston: Little Brown, 1994).

36. Jeffrey L. Covell, "Peter Kiewit Sons', Inc.," *International Directory of Company Histories*, ed. Paula Kepos, vol. 8 (Detroit: St. James Press, 1994), 422; "Honoring the Past, Focused on the Future," timeline, *Kieways* 64 (October 2008): 10–13; Harold B. Meyers, "The Biggest Invisible Builder in the World," *Fortune* 73 (April 1966): 149.

37. "Honoring the Past, Focused on the Future"; Hollis Limprecht, *The Kiewit Story: Remarkable Man, Remarkable Company* (Omaha, NB: Omaha World-Herald, 1981): 59–60; Meyers, "The Biggest Invisible Builder in the World," 148–51. On the collapse in construction, see, "Building Activity in Los Angeles: The Year 1929," *Eberle Economic Service* 7, no. 7 (17 January 1930); "Building Activity in Los Angeles: The Year 1930," *Eberle Economic Service* 8, no. 3 (19 January 1931); "Building by Districts and Types," *Eberle Economic Service* 8, no. 12 (23 March 1931); "Building Activity in Los Angeles: The Year 1931," *Eberle Economic Service* 9, no. 7 (15 February 1932); Fogelson, *Downtown*, 218–21. Copies of *Eberle Economic Service*, a weekly newsletter, may be read at the Southern Regional Library Facility, University of California, Los Angeles.

38. Meyers, "The Biggest Invisible Builder in the World," 150; Norman L. "Red" Metcalf interview.

39. Cf. Jill Bettner, "Only the Best," *Forbes*, 1 August 1983, 106; Bruce Grewcock, "Growing the Team," president's message, *Kieways* 63 (April 2007).

40. "Honoring the Past, Focused on the Future"; Meyers, "The Biggest Invisible Builder in the World," 150–51. On defining groups within organizations in terms of real or invented enemies, see Warren Bennis, "The Secret of Great Groups," *Leader to Leader* 3 (Winter 1997): 29-33; Bennis and Biederman, *Organizing Genius*, 23–5. In the cases cited by Bennis and Biederman, the perceived enemy was invariably external to the organization.

41. Meyers, "The Biggest Invisible Builder in the World," 150; Michael R. Adamson, Memorandum of Conversation with Russell J. Osterman, Pasadena, CA, 29 July 2009; Alan D. Murk interview, 15 December 2008.

42. Bennis and Biederman, *Organizing Genius*, 21.

43. Limprecht, *The Kiewit Story*, 72–7; Meyers, "The Biggest Invisible Builder in the World," 147–51, 198 (Peter Kiewit quoted); Peter Kiewit quoted in Bruce Grewcock, "Pleased, But Not Satisfied," president's message, *Kieways* 63 (January 2007).

44. Covell, "Peter Kiewit Sons', Inc.," 422; Limprecht, *The Kiewit Story*, 47–8; Meyers, "The Biggest Invisible Builder in the World," 151.

45. "86 Firms Rank as Top Contractors," *ENR*, 2 July 1964, 53–4; Grewcock, "Growing the Team"; Limprecht, *The Kiewit Story*, 47–8; Meyers, "The Biggest Invisible Builder in the World," 148. On Sloan and General Motors, see Chandler, *The Visible Hand*, chap. 14.

46. Alan D. Murk interview, 15 December 2008; Dean E. Stephan interview (quoted). The author's review of bid and contract files in the records of the Department of Public Works uncovered no contracts awarded to Peter Kiewit Sons' for the construction of any state building in Los Angeles from 1949 to 1954, but records do not cover all projects for the period. Murk suggests that it may have been an exhibition building. If so, the building may have been one of several agricultural exhibit halls erected in the area at this time, other than the projects associated with the renovation and expansion of the California State Exposition— renamed the Museum of Science and Industry of California in 1953 ("Summary of Activities," reports for June, July, and November 1952, February, May, and September 1953, and March 1954; Contract Architects Section, Progress Reports: No. 14, 1 January 1951; No. 19, 1 June 1951 [all found in DPW, box 13]; Contract Architects Section, Progress Reports: No. 26, supplemental, 1 January 1952, No. 29, 1 April 1952; No. 30, 1 May 1952, No. 32, 1 July 1952, No. 38, supplemental, 1 January 1953 [all found in DPW, box 14]; Boyd to Durkee, 24 January 1952; Boyd to Durkee, 27 February 1952; Boyd to Venator, 18 April 1952; Boyd to Durkee, 18 June 1952 [all found in DPW, box 15]; California Department of Public Works, Division of Architecture, *Construction Program: January 1, 1946 to June 30, 1954* [Sacramento, 1955]; idem, *Construction Program: July 1, 1955 to June 30, 1960* [Sacramento, 1960]; "State Exposition Name Made 'More Fitting,'" *LAT*, 24 May 1953, A5; "New $268,700 Wing Set for State Museum," *LAT*, 22 January 1956,

E15; "10-Year Plan on Exposition Park Offered," *LAT*, 28 February 1957, B1; "Big Addition Planned for L.A. Museum," *LAT*, 24 March 1957, F1; "Bids Received for Museum," *LAT*, 8 May 1958, B7.)

47. Albert W. Fink interview; Thomas D. Verti interview; William "Red" Ward interview.

48. Peterson, "History and Development of Precast Concrete in the United States," 487 (quoted), 488; Alan D. Murk interview, 15 December 2008 (quoted).

49. Charles J. Pankow, Raymond E. Davis Lecture. Pankow stated that on-site precasting took place in 1954, but note that Local 78 of the United Association of Plumbers and Pipefitters went on strike in July 1955: the only such labor action in Southern California between 1952 and 1955. It lasted less than a week ("Plumbers Strike; Shutdown Slated," *LAT*, 2 July 1955, I:2; "Plumbers Set to Go Back to Work Today," *LAT*, 8 July 1955).

50. Pankow quoted in Kinchen, "A Love Affair with Concrete"; Alan D. Murk interview, 15 December 2008; Peterson, "History and Development of Precast Concrete in the United States."

51. Steve Pankow interview; Lee Sandahl interview; Dean E. Stephan interview; "Top Projects of the Century in Washington State," *Daily Journal of Commerce* (Seattle, WA), special issue, 9 December 1999, www.djc.com/special/century.

52. Dean Browning interview; Conan "Doug" Craker interview; Albert W. Fink interview, 13 February 2009; Robert Heisler interview; George F. Hutton interview; Brad Inman interview; Richard M. Kunnath interview, 21 October 2008; Robert Law interview; Norman L. "Red" Metcalf interview (quoted); Alan D. Murk interview, 15 December 2008 (quoted); Mark J. Perniconi interview; Joseph Sanders interview; Kevin Smith interview; Thomas D. Verti interview; Russell L. Wahl interview (quoted); Dick Walterhouse interview.

53. On the group of tinkerers like himself that Henry Ford gathered around him, see Zunz, *Making America Corporate*, 79–90. On the culture of Silicon Valley start-ups, see AnnaLee Saxenian, *Regional Advantage: Culture and Competition in Silicon Valley and Route 128* (Cambridge, MA: Harvard University Press, 1994), chap. 2. On the characteristics that "Great Groups" share, see Bennis and Biederman, *Organizing Genius*, chap. 1.

54. Conan "Doug" Craker interview.

55. Brad Inman interview; "New Large Construction Firm Based in Altadena," *Pasadena Independent Star-News*, 4 August 1963.

56. Michael R. Adamson, Memorandum of Conversation with Russell J. Osterman, Pasadena, CA, 29 July 2009.

57. George F. Hutton interview; Alan D. Murk interview, 15 December 2008.

58. Norman L. "Red" Metcalf interview, "Employee in Focus: Jack Grieger," *CPI News* 2 (Summer/Fall 1984); "Employee in Focus: Norman 'Red' Metcalf," *CPI News* 3 (Fall 1985).

59. "Civil Engineering Alumni Achievement Awards," *Transitions* 4 (Summer 1999): 15; "Alumni Features," *Transitions* 5 (Spring/Summer 2000): 12; Thomas

D. Verti interview (quoted). *Transitions* is the newsletter of the School of Civil Engineering at Purdue University.

60. Dean E. Stephan interview.

61. Pankow quoted in "New Large Construction Firm Based in Altadena."

62. "Cement Firm Plans 13-Story Office Building," *LAT*, 26 February 1959, II:1; "The Image of the American Cement Corporation," in advertising supplement, "The American Cement Building . . . Concrete Symbol of the Future," *LAT*, 11 July 1961 (quoted). On the use of the skyscraper as a tool of public relations in early twentieth century America, see the references in note 17, page 390. On the Alcoa Building, see Stuart W. Leslie, "The Strategy of Structure: Architectural and Managerial Style at Alcoa and Owens-Corning," *Enterprise & Society* 12 (December 2011): 863–902.

63. Charles J. Pankow, Raymond E. Davis Lecture (quoted); "The American Cement Building . . . Concrete Symbol of the Future" (quoted); "Cement Firm Plans 13-Story Office Building"; "Ground Broken for Wilshire Structure," *LAT*, 2 August 1959, VI:16; James Kesler and Albert Salibian, "Decorative Concrete Grillwork Doubles as Load-Bearing Wall," *ENR*, 17 August 1961, 46. For a profile of DMJM, see, "Around the World in 14 Years," *ENR*, 11 February 1960, 53–6. DMJM Design is now part of AECOM.

64. "Kiewit executive" quoted in "American Cement Building's New Ideas Began with New Bidding," in advertising supplement, "The American Cement Building . . . Concrete Symbol of the Future"; Charles J. Pankow, Raymond E. Davis Lecture (quoted).

65. "13-Story Office Building On Wilshire Blvd.," *Kieways* 16 (March-April 1961): 8–9; Kesler and Salibian, "Decorative Concrete Grillwork Doubles as Load-Bearing Wall," 46–8; "Reinforced-Concrete Office Building Rises," *LAT*, 21 August 1960, J2; "New Building Illustrates Varied Uses of Concrete," *LAT*, 16 July 1961, J13.

66. Kesler and Salibian, "Decorative Concrete Grillwork Doubles as Load-Bearing Wall," 48 (quoted).

67. Ibid., 48 (quoted); "American Cement Building's New Ideas Began with New Bidding" (quoted). On the contemporary use of slipforming elsewhere in the state, see "How to Build High-Rise Apartments for Less," *Buildings* 57 (June 1963); August E. Waegemann, "Unique Design Approach Combines Three Structural Concepts to Achieve Economy," *Western Architect and Engineer* 219 (May 1960): 42–4.

68. "13-Story Office Bldg. On Wilshire Blvd.," 9; Charles J. Pankow, Raymond E. Davis Lecture.

69. Kesler and Salibian, "Decorative Concrete Grillwork Doubles as Load-Bearing Wall," 48.

70. Robert Heisler interview; Lee Sandahl interview.

71. Lee Sandahl interview; "Air-Conditioning with Low-Cost Gas," in advertising supplement, "The American Cement Building . . . Concrete Symbol of the Future."

72. "First and C Building," technical report, CPBL, Pasadena, CA, n.d. (quoted); "Concrete Frame Made Quakeproof," *ENR*, 23 May 1963, 71–2.

73. "Concrete Frame Made Quakeproof." The UBC classified earthquake hazard on a scale from 0 (least hazardous) to 4 (most hazardous). Building officials used the values to determine the strengths of structural components needed to resist seismic forces. The UBC was first published in 1927 by the International Council of Building Officials and updated periodically until 1997. It was superseded in 2000 by the International Building Code, issued by the International Code Council, which was formed by the merger of three organizations, including the ICBO.

The First and C Building's height would stand as a record for a concrete structure for a quarter of a century, owing to the preference in the UBC and local building codes for steel over concrete in tall structures and the need for consensus within the engineering community to achieve code changes.

Coincidentally, the 13-story limit was the same as Los Angeles's height limit, which had been repealed in 1957. That limit, imposed in 1904 and updated in 1911, had nothing to do with seismic safety, however. Rather, it was adopted to limit urban density.

74. Ibid.; "24-Story San Diego Office Renamed for Major Tenant," *LAT*, 24 February 1963, J17.

75. Starr, *Golden Dreams*, 70–3.

76. "First and C Building," technical report; George F. Hutton interview; Norman L. "Red" Metcalf interview; Alan D. Murk interview.

77. "First and C Building," technical report; "Concrete Frame Made Quakeproof," 72; George F. Hutton interview.

78. "Concrete Frame Made Quakeproof," 71 (quoted), 72.

79. George F. Hutton interview; Covell, "Peter Kiewit Sons' Inc.," 423; Meyers, "The Biggest Invisible Builder in the World," 151.

80. "A Story of Problems, Experimentations, and Success," *SB&C*, 26 April 1963, 8; Tom Cameron, "New Techniques Expand Uses of Concrete," *LAT*, 10 March 1963, P1, P23.

81. Robert L. Heisler interview; Lee Sandahl interview; Norman L. "Red" Metcalf interview.

82. Ibid.; Cory Buckner, "Architect A. Quincy Jones," Eichler Network, www.eichlernetwork.com/ENStry20.html; Starr, *Golden Dreams*, 42–9. On Jones and Emmons, see also, Cory Buckner, *A. Quincy Jones* (London: Phaidon, 2002). On the Case Study House project, see, Reyner Banham, *Los Angeles: The Architecture of Four Ecologies* (Berkeley: University of California Press, 1971), 205–15; Esther McCoy, *Modern California Houses: Case Study Houses, 1945–1962* (New York: Reinhold, 1962); Elizabeth A. T. Smith, *Blueprints for Modern Living: History and Legacy of the Case Study Houses* (Los Angeles and Cambridge, MA: Museum of Contemporary Art and MIT Press, 1989). The Case Study house was not built because the Los Angeles City Council's zoning committee rejected it.

83. "A Story of Problems, Experimentations, and Success," 8 (quoted), 9, 76–8; Buckner, "Architect A. Quincy Jones"; Cameron, "New Techniques Expand Uses of Concrete"; Norman L. "Red" Metcalf interview.

84. "Concrete Awards to Southland Structures," *SB&C*, 25 February 1966, 18–19.

85. Alan D. Murk interview, 15 December 2008; "Hillcrest North Medical Center in San Diego Utilizes Over 800 Channel-shaped Concrete Panels," *SB&C*, 25 October 1963, 46–50. In April 1959, Ward Deems launched Deems-Martin Associates as a branch office of AC Martin Partners, a Los Angeles architectural firm. Bill Lewis, who, like Deems, earned his degree from the School of Architecture at the University of Southern California and worked for AC Martin, joined Deems in San Diego shortly thereafter. The firm changed its name several times, eventually becoming Deems Lewis McKinley Architects (Modern San Diego, www.modernsandiego.com/DeemsLewis.html).

86. Hillcrest North Medical Center in San Diego Utilizes Over 800 Channel-shaped Concrete Panels," 46.

87. George F. Hutton interview; Norman L. "Red" Metcalf interview.

88. Margaret B. W. Graham, "Entrepreneurship in the United States, 1920–2000," in *The Innovation of Enterprise: Entrepreneurship from Ancient Mesopotamia to Modern Times*, ed. David S. Landes, Joel Mokyr, and William J. Baumol (Princeton, NJ: Princeton University Press, 2010), 401–25.

89. Knight quoted in Ray Hebert, "Work on Music Center to Begin in December," *LAT*, 27 June 1961, III:2; "Los Angeles Hoists Flag of Culture," *Business Week*, 21 November 1964, 32; "Music Center Construction Will Start Next Summer," *LAT*, 30 December 1960, III:1.

90. Ray Hebert, "Plans for Expanded Music Center Given Approval by Supervisors," *LAT*, 8 March 1961, III:1, 3; "Los Angeles Hoists Flag of Culture"; "Music Center Construction Will Start Next Summer"; "Supervisors OK Financing of $15 Million Music Center," *LAT*, 4 November 1960, II:1; John Anson Ford, *Thirty Explosive Years in Los Angeles County* (San Marino, CA: The Huntington Library Press, 2010 [1961]), chap. 22; Starr, *Golden Dreams*, 153–60.

91. Leslie, "The Strategy of Structure," 889.

92. "What Makes an Internationally Famous Architect's Firm Tick?" *SB&C*, 23 June 1961, 8–9, 42–4, 43 (quoted); "Music Center Architect Nursemaids Handiwork," *LAT*, 4 August 1963, III:1 (quoted); Welton Becket and Associates, *Vision . . . Through Supervision* (Los Angeles: the firm, 1964); Arthur Love interview (quoted); William Dudley Hunt, *Total Design: Architecture of Welton Becket and Associates* (New York: McGraw-Hill, 1972). The firm's shopping centers in California included Broadway (Anaheim), Hillsdale (San Mateo), Los Altos (Long Beach), Panorama (San Fernando), Stonestown (San Francisco), and Stanford (Palo Alto).

93. "Music Center Contract Finally Awarded," *ENR*, 22 March 1962, 66; "Work on Music Center to Begin in December" (quoted).

94. Ray Hebert, "Lease and Contract Signed for Start of Music Center," *LAT*,

28 February 1962, II:1; idem, "Work on Music Center to Begin in December"; "Mayor Favors Go-Ahead on New Music Center," *LAT*, 17 October 1961, II:1; "Music Center Bid Call Issued by Supervisors," *LAT*, 21 October 1961, I:3.

95. George F. Hutton interview; Norman L. "Red" Metcalf interview.

96. "Bids Made on Music Center Units," *LAT*, 29 November 1961, II:1; Hebert, "Lease and Contract Signed for Start of Music Center"; idem, "Music Center Work Will Start Today," *LAT*, 12 March 1962, II:1; idem, "Way Clear for Quick Start on Music Center," *LAT*, 1 February 1962, II:1; "Music Center Contract Finally Awarded."

97. Hebert, "Music Center Work Will Start Today"; "Music Center Architect Nursemaids Handiwork," III:1 (quoted); "Music Center for the Performing Arts," *Kieways* 18 (March-April 1963): 6–8.

98. Conan "Doug" Craker interview; George F. Hutton interview; Norman L. "Red" Metcalf interview; "New Large Construction Firm Based in Altadena." Volke began working for Kiewit in 1952 on the Willow Creek Dam, and was experienced in both heavy construction and building work ("13 Stories of Slipform Construction," *Kieways* 18 [May-June 1963]: 5).

99. "Construction Progressing on Largest of Three Buildings," *SB&C*, 26 July 1963, 50; Ray Hebert, "Music Center Makes Giant Strides in Plans, Building, and Donations," *LAT*, 2 December 1962, III:1; "Music Center Parking Will Begin April 1," *LAT*, 13 March 1963, II:1.

100. George F. Hutton interview; "Construction Progressing on Largest of Three Buildings"; "Giant Crane Speeds Music Center Work," *LAT*, 12 August 1962, II:1; Ray Hebert, "Music Center's Classic Beauty Taking Shape," *LAT*, 7 April 1963, C1; idem, "Music Center Makes Giant Strides in Plans, Building, and Donations"; "Music Center Architect Nursemaids Handiwork."

101. Arthur Love interview.

102. Norman L. "Red" Metcalf interview; Alan D. Murk interview, 15 December 2008; Lee Sandahl interview. Though its employees may have felt otherwise, the building division was not ignored entirely. The American Cement Building, for one, was featured on the cover of the March-April 1961 issue of *Kieways*, the company's magazine.

103. Conan "Doug" Craker interview.

104. George F. Hutton interview; Michael R. Adamson, Memorandum of Conversation with Russell J. Osterman, Pasadena, CA, 29 July 2009.

105. Conan "Doug" Craker interview; Robert L. Heisler interview; Norman L. "Red" Metcalf interview; Steve Pankow interview; Charles Pankow, Inc., articles of incorporation, 17 May 1963, State of California, Secretary of State, Sacramento; "MacArthur Shops Let Contract," *Oakland Tribune*, 6 September 1963, I:15; "Unique Features Planned for $10 Million Center," *Oakland Tribune*, 26 May 1963, R1.

106. George F. Hutton interview; Norman L. "Red" Metcalf interview; Alan D. Murk interview, 15 December 2008. David Boyd replaced Murk. It is not clear whether Pankow assigned Boyd to the project or if Murk nominated him as his

replacement ("Hillcrest North Medical Center in San Diego Utilizes Over 800 Channel-shaped Concrete Panels").

107. George F. Hutton interview; Norman L. "Red" Metcalf interview; Alan D. Murk interview, 15 December 2008; "New Large Construction Firm Based in Altadena"; "Construction Firm Formed," *LAT*, 25 August 1963, Q10.

108. Norman L. "Red" Metcalf interview.

109. Ibid.

110. Ibid.; "Contract Awarded for Valley Music Theater," *LAT*, 19 December 1963, F1; "Music Theater Dome Formed on Earth," *SB&C*, 24 April 1964, 44–9.

Chapter 2

1. Rick Pankow interview; Steve Pankow interview. For three years, Doris Pankow was CPI's only administrative assistant.

2. Pankow quoted in "New Large Construction Firm Based in Altadena."

3. Michael R. Adamson, Memorandum of Conversation with Richard M. Kunnath, San Francisco, 9 July 2009; Albert W. Fink interview, 25 March 2009; Robert Heisler interview.

4. Michael R. Adamson, Memorandum of Conversation with Russell J. Osterman, Pasadena, CA, 29 July 2009. See, also, Dean E. Stephan interview.

5. Alan D. Murk interview, 15 December 2008; Dean E. Stephan interview. On the difficulties faced by Perini and Turner as public companies, cf. April Dougal Gasbarre and Dorothy Kroll, "The Turner Corporation," in *International Directory of Company Histories*, ed. Tina Grant, vol. 23 (Detroit: St. James Press, 1998), 485–8; Sara Pendergast and Tom Pendergast, "Perini Corporation," in *International Directory of Company Histories*, ed. Paula Kepos, vol. 8 (Detroit: St. James Press, 1994), 418–21.

6. Conan "Doug" Craker interview (quoted); Robert Heisler interview; Timothy P. Murphy interview; Pankow quoted in Minutes, Offsite Managers' Meeting, Monterey, CA, 2–5 May 1976.

7. Conan "Doug" Craker interview; finding aid to the Sollitt Construction Company collection; George Sollitt Construction website, www.sollitt.com. Ralph Shannon Sollitt had three children with wife, Ellen Anita, daughter of Thomas Bishop of Chicago: Gwen Charlotte (born 7 July 1918), Ralph Bishop (born 3 March 1926), and Gloria Sharillon (born 30 December 1929) (Roll, *Indiana*). Sollitt's brother, George Thomas Sollitt, assumed the leadership of the Chicago branch of the firm after the death of Ralph Sollitt, the father, in 1940. The name of that business changed to George Sollitt Construction. George died in 1955, but there were no further name changes to the business, which traces its origins to 1838, and remains active today. By 1993, Sollitt Construction Company had ceased doing business.

8. Timothy P. Murphy interview; Steve Pankow interview; Rick Pankow interview; Michael R. Adamson, Memorandum of Conversation with Doris Pankow, Steve Pankow, and Rick Pankow, Altadena, CA, 12 February 2009;

Michael Stremfel, "All for One," *Los Angeles Business Journal*, 3 June 1991, 40 (quoted).

9. Conan "Doug" Craker interview; Timothy P. Murphy interview; Dean E. Stephan interview.

10. Dean Browning interview; Dean E. Stephan interview; Cohan, "8 Billion Dollar Decade," D33; Fisher, *The Boom in Office Buildings*; Daniel Seligman, "The Future Office-Building Boom," *Fortune* (March 1963): 84–6, 216 (quote), 218; Royal Shipp and Robert Moore Fisher, "The Postwar Boom in Hotels and Motels," Federal Reserve Staff Economic Study (Washington, DC: Board of Governors of the Federal Reserve System, 1965).

11. Conan "Doug" Craker interview; Richard M. Kunnath interview, 21 October 2008 (quoted); Charles Pankow Builders, Ltd., *40 Years of Building Innovation, 1963–2003*, marketing brochure, CPBL, Pasadena, CA.

12. *United States v. Paramount Pictures, Inc. et al.*, 334 U.S. 131 (1948). The antitrust case involved eight studios. Only the five studios with motion picture theater chains signed the consent decree. With their vertically integrated stranglehold on distribution removed, the Big Five studios experimented with new revenue models, with mixed results.

13. "E. E. Herrscher Dies; Built Century City," *LAT*, 10 June 1983, C4; Stephanie Frank, "Building Century City: Twentieth Century Fox and Urban Development in Los Angeles, 1920–1975," a paper presented at the annual meeting of the Urban History Association, Las Vegas, October 2010; Richard Longstreth, *The American Department Store Transformed, 1920–1960* (New Haven, CT: Yale University Press, 2010), 243–5; "MacArthur/Broadway 'Go' Tomorrow," *Oakland Tribune*, 8 September 1965, A2; "Mayfair Executive: Founder of Century City," *LAT*, 6 June 1965, A2; William H. Nichols, *Seabiscuit: The Rest of the Story* (Mustang, OK: Tate, 2007), 72; H. May Spitz, "Upscale Living in Tinseltown's Back Lot," *LAT*, 11 January 2004, K2; *Walter N. Shorenstein: An Oral History*, conducted by Richard Cándida Smith and Laura McCreery, 2002 (Regional Oral History Office, Bancroft Library, University of California, Berkeley, 2010), 30–1.

The idea of the "new city" was architect Victor Gruen's. Even as he designed regional shopping centers that drained the life out of downtown shopping districts, Gruen was confidently prescribing the new building type as the antidote to their deterioration. He applied the term "new city" to the development of greenfield sites as communities composed of clusters of neighborhood units. Separated by greenbelts and linked by ring roads, clusters of communities anchored by regional shopping centers would constitute "a new type of metropolitan organization," which Gruen labeled "the 'cellular' form of urban planning'" (Gruen quoted in "Cities in Trouble—What Can Be Done," interview with Victor Gruen, *U.S. News & World Report*, 20 June 1960, 86; M. Jeffrey Hardwick, *Mall Maker: Victor Gruen, Architect of an American Dream* (Philadelphia: University of Pennsylvania Press, 2004), chaps. 7–8; Alex Wall, *Victor Gruen: From Urban Shop to New City* (Barcelona: Actar, 2005), 116–97.

14. Edmond E. Herrscher, "A Warm Welcome," *Oakland Tribune*, 8 September 1965, A3; "Key Plopped into Bay to Start Center," *Oakland Tribune*, 8 September 1965, A4; "Mayfair Executive: Founder of Century City."

15. Herrscher quoted in "Unique Features Planned for $10 Million Center," R1; "Key Plopped into Bay to Start Center." On the overwhelmingly suburban location of first generation regional shopping centers, see Lizabeth Cohen, "From Town Center to Shopping Center: The Reconfiguration of Community Marketplaces in Postwar America," *American Historical Review* 101 (October 1996): 1050–81; Victor Gruen and Larry Smith, *Shopping Towns USA: The Planning of Shopping Centers* (New York: Reinhold, 1960), chap. 12; Hardwick, *Mall Maker*, chaps. 4–6; Kenneth T. Jackson, "All the World's a Mall: Reflections on the Social and Economic Consequences of the American Shopping Center," *American Historical Review* 101 (October 1996): 1113–4; Witold Rybczynski, "The New Downtowns," *Atlantic Monthly* (May 1993): 98–103.

16. Richard Longstreth, "Sears, Roebuck and the Remaking of the Department Store, 1924–42," *Journal of the Society of Architectural Historians* 65 (June 2006): 281–3; "Parking on the Roof," *ENR*, 9 December 1937, 939–41; "Store Building for Sears, Roebuck and Company, Los Angeles, Calif.," *AF* 72 (February 1940): 70–6.

17. "Unique Features Planned for $10 Million Center"; "Prestressed Concrete Institute Makes Awards for 1965," *SB&C*, 27 August 1965, 61; "Some Incidental Facts and Figures," *Oakland Tribune*, 8 September 1965, A5; Howard Gillette Jr., "The Evolution of the Planned Shopping Center in Suburb and City," *Journal of the American Planning Association* 51 (Autumn 1985): 449–51. Figures on the leasable areas of early regional shopping centers (at the time of their opening) may be found in Gruen and Smith, *Shopping Towns USA*, table on 216–8.

18. Herrscher, "A Warm Welcome" (quoted); Shapiro quoted in "Unique Features Planned for $10 Million Center," R3; "MacArthur/Broadway 'Go' Tomorrow"; "Some Incidental Facts and Figures."

19. Conan "Doug" Craker interview; George F. Hutton interview; Richard M. Kunnath interview, 21 October 2008; Norman L. "Red" Metcalf interview; Steve Pankow interview; Charles Pankow, Inc., articles of incorporation, 17 May 1963, State of California, Secretary of State, Sacramento; "Home Office Planned for Savings, Loan Firm," *LAT*, 16 June 1963, N32; "MacArthur Shops Let Contract"; "Several New Buildings in Area Planned," *LAT*, 9 December 1963; "Start," photograph, *Oakland Tribune*, 17 May 1963, I:19; "Stained Glass Skylight Feature of New Office," *LAT*, 30 August 1964, L3. Shapiro's award-winning design for the 3-story Columbia Savings and Loan Association Building featured a 1,300-square-foot *dalle-de-verre* skylight by Roger Darricarrere and a 45-foot-long brass screen-waterfall structural fountain by Taki. Richard Bradshaw was structural engineer. In December 2009, the Los Angeles City Council approved an environmental impact report for the redevelopment of the block on which Columbia Savings Building was located. Demolition of the structure began immediately.

20. Conan "Doug" Craker interview; George F. Hutton interview; Alan D. Murk interview, 15 December 2008; "Some Incidental Facts and Figures." Henry J. Kaiser set up Permanente Cement in 1939 to supply his Shasta Dam project. During World War II, it became the largest cement production operation in the world (Foster, *Henry J. Kaiser*, 64–6, 165–6).

21. George F. Hutton interview; Lin, "Tall Buildings in Prestressed Concrete," 168; "Prestressed Concrete Institute Makes Awards for 1965," 61; "Some Incidental Facts and Figures" (quoted); "Unique Features Planned for $10 Million Center." The cost to construct the office building was less than $9 per square foot.

22. "Key Plopped into Bay to Start Center"; "MacArthur/Broadway 'Go' Tomorrow"; "Unique Features Planned for $10 Million Center."

23. George F. Hutton interview; Lin, "Tall Buildings in Prestressed Concrete," 166–8; "Prestressed Concrete Institute Makes Awards for 1965."

24. Michael R. Adamson, Memorandum of Conversation with Doris Pankow, Steve Pankow, and Rick Pankow, Altadena, CA, 12 February 2009; "New Large Construction Firm Based in Altadena."

25. Conan "Doug" Craker interview; A. J. Dell'Isola, "Value Engineering in Construction," *Civil Engineering* 36 (September 1966): 59–61.

26. Conan "Doug" Craker interview.

27. Ibid.

28. Robert P. Kessler and Chester W. Hartman, "The Illusion and the Reality of Urban Renewal: A Case Study of San Francisco's Yerba Buena Center," *Land Economics* 49 (November 1973): 441; Chester Hartman, *City for Sale: The Transformation of San Francisco*, with Sarah Carnochan, rev. ed. (Berkeley: University of California Press, 2002), 64, 217–9. *City for Sale* remains the most comprehensive review and critique of redevelopment in San Francisco. It focuses on the Yerba Buena Center project to illustrate the dominance of downtown business interests in the city's growth trajectory. Brian T. Godfrey sets San Francisco's urban renewal in geographical and historical context in, "Urban Development and Redevelopment in San Francisco," *Geographical Review* 87 (July 1997): 309–33.

29. Fogelson, *Downtown*, chap. 7; Leslie Fulbright, "Sad Chapter in Western Addition History Ending," *San Francisco Chronicle*, 21 September 2008, B1; Gregory Paul Williams, *The Story of Hollywood: An Illustrated History* (Los Angeles: BL Press, 2006), 335–73; Victor Valle, *City of Industry: Genealogies of Power in Southern California* (New Brunswick, NJ: Rutgers University Press, 2009), 97–102, 119–28.

30. Fogelson, *Downtown*, 146–52; Hartman, *City for Sale*, 8; Alison Isenberg, *Downtown America: A History of the Place and the People Who Made It* (Chicago: University of Chicago Press, 2004), chap. 5.

31. Tom Sitton, *Los Angeles Transformed: Fletcher Bowron's Urban Reform Revival, 1938–1953* (Albuquerque: University of New Mexico Press, 2005), chap. 8; Gwendolyn Wright, *Building the Dream: A Social History of Housing in America* (New York: Pantheon, 1981), chap. 12.

32. Kessler and Hartman, "The Illusion and the Reality of Urban Renewal," 441–4; Valle, *City of Industry*, 97–102.

33. Fulbright, "Sad Chapter in Western Addition History Ending"; Hartman, *City for Sale*, 25, 62–5; idem, "The Housing of Relocated Families," in *Urban Renewal: The Record and the Controversy*, ed. James Q. Wilson (Cambridge, MA: MIT Press, 1966), 292–335.

34. City of San Francisco, Planning Commission, Minutes of Special Meeting, 11 October 1961; Joseph Eichler, remarks, Geneva Towers Dedication, 4 December 1965; Dick Girvin, press release, Geneva Towers Dedication, 4 December 1965 (all found in Oakland and Imada Collection (2002-03), EDA, folder: Geneva Terrace Correspondence); Eichler quoted in "Joseph Eichler's Success Formula: 'I try everything—I'm never satisfied,'" *American Builder* 85 (August 1963): 67; Paul Adamson and Marty Arbunich, *Eichler: Modernism Rebuilds the American Dream* (Salt Lake City, UT: Gibbs Smith, 2002), 205–15.

35. Kulka to Oakland, 16 August 1961; Claude Oakland, "Proposed Visitacion Valley Development for Eichler Homes, Inc., 10 October 1961; idem, "Eichler Homes Development in Visitacion Valley, 3 May 1962; Joseph Eichler, remarks, Geneva Towers Dedication, 4 December 1965 (quoted); Dick Girvin, press release, Geneva Towers Dedication, 4 December 1965 (all found in Oakland and Imada Collection (2002-03), EDA, folder: Geneva Terrace Correspondence); Claude Oakland, "Specifications for Western Addition Garden Apartments for Eichler Homes," 12 May 1961, Oakland and Imada Collection (2002-03), EDA, folder: Eichler Homes—FHA—Laguna Heights Project; "Joseph Eichler's Success Formula," 67–8; Adamson and Arbunich, *Eichler*, 205–11; Richard Brandi, "Diamond Heights: Combining Suburban Ideals with San Francisco's Urban Amenities," paper presented at the annual meeting of the Society for American City and Regional Planning History, Oakland, CA, October 2009; Annie Nakao, "Time to 'Blow that Sucker Up,'" *San Francisco Examiner*, 14 May 1998.

36. Adamson and Arbunich, *Eichler*, 212–5; Dave Weinstein, "Claude Oakland: Modern Homes for the Masses," *San Francisco Chronicle*, 1 January 2005, F1.

37. David Boyd, "Eichler Home Apartments," technical report, n.d. [1964], CPBL, Pasadena, CA; "Slipforms Mold All Bearing Walls for Apartment Towers," *Construction Methods and Equipment* (August 1964).

38. Ibid.

39. Ibid.; Pankow, "Slipform Construction of Buildings," 34–4; Robert Law interview.

40. Dick Girvin, press release, Geneva Towers Dedication, 4 December 1965, Oakland and Imada Collection (2002-03), EDA, folder: Geneva Terrace Correspondence; Paul Adamson, "Last Days of Eichler Homes," Eichler Network, www.eichlernetwork.com/article/last-days-eichler-homes, 9 November 2011 (quoted); Adamson and Arbunich, *Eichler*, 216–22. In May 1998, Geneva Towers was demolished. By the time that HUD took over the project in 1991, it was ridden with crime and gang violence. The agency foreclosed on the property for safety

reasons. The townhouses remain standing (Jason B. Johnson, "Geneva Towers to Tumble," *San Francisco Chronicle*, 16 May 1998; Nakao, "Time to 'Blow that Sucker Up'").

41. Dick Girvin, press release, Geneva Towers Dedication, 4 December 1965, Oakland & Imada Collection (2002–03), EDA, folder: Geneva Terrace Correspondence; "Joseph Eichler's Success Formula," 66–8, 67 (Eichler quoted). For discussion of the impact of setback laws and zoning regulations on postwar urban environments see, for instance, Carl Abbott, *Portland: Planning, Politics, and Growth in a Twentieth-Century City* (Lincoln: University of Nebraska Press, 1983), chaps. 10–12; Robert Bruegmann, *Sprawl: A Compact History* (Chicago: University of Chicago Press, 2005), 42–50, 101–7; William A. Fischel, *The Homevoter Hypothesis: How Home Values Influence Local Government Taxation, School Finance, and Land-Use Policies* (Cambridge, MA: Harvard University Press, 2001), chap. 10; James Howard Kunstler, *The Geography of Nowhere: The Rise and Decline of America's Man-Made Landscape* (New York: Simon & Schuster, 1993), chap. 8.

42. Hartman, *City for Sale*, chap. 1; Mitchell Schwarzer, "Yerba Buena as an Incubator for Architectural Innovation," paper presented at the annual meeting of the Society for American City and Regional Planning History, Oakland, CA, October 2009.

43. Gray Brechin, *Imperial San Francisco: Urban Power, Earthly Ruin* (Berkeley: University of California Press, 1999), 44–53, 125–9, 142–4, 169–70; John Bowden Martyn, "The Dynamics of City Growth: An Historical Geography of the San Francisco Central District, 1850–1931," PhD diss., Department of Geography, University of California, Berkeley, 1967.

44. Hartman, *City for Sale*, chap. 3. Redevelopment agencies typically assemble land for resale at reduced prices to developers who agree to construct projects according the agency's plans.

45. Hartman, *City for Sale*, chap. 5, 112–16, 216–25.

46. Wright, *Building the Dream*, 237–9; idem, "The Evolution of Public Housing and Design in the San Francisco Bay Area," PhD exam, Department of Architecture, University of California, Berkeley, 22 November 1976, 44–9; "U.S. Department of Housing & Urban Development, 1930 to 2010," timeline, www.huduser.org/hud_timeline/index.html.

47. Lawrence M. Friedman, "Public Housing and the Poor: An Overview," *California Law Review* 54 (1966): 624–69; Wright, "The Evolution of Public Housing and Design in the San Francisco Bay Area," 45 (quoted), 46.

48. David Parry, "John Savage Bolles," *Encyclopedia of San Francisco*, www.sfhistoryencyclopedia.com/articles/b/bollesJohn.html; Wright, "The Evolution of Public Housing and Design in the San Francisco Bay Area," 43, 46–7.

49. Harry McHugh, "Yerba Buena Senior Community," technical report, 28 April 1971, CPBL, Pasadena, CA; Twin-Tower Apartments Started in S.F. for Low-Income Elderly," *Oakland Tribune*, 8 March 1970, C7; "Yerba Buena Turnkey

Apartment House Project Goes Up Fast With Slipforming," *Western Construction* (January 1971): 33–4. McHugh and Wayne Brown were the project engineers on Clementina Towers.

50. McHugh, "Yerba Buena Senior Community"; "Yerba Buena Turnkey Apartment House Project Goes Up Fast With Slipforming." Henderson died before the completion of the project. Jack Grieger finished the work.

51. Wai-Fah Chen and Charles Scawthorn, eds., *Earthquake Engineering Handbook* (Boca Raton, FL: CRC Press 2003), chap. 11, pp. 9–10.

52. Felix Kulka, Y. C. Yang, and T. Y. Lin, "Prestressed Concrete Building Construction Using Precast Wall Panels," paper presented at the VII Congress, Federation International de la Precontrainte, New York, May 1974, T. Y. Lin Papers, Bancroft Library, University of California, Berkeley, carton 1; Pankow, "Automation on the Job Site," 281–7, 287 (quoted).

53. John M. Findlay, *Magic Lands: Western Cityscapes and American Culture After 1940* (Berkeley: University of California Press, 1992), 129–42.

54. Hunter Properties website, www.hunterproperties.com/about.html. CPI also completed two smaller office buildings for Demmon and Hunter: the Dean Witter Building and the IBM Building, in San José and Menlo Park, respectively (see appendix A).

55. "Construction Moves Fast with Site Precasting," *ENR*, 3 June 1965, 30–1.

56. Kulka, Yang, and Lin, "Prestressed Concrete Building Construction Using Precast Wall Panels"; Pankow, "Automation on the Job Site," 281–2.

57. "Construction Moves Fast with Site Precasting," 31; Pankow, "Automation on the Job Site," 283–7.

58. Thomas J. Branson, "San José State College Residence Hall and Dining Facility," technical report, n.d. [1967], CPBL, Pasadena, CA; "Zig-Zag Dormitory Is Slipformed for Economy," *ENR*, 29 February 1968, 26–8; "Joe West Hall," Emporis.com, www.emporis.com/building/joewesthall-sanjose-ca-usa.

59. Pankow, "Automation on the Job Site," 287 (quoted); Kim Lum interview; Alan D. Murk interview, 15 December 2008; Thomas D. Verti interview; Dick Walterhouse interview; William "Red" Ward interview. In April 1966, CPI signed a $2.58 million contract to construct the First American Building for Demmon and Hunter (see appendix A).

60. One of the Southern California projects, the Sixth & Harvard Building, provided summer work for Rick Pankow, Charlie and Doris Pankow's second child, between one of his semesters at the University of California, Los Angeles. The building was constructed between December 1968 and December 1969 at the corner of West Sixth Street and Harvard Boulevard in the Mid-Wilshire district of Los Angeles (see appendix A).

61. "Borel Shopping Center Expected to Become a Peninsula Focal Point," *San Mateo Times*, 19 October 1967, 11A; E. J. Halcrow, "Swiss Echoes—A Vision Transplanted," report on Antoine Borel and the Borel Estate, June 1955, California and Western Manuscript Collection, M0119, Department of Special Collections,

Stanford University Libraries, Stanford, CA, box 1, folder 3. On the formation and investment activities of the Bank of California and the role of Spring Valley Water Company in the early development of San Mateo County, see Brechin, *Imperial San Francisco*, 38–44, 71–90. On California's ranchos—large land grants from the Spanish and Mexican eras—and their transfer to Yankee owners after the Mexican War, see Robert Glass Cleland, *The Cattle on a Thousand Hills: Southern California, 1850–1870* (San Marino, CA: The Huntington Library Press, 1941), 135–83; W. H. Hutchinson, *Oil, Land, and Politics: The California Career of Thomas Robert Bard*, vol. 1 (Norman: University of Oklahoma Press, 1965), 65–75; Leonard Pitt and Ramon A. Gutierrez, *Decline of the Californios: A Social History of the Spanish-Speaking Californias, 1846–1890* (Berkeley: University of California Press, 1999), esp. chaps. 5–6; W. W. Robinson, *Land in California: The Story of Mission Lands, Ranchos, Squatters, Mining Claims, Railroad Grants, Land Scrip, Homesteads* (Berkeley: University of California Press, 1979).

62. Fogelson, *Downtown*, 383–4.

63. "Borel Shopping Center Expected to Become a Peninsula Focal Point." On the problem of retail decentralization from the point of view of downtown interests generally, see Fogelson, *Downtown*, chap. 8; Isenberg, *Downtown America*, chap. 5. Peter Kiewit Sons' constructed the concrete trestle of the San Mateo-Hayward Bridge of precast concrete slabs placed on bents made of precast, prestressed concrete piles and a precast concrete pile cap. To create the 24-inch-diameter piles, Kiewit used a horizontal slipform method of precasting that Charlie Pankow developed while he was with the company (E. R. Foley and J. Philip Murphy, "World's Longest Orthotropic Section Feature of San Mateo-Hayward Bridge," *Civil Engineering* 38 [April 1968]: 54–8; Charles J. Pankow, Raymond E. Davis Lecture).

64. Michael R. Adamson, Memorandum of Telephone Conversation with William Wilson III, 29 October 2009.

65. Ibid.; Rick Pankow interview; "Owens-Illinois Is Moving to Mateo," *San Mateo Times*, 19 May 1966, 23; "Borel Shopping Center Expected to Become a Peninsula Focal Point"; "Petroleum Firm Moving to Borel," *San Mateo Times*, 1 April 1969, 25; "Fluor Utah in Move to Borel," *San Mateo Times*, 12 March 1970, 27. The "town and country" design of the 80,000-square-foot center did not offer CPI the opportunity to utilize precast or slipform operations, but Wilson selected CPI to build it because the contractor had performed well on the first project, and he had become familiar with the company's personnel.

66. "Borel Shopping Center Expected to Become a Peninsula Focal Point"; Kulka, Yang, and Lin, "Prestressed Concrete Building Construction Using Precast Wall Panels"; Pankow, "Automation on the Job Site," 287.

67. "Owens-Illinois Is Moving to Mateo"; "Petroleum Firm Moving to Borel"; "Fluor Utah in Move to Borel."

68. Michael R. Adamson, Memorandum of Telephone Conversation with William Wilson III, 29 October 2009; Webcor Builders, Inc., articles of incorporation, 19 January 1971, State of California, Secretary of State, Sacramento.

Duties of the project engineer quoted from "Project Organization," memorandum, 28 February 1998, CPBL, Pasadena, CA. Field engineers were responsible "for the technical execution, inspection, and quality control for structural concrete" and for helping project engineers write technical reports.

69. Dean Browning interview; Conan "Doug" Craker interview; Alan D. Murk interview, 15 December 2008; Lee Sandahl interview.

70. "New Office Plans Told," *San Mateo Times*, 13 February 1970, 4.

71. "Project Organization," memorandum, 28 February 1998, CPBL, Pasadena, CA.

72. Conan "Doug" Craker interview.

73. "Webcor Builders," *California Construction* (May 2000): 36; Webcor Builders, www.webcorbuilders.com.

74. Larry E. Greiner, "Evolution and Revolution as Organizations Grow," *Harvard Business Review* 50 (July-August 1972): 37–46, 42 (quoted). The article was reprinted with minor revisions as an *HBR* "classic" in the May-June 1998 issue (vol. 76, pp. 55–67).

75. Conan "Doug" Craker interview; George F. Hutton interview; Alan D. Murk interview, 15 December 2008.

Chapter 3

1. George F. Hutton interview; Robert Law interview; Timothy P. Murphy interview; Dean E. Stephan interview.

2. George F. Hutton interview.

3. *Construction in Hawaii, 1989* (Honolulu: Bank of Hawaii, 1989); *Hawaii '68: Annual Economic Review* (Honolulu: Bank of Hawaii, 1968), 22–3; *Hawaii '71: Annual Economic Review* (Honolulu: Bank of Hawaii, 1971), 26–7.

4. Dean E. Stephan interview.

5. George F. Hutton interview.

6. "Office-Parking Building Planned by Campbell Estate on Fort St.," *Sunday Star-Bulletin and Advertiser*, 11 October 1964, A1; Robyn Rickard, "New Campbell Block Plans Are Complete," *Honolulu Star-Bulletin*, 12 October 1964, A4; "Estate Plans Larger Downtown Building," *Honolulu Advertiser*, 31 October 1966, B4. Born in Ireland in 1826, James Campbell came to Maui by way of Tahiti at age 24. Ten years later he established a sugar processing mill that generated the wealth that he used to become one of Oahu's largest private landowners. In 1877 Campbell sold his interest in the mill, married, and purchased some 41,000 acres on Oahu's Ewa plain. Two years later, he drilled Oahu's first artesian well, initiating an irrigation program that converted dry, barren land into sugar cane fields (Estate of James Campbell, *The Estate of James Campbell* [Honolulu: the trust, 1978]; Rob Perez, "James Campbell Company Taking Shape," *Honolulu Advertiser*, 22 May 2005, A1; "Campbell Estate now James Campbell Co.," *Pacific Business News*, 8 October 2004).

7. James T. Burns Jr., "Financial Plaza of the Pacific," *Progressive Architecture* 50 (July 1969): 87–8; Downtown Improvement Association, *Honolulu: A Master*

Plan for the Central Business District: An Interim Planning Report (Honolulu: the association, July 1962) (quoted); Hardwick, *Mall Maker*, 191–7; "Redevelopment Proposed for Downtown Honolulu," *Progressive Architecture* 45 (April 1964); Victor Gruen Associates, *Report of the Studies and Recommendations for a Program of Revitalization of the Central Business District of Downtown Honolulu*, April 1968, 6–9; Leo S. Wou, "Designing the Facility," *Journal of Property Management* 38 (January/February 1973): 36.

8. Warren G. Haight, "Deciding the Need," *Journal of Property Management* 38 (January/February 1973): 30 (quoted); Shurei Hirozawa, "Financial Plaza— Success Monument," *Honolulu Star-Bulletin*, 11 July 1968, C1; "6th Headquarters for C & C," *Sunday Star-Bulletin and Advertiser*, 14 July 1968, F7; "Old C & C Building To Be Razed," *Sunday Star-Bulletin and Advertiser*, 11 October 1964, A1A. The Big Five included C. Brewer & Co., founded in 1826, Theo. H. Davies & Co., founded in 1845, Amfac, founded in 1849, Castle & Cooke, founded in 1851, and Alexander & Baldwin, founded in 1870. Castle & Cooke was founded by missionaries Samuel Northrup Castle and Amos Starr Cooke as a merchandising operation. In 1907 it became interested in Matson Navigation. It later acquired James Dole's Hawaiian Pineapple, which became the food division of the company (Lyn Danninger, "Isle Institutions' Economic Impact Endures," *Honolulu Star-Bulletin*, 29 September 2002).

9. Burns, "Financial Plaza of the Pacific," 87–8; Hardwick, *Mall Maker*, chap. 7–8; Hirozawa, "Financial Plaza—Success Monument"; Victor Gruen Associates, *Report of the Studies and Recommendations for a Program of Revitalization of the Central Business District of Downtown Honolulu*, 7. On Gruen, see, also, chap. 2, fn13.

10. Wou, "Designing the Facility," 36.

11. "Big Step Toward Redevelopment," *Sunday Star-Bulletin and Advertiser*, 14 July 1968, F6; Burns, "Financial Plaza of the Pacific," 88–93; William Curlett, "Constructing the Complex," *Journal of Property Management* 38 (January/February 1973): 39; Haight, "Deciding the Need," 30–1; "Individuality Is Expressed," *Sunday Star-Bulletin and Advertiser*, 14 July 1968, F8; Bob Jones, "$15 Million Financial Plaza Planned by 5 Island Firms," *Sunday Star-Bulletin and Advertiser*, 11 October 1964, A1; H. Baird Kidwell, *Journal of Property Management* 38 (January/February 1973): 32–5; "Pacific Superblock," *Progressive Architecture* 47 (February 1966): 192–3.

12. Founded in 1893 by Peter Janss as Janss Investment Company, Janss Corporation developed almost 100,000 acres of residential real estate in Southern California by 1929. It eventually developed much of the San Fernando Valley. Perhaps most significantly, it also developed Westwood Village, a major business center that Peter Janss and his two sons, Edwin and Harold, planned in the wake of the decision of the Regents of the University of California, in 1926, to relocate the recently created Los Angeles campus to a 384-acre site in the center of Westwood Hills, a tract that Janss had been marketing to affluent Angelenos

(Richard Longstreth, *City Center to Regional Mall: Architecture, the Automobile, and Retailing in Los Angeles, 1920–1950* [Cambridge, MΛ: MIT Press, 1997], 159–75).

13. Curlett, "Constructing the Complex," 39–42, 39 (quoted); Hirozawa, "Financial Plaza—Success Monument."

14. Davis quoted in "Estate Plans Larger Downtown Building"; Rickard, "New Campbell Block Plans Are Complete"; "Big Step Toward Redevelopment."

15. George F. Hutton interview; Wou, "Designing the Facility," *Journal of Property Management*, 36 (quoted).

16. George F. Hutton interview.

17. Ibid.; *Construction in Hawaii, 1968* (Honolulu: Bank of Hawaii, 1968), 11–12.

18. Conan "Doug" Craker interview; George F. Hutton interview (quoted); Norman "Red" Metcalf interview; *Construction in Hawaii, 1968*, 13–18. Al Fink, who worked with Pankow in Hawaii for almost four decades, notes that Wou apparently preferred conveying his ideas on constructing his designs in person to detailing them thoroughly in working drawings. Fink drew his conclusions from working with Wou on the Esplanade, discussed in this chapter (Albert W. Fink interview, 13 February 2009). A brief article, noting that Wou made his initial sketches of the Financial Plaza buildings on stationery from the Beverly Hills Hilton while on a flight to Hawaii, is suggestive ("Building Designed In Flight," *Sunday Star-Bulletin and Advertiser*, 14 July 1968, F7). On the lack of a large, skilled labor force in Hawaii, see, for instance, the comments of B. L. Snow, a builder of townhouses in Hawaii Kai, to Senate and House committees in, Janos Gereben and Beverly Creamer, "Housing Goals Are Failing in Low-Income Field," *Honolulu Star-Bulletin*, 12 April 1969, B16.

19. "Project Organization," memorandum, 28 February 1998, CPBL, Pasadena, CA.

20. Conan "Doug" Craker interview; George F. Hutton interview; "Employee in Focus: Bob Crawford," *CPI News* 1 (Fall 1983).

21. Conan "Doug" Craker interview.

22. George F. Hutton interview (quoted). The joint venture also included Honolulu-based Pacific Construction Company and San Francisco–based Swinerton & Walberg ("Project a Joint Venture," *Sunday Star-Bulletin and Advertiser*, 14 July 1968, F9. Dredging's parent was Dillingham Corporation, a top-25 contractor on *ENR's* Top 400 list whose local projects included the Ala Moana Center that Pankow would renovate and expand ("Dillingham Deals from Strength in Hawaii," *ENR*, 5 September 1968, 28–32).

23. George F. Hutton interview; Richard M. Kunnath interview, 1 October 2008.

24. Conan "Doug" Craker interview; George F. Hutton interview; Timothy P. Murphy interview; Dean E. Stephan interview.

25. George F. Hutton interview.

26. Foster, *Henry J. Kaiser*, 132–4, 254–61; Jack Smith, "Industrialist, 77 Guides Big Empire Personally," *LAT*, 2 August 1959, A2; idem, "Restless Kaiser Invades Hawaiian Island Industry," *LAT*, 6 August 1959, A4.

27. Foster, *Henry J. Kaiser*, 254 (Kaiser quoted), 262.

28. Ibid., 262–3.

29. Ibid., 263–4.

30. Ibid., 264–6.

31. Lambreth Hancock, *Hawaii Kai: The First 20 Years* (Hawaii Kai: Rotary Club of Hawaii Kai, 1983), 39–41; *Construction in Hawaii, 1970*, 16–17; *Construction in Hawaii, 1972* (Honolulu: Bank of Hawaii, 1972), 20–1. Hancock worked for various Kaiser companies for 33 years until his retirement in 1974. From 1967 to 1974, he was vice-president of marketing for Hawaii Kai Development Company.

32. George F. Hutton interview; Charles Pankow Associates, "Projects on Oahu," September 1977, spreadsheet attached to memorandum, Al Fink, "Tour of Charles Pankow Associates Projects in Hawaii," *Annual Meeting 1977*, binder, CPBL, Pasadena, CA.

33. Ibid. Hogan and Chapman established their practice in 1958. They later designed Hawaii Kai Corporate Plaza, which included a 2-story building and a 3-story building connected by walkways to compose a central courtyard. The project won honorable mention in the Project of the Year Awards of the City and County of Honolulu for 1988 (Erika Engle, "Davies, Pan Am Buildings Sold," *Honolulu Star-Bulletin*, 30 December 2003; CDS International website, www. cdsintl.com/projects_hawaiikai.php).

34. George F. Hutton interview; Lybrand, Ross Bros. & Montgomery, "Tudor Engineering Company: Report on Examination of Financial Statements for the years ended December 31, 1969 and 1968," carton 2; Tudor Engineering Company, newsletters of September 1968, Winter 1969, and Spring 1970, carton 3 (all found in Tudor Engineering Company Records, 1933–2006, BANC MSS 2007/134, Bancroft Library, University of California, Berkeley); Finding Aid to the Tudor Engineering Company Records, 1933–2006, Online Archive of California, oac. cdlib.org/findaid/ark:/13030/kt28702559; "The Statehood Decade: Patterns of Development," *Hawaii '69* (Honolulu: Bank of Hawaii. 1969), 16; Benson Bobrick, *Parsons Brinckerhoff: The First 100 Years* (New York: Van Nostrand Reinhold, 1985).

35. *Construction in Hawaii, 1968*, 14–15.

36. Ibid., 14–16; *Construction in Hawaii, 1978* (Honolulu: Bank of Hawaii, 1978), 6–7.

37. Lybrand, Ross Bros. & Montgomery, "Tudor Engineering Company: Report on Examination of Financial Statements for the years ended December 31, 1969 and 1968," carton 2; Lybrand, Ross Bros. & Montgomery, "Tudor Engineering Company: Report on Examination of Financial Statements for the years ended December 31, 1971 and 1970," carton 2; Tudor Engineering Company, newsletter,

Fall 1970, carton 3 (all found in Tudor Engineering Company Records, 1933–2006, BANC MSS 2007/134, Bancroft Library, University of California, Berkeley); *Construction in Hawaii, 1970*, 12.

38. George F. Hutton interview; Charles Pankow Associates, "Projects on Oahu," September 1977, spreadsheet attached to Al Fink, "Tour of Charles Pankow Associates Projects in Hawaii," *Annual Meeting 1977*, binder, CPBL, Pasadena, CA; "Employee in Focus: Don Kimball," *CPI News* 3 (Winter 1985); *Construction in Hawaii, 1972*, 20–1.

39. George F. Hutton interview (quoted).

40. Kelli Abe Trifonovitch, "Victoria Ward, Ltd.," *Hawaii Business* (November 2001); Jacy L. Youn, "A Future for Kakaako," *Hawaii Business* (February 2002). (Both articles accessed from the *Hawaii Business* website, www.hawaiibusiness.com.)

41. Hustace quoted in "Low-Rise Plaza Blooms—'Esthetically,'" *Honolulu Advertiser*, 19 September 1968, D3; Phil Mayer, "Well, Not Every New Building Is a High Rise," *Honolulu Star-Bulletin*, 6 July 1970, E2.

42. Mayer, "Well, Not Every New Building Is a High Rise."

43. George F. Hutton interview; Mayer, "Well, Not Every New Building Is a High Rise."

44. George F. Hutton interview.

45. Ibid.

46. Among quoted in Buck Donham, "City Housing Problem Called Worst in U.S.," *Honolulu Star-Bulletin*, 31 July 1969, E4.

47. Janos Gereben, "Gill Poses Triple-Threat Plan to Ease Isle Housing Crisis," *Honolulu Star-Bulletin*, 26 March 1969, A1; "Gill: 'Social Disaster' in Housing," *Honolulu Advertiser*, 15 February 1969, A4; Hawaii Housing Authority, *Annual Report for 1968–1969* (Honolulu: Hawaii Housing Authority, 1969). The reports on housing issued by the Office of the Lieutenant Governor included *Housing Costs in Hawaii: Report to the Legislature of the State of Hawaii* (Honolulu, 1969) and *Hawaii's Crisis in Housing* (Honolulu, 1970).

48. Yano quoted in "Urgent and Monstrous Problem," *Honolulu Star-Bulletin*, 26 March 1969, A1.

49. *Hawaii, '71: Annual Economic Review* (Honolulu: Bank of Hawaii, 1971), 26; *Hawaii, '80: Annual Economic Review* (Honolulu: Bank of Hawaii, 1980), table on page 24. All population figures are US Census figures.

50. *Hawaii, '71*, 27; "Home Prices Here Far Above U.S. Average," *Honolulu Star-Bulletin*, 26 March 1969, A1; Russ Lynch, "Low-Income Housing Is Lagging," *Honolulu Star-Bulletin*, 5 March 1969, A10; Robert E. Udick, "Some Facts, Figures About Island Building," *Honolulu Star-Bulletin*, 2 September 1969, D6.

51. Donham, "City Housing Problem Called Worst in U.S."; Gereben and Creamer, "Housing Goals Are Failing in Low-Income Field"; "Home Prices Here Far Above U.S. Average"; Robert E. Udick, "Architects Give Housing Views," *Honolulu Star-Bulletin*, 14 August 1969, A2; Toni Withington, "HUD

Aide Issues Ultimatum to Fasi, Home Builders," *Honolulu Star-Bulletin*, 31 July 1970, D16.

52. Gill quoted in "Gill: 'Social Disaster' in Housing"; Bank of Hawaii, "The Statehood Decade," 15; Governor's Committee on Hawaii's Economic Future, *Hawaii's Economic Future: Final Report* (Honolulu: Hawaii Department of Planning and Economic Development, 1985), 80–4; *Hawaii, '80*, table on page 24.

53. Thomas H. Creighton, "What Is the Housing Crisis? What Can Be Done About It?" *Sunday Star-Bulletin and Advertiser*, 19 September 1971, A19; Gill quoted in Gereben, "Gill Poses Triple-Threat Plan to Ease Isle Housing Crisis," A4; Governor's Committee on Hawaii's Economic Future, *Hawaii's Economic Future*, 80; *Hawaii, 1980*, table on page 24.

54. *Affordable Housing: The Years Ahead*, Ford Foundation Program Paper (New York: Ford Foundation, 1989), 25–30; B. T. Fitzpatrick, "FHA and FNMA Assistance for Multifamily Housing," *Law and Contemporary Problems* 32 (Summer 1967): 439–64.

55. Yanagwa quoted in Helen Altann, "Shortage of Low-Income Housing for Elderly," *Honolulu Star-Bulletin*, 28 November 1968, I8; Lynch, "Low-Income Housing Is Lagging."

56. Gill quoted in "Gill Cites Isle Needs in Housing," *Honolulu Advertiser*, 10 July 1968, D3; Altann, "Shortage of Low-Income Housing for Elderly"; Joe Arakai, "Fasi, Anderson Differ on City's Housing Solutions," *Honolulu Star-Bulletin*, 9 October 1968; Gereben, "Gill Poses Triple-Threat Plan to Ease Isle Housing Crisis"; Janos Gereben, "More Action, Less Study Asked on Housing Woes," *Honolulu Star-Bulletin*, 22 April 1969, A6; "Rutledge Wants Action Now on Housing Crisis," *Honolulu Star-Bulletin*, 12 September 1969, A15.

57. Alfred L. Castle, *A Century of Philanthropy: A History of the Samuel N. and Mary Castle Foundation* (Honolulu: Hawaii Historical Society, 1992), 181–2.

58. Neill quoted in Gereben, "More Action, Less Study Asked on Housing Woes"; "New Column Will Help Consumer in Housing Area," *Honolulu Star-Bulletin*, 18 February 1970, A18.

59. Norman "Red" Metcalf interview.

60. Neill quoted in "Much Work Preceded Housing Project Start," *Honolulu Star-Bulletin*, 1 November 1969, A8; Douglas Boswell, "Nonprofit Housing Job Begins," *Honolulu Advertiser*, 4 November 1969, B1; "New Project Slated," *Honolulu Advertiser*, 24 December 1970, B7; "Variance Sought in Building Project for Elderly," *Honolulu Star-Bulletin*, 5 June 1968, A9; "Projects on Oahu," September 1977, spreadsheet attached to Al Fink, "Tour of Charles Pankow Associates Projects in Hawaii," *Annual Meeting 1977*, binder, CPBL, Pasadena, CA. The value of the Kauluwela Elderly construction contract was $2.87 million.

61. Boswell, "Nonprofit Housing Job Begins"; "Much Work Preceded Housing Project Start"; "Variance Sought in Building Project for Elderly."

62. Neill quoted in Altann, "Shortage of Low-Income Housing for Elderly."

63. George F. Hutton interview; Neill quoted in "Much Work Preceded Housing

Project Start"; Pang quoted in Withington, "HUD Aide Issues Ultimatum to Fasi, Home Builders"; Slavsky quoted in "Variance Sought in Building Project for Elderly."

64. *Affordable Housing*, 30; Chris Bonastia, "Hedging His Best: Why Nixon Killed HUD's Desegregation Efforts," *Social Science History* 28 (Spring 2004): 19–52; Castle, *A Century of Philanthropy*, 182; "Much Work Preceded Housing Project Start"; "New Column Will Help Consumer in Housing Area"; Nadine Wharton, "'Banyan Tree' Housing Project Explained to Waimanalo Group," *Honolulu Star-Bulletin*, 23 October 1970, A8. The Kauluwela Elderly project broke ground in December 1970, one week before crews finished the Kauluwela Co-op building.

65. Minutes, Offsite Managers' Meeting, Monterey, CA, 2–5 May 1976.

66. "Bank Says Building Boom Over, Isles More 'Normal,'" *Honolulu Star-Bulletin*, 28 May 1976, D7; *Construction in Hawaii, 1985* (Honolulu: Bank of Hawaii, 1985), 1 (quoted); Governor's Committee on Hawaii's Economic Future, *Hawaii's Economic Future*, 61–2; *Hawaii '76: Annual Economic Review* (Honolulu: Bank of Hawaii, 1976), 20–1; *Hawaii '82: Annual Economic Review* (Honolulu: Bank of Hawaii, 1982), 25–6; *Hawaii '83: Annual Economic Review* (Honolulu: Bank of Hawaii, 1983), 14. Inflation figures derived from CPI-U, the Consumer Price Index for All Urban Consumers, US City Average (US Bureau of Labor Statistics). From 1974 to 1982, the CPI slightly more than doubled. At the same time, median household income almost doubled, from $12,902 to $23,433 (US Bureau of the Census, Historical Income Tables—Families, Table F-7, www. census.gov/hhes/www/income/histinc/F07ar.html). Housing prices in Hawaii tripled, however, as an Oceanic Properties executive vice president noted ruefully (Randolph G. Moore, "Housing Supply in Hawaii," *Construction in Hawaii, 1981* [Honolulu: Bank of Hawaii, 1981], special supplement, n.p.).

67. Jon Eicholtz interview; Hutton quoted in Bill Wood, "The Secret to Pankow's Success," *Hawaii Investor* 10 (July 1990): 48; David Walker, "The Ups and Downs of Construction," *Honolulu Star-Bulletin*, Progress Section, 22 February 1977, I:2. On the plight of contractors in Hawaii, cf. "Former THC Company Sold," *Honolulu Star-Bulletin*, 14 July 1977, B6; Kit Smith, "Dillingham Easing Off Construction," *Sunday Star-Bulletin and Advertiser*, 11 May 1975, B6; Jerry Tune, "Pacific Construction Co., Ltd. Looks to New Growth Areas," *Sunday Star-Bulletin and Advertiser*, 3 July 1983, G2; "Two Island Construction Firms Merge," *Honolulu Star-Bulletin*, 6 April 1978, D3.

68. Albert W. Fink interview, 13 February 2009; George F. Hutton interview; Timothy P. Murphy interview; Board of Governors of the Federal Reserve System, *Bank Prime Loan Rate Changes: Historical Dates of Changes and Rates*, research.stlouisfed.org/fred2/data/PRIME.txt; Hawaii Real Estate Commission, Supplementary Horizontal Property Regimes (Condominium) Public Report, Hobron in Waikiki, Registration No. 1001, issued 31 October 1983, accessed on the Department of Commerce and Consumer Affairs website, hawaii.gov/dcca_condo; "Hobron Backers Scramble to Find $18.7 Million Loan," *Honolulu*

Star-Bulletin, 5 April 1979, A2; "Judge Approves Plans for Hobron Development," Honolulu Star-Bulletin, 1 August 1979, A2. The prime rate had increased from 4.75 percent in January 1972 to 12 percent in 1974, before falling back to 6.25 percent during 1976.

Construction on the Hobron condominium halted after foundation work had been completed. It resumed in 1984 after the Bankruptcy Court ruled that the original building permits were still valid. The City had sought to revoke them, contending that they did not conform to the Waikiki Special Design District Ordinance, passed in 1976. The City won a ruling to "restart the clock" on the time allowed to complete the structural work, leaving CPA six months to do so ("Projects of the Quarter: Hobron and Mandarin," CPI News 2 [Winter 1984]).

69. Construction in Hawaii, 1977 (Honolulu: Bank of Hawaii, 1977), 15; Hawaii Real Estate Commission, Preliminary Horizontal Property Regimes (Condominium) Public Report, Waikiki Tusitala, Registration No. 754, issued 12 February 1975, accessed online at the Department of Commerce and Consumer Affairs website, hawaii.gov/dcca_condo.

70. George F. Hutton, "Waikiki Tusitala, A Salvage Job," a paper delivered at the annual meeting of CPI, San Francisco, 25 September 1976, Annual Meeting 1976, binder, CPBL, Pasadena, CA; Hawaii Real Estate Commission, Preliminary Horizontal Property Regimes (Condominium) Public Report, Waikiki Tusitala.

71. Construction in Hawaii, 1978, 6, 13; Hutton, "Waikiki Tusitala, A Salvage Job."

72. Hutton, "Waikiki Tusitala, A Salvage Job."

73. Albert W. Fink interview, 13 February 2009 (quoted); Hutton, "Waikiki Tusitala, A Salvage Job" (quoted).

74. Hutton, "Waikiki Tusitala, A Salvage Job."

75. Albert W. Fink interview, 13 February 2009; Norman "Red" Metcalf interview; Leila Fujimori, "Real Estate Developer Ralph D. Cornuelle, 69," obituary, Honolulu Star-Bulletin, 30 August 2000; Hutton, "Waikiki Tusitala, A Salvage Job" (quoted); Hawaii Real Estate Commission, "Preliminary Horizontal Property Regimes (Condominium) Public Report, Waikiki Lanais, Registration No. 892, issued 10 January 1977, accessed online at the Department of Commerce and Consumer Affairs website, hawaii.gov/dcca_condo.

76. Hutton, "Waikiki Tusitala, A Salvage Job" (quoted); Hawaii Real Estate Commission, Final Horizontal Property Regimes (Condominium) Public Report, Waikiki Lanais, Registration No. 892, issued 4 October 1977, accessed online at the Department of Commerce and Consumer Affairs website, hawaii.gov/dcca_condo.

77. George F. Hutton interview; Hawaii Real Estate Commission, Final Horizontal Property Regimes (Condominium) Public Report, Wilder at Piikoi, Registration No. 937, issued 10 November 1977, accessed online at the Department of Commerce and Consumer Affairs website, hawaii.gov/dcca_condo; Jerry Tune, "Garden Offices Planned," Honolulu Star-Bulletin, 25 September 1975, A6.

78. George F. Hutton interview; Norman "Red Metcalf" interview; "Employee in Focus: Norman 'Red' Metcalf," *CPI News* 3 (Fall 1985).

79. Albert W. Fink interview, 13 February 2009; "Project Organization," memorandum, 28 February 1998, CPBL, Pasadena, CA.

80. Ibid.; George F. Hutton interview; Norman "Red Metcalf" interview; "Employee in Focus: Jack Grieger," *CPI News* 2 (Summer/Fall 1984); "Employee in Focus: Don Kimball," *CPI News* 3 (Winter 1985).

81. George F. Hutton interview; Richard M. Kunnath interview, 1 October 2008 (quoted); Kim Lum interview (quoted).

82. Albert W. Fink interview, 13 February 2009; George F. Hutton interview.

83. Albert W. Fink interview, 13 February 2009; Norman "Red" Metcalf interview; "Hawaii Sports Report," *CPI News* 2 (Winter 1984).

84. Kim Lum interview (quoted).

85. Fink quoted in "Forming Strategy Slips Job a Month Ahead of Schedule," *ENR*, 28 October 1982, 28.

86. George F. Hutton, "Contractor's Contribution to Developers' Projects," a paper delivered at the annual meeting of CPI, San Francisco, 12 September 1975, *Annual Meeting 1975*, binder, CPBL, Pasadena, CA; Downtown Improvement Association, *Honolulu: A Master Plan for the Central Business District* (quoted); "Low-Priced Homes Aim of Firm," *Sunday Star-Bulletin and Advertiser*, 19 October 1969, A13; "One of Many Oceanic Developments," *Sunday Star-Bulletin and Advertiser*, 14 July 1968, F8. On the Sea Ranch project, see Charles Moore, Gerald Allen, and Donlyn Lyndon, *The Place of Houses* (New York: Holt, Rinehart, and Winston, 1974), 31–48.

87. Hutton, "Contractor's Contribution to Developers' Projects"; Albert W. Fink interview, 13 February 2009; Norman "Red Metcalf" interview.

88. Hutton, "Contractor's Contribution to Developers' Projects."

89. Ibid. (quoted); Dean E. Stephan interview (quoted); "Fancy Forms: How They Cut Concrete Building Costs," *Civil Engineering* 48 (March 1978): 84–9; "Team Cooperation Slashes Construction Time 35 Percent," *Journal of the American Concrete Institute* 74 (August 1977): N5–N6. According to Clyde Sisson, a superintendent with Sweet Associates, a Schenectady, New York–based contractor, flying forms were cost effective if used eight or more times on a job.

90. Hogan to Hutton, letter, 9 April 1976 (quoted); "Team Cooperation Slashes Construction Time 35 Percent."

91. Albert W. Fink interview, 13 February 2009 (quoted).

92. Ibid. (quoted); Norman "Red" Metcalf interview; Minutes, Offsite Managers' Meeting, Monterey, 2–5 May 1976; "Team Cooperation Slashes Construction Time 35 Percent." Pacific Concrete and Rock attempted unsuccessfully to recover its loss from Travelers Insurance.

93. L. Robert Allen, "Operating in a Changing Marketplace," *Construction in Hawaii, 1981* (Honolulu: Bank of Hawaii, 1981), special supplement, n.p.; "'Local Boy' Thrives on Urban Projects," *Honolulu Star-Bulletin*, "On the Run in

'81," special section, 17 February 1981, I:16; "Self-Made Man Shapes Honolulu," *ENR*, 28 October 1982, 33; Kit Smith, "Brokers Say Downtown Office Space Glut Fading," *Sunday Star-Bulletin and Advertiser*, 19 February 1984, B1.

94. "Why the Growing Popularity of Post-Tensioned Floors?" *Civil Engineering* 48 (March 1978): 89–91.

95. Steven M. Baldridge, "Tall Structural Sustainability in an Island Context: The Hawaii Experience," a paper delivered at the 8th World Congress, Council on Tall Buildings and Urban Habitat (CTBUH), Dubai, March 2008, 3 (quoted), 4 (quoted); "$120 Million Plan Unveiled," *Honolulu Star-Bulletin*, 4 February 1980, A2; "Forming Strategy Slips Job a Month Ahead of Schedule" 28; "Forming System for Core Walls Saves Time on 30-Story Towers and Casting Methods Speed Nearby 41-Story Job," *ENR*, 23 February 1978, 26.

96. Allen, "Operating in a Changing Marketplace"; "Forming Strategy Slips Job a Month Ahead of Schedule," 28.

97. "Forming System for Core Walls Saves Time on 30-Story Towers and Casting Methods Speed Nearby 41-Story Job."

98. "Forming Strategy Slips Job a Month Ahead of Schedule," 28–29, 33; Stu Glauberman, "Builder Opts for Higher Fee," *Honolulu Star-Bulletin*, 9 July 1981, A3. By this time, CPA also was saving money on slipform operations because its volume of work made it economical to purchase the equipment needed for the technique.

99. Fink quoted in "Forming Strategy Slips Job a Month Ahead of Schedule," 29.

100. The Craigside condominium consisted of twin 27-story towers designed by Norman Lacayo, who conceived of the three-level, 6,000-square-foot-plus penthouse that he designed for Hutton in the second tower as a "villa" and "the last thing in the world you would expect when you get out of the elevator." By including planters, a greenhouse, and a rooftop garden, Lacayo "tried to create a feeling of living in something other than a high-rise condominium" (Jerry Tune, "Norman Lacayo Designs Spacious Villa in the Sky," *Sunday Star-Bulletin and Advertiser*, 18 October 1987, I4).

101. Bill Bramschreiber interview; "Forming Strategy Slips Job a Month Ahead of Schedule," 28–9.

102. *Construction in Hawaii, 1973* (Honolulu: Bank of Hawaii, 1973), 12 (quoted); Albert W. Fink interview, 13 February 2009 (quoted). The City of Honolulu issued the building permit, valued at $7,715,000, in December 1971. The construction contract was worth $9,459,799.

103. On the cultural belief in the capacity of technology to underpin social and economic progress, see Thomas P. Hughes, *American Genesis: A Century of Invention and Technological Enthusiasm, 1870–1970* (New York: Viking, 1989). On civic boosterism theory, cf. Kevin R. Cox and Andrew Mair, "Locality and Community in the Politics of Local Economic Development," *Annals of the Association of American Geographers* 78 (1988): 307–25; John R. Logan and Harvey L. Molotch, *Urban Fortunes: The Political Economy of Place* (Berkeley: University of California Press, 1987), chap. 3; Harvey L. Molotch, "The City as a Growth

Machine: Toward a Political Economy of Place," *American Journal of Sociology* 82 (1976): 309–31. For a recent assessment of civic boosterism theory, see Mark Boyle, "Growth Machines and Propaganda Projects: A Review of Readings of the Role of Civic Boosterism in the Politics of Local Economic Development," in *The Urban Growth Machine: Critical Perspectives Two Decades Later*, ed. Andrew E. G. Jones and David Wilson (Albany, NY: SUNY Press, 1999), 55–70.

104. Carl Hebenstreit, "Publisher's Notebook," *Building Industry* 36 (June 1990): 5 (quoted).

105. Russ Lynch, "'Isle' Construction Fell 8% in 1984," *Honolulu Star-Bulletin*, 26 June 1983, B1, citing *Construction in Hawaii, 1985* (Honolulu: Bank of Hawaii, 1985). As the value of commercial construction authorizations slid 7.6 percent in nominal terms in 1984, the value of authorized government contracts soared 49 percent. Charlie Pankow's business model, however, precluded the company from competing for work in this sector.

Chapter 4

1. Dean E. Stephan interview.

2. Don Nash, "Condition of the Surety Market," n.d. [1976] (quoted); Russell J. Osterman, "Condition of the Surety Market," cover letter, n.d. (quoted) (both enclosed in binder, Offsite Managers' Meeting, Monterey, CA, 2–5 May 1976).

3. Conan "Doug" Craker interview; Richard M. Kunnath interview, 21 October 2008 (quoted); Dean E. Stephan interview (quoted).

4. "He's Built More Than 75 Malls," *Syracuse Post-Standard*, 25 March 1976, special section. On the relationship between Smith and Gruen, see Hardwick, *Mall Maker*, 114–6.

5. Sid Copeland, *The SAFECO Story, 1923–1980* (Seattle, WA: Safeco Corporation, 1981), chap. 15. In 2008 Safeco was acquired by Liberty Mutual Insurance. As of 2011, it operates as a subsidiary of the parent organization.

6. Robert Heisler interview (quoted); Lee Sandahl interview; Longstreth, *City Center to Regional Mall*, 335–44. Beginning with Lakewood Center (1950–1952), the May Company—founded in Cleveland, Ohio in 1877— had embarked on a series of postwar shopping center developments in Southern California. The earliest project that involved Heisler's firm was most likely Eastland Shopping Center, in West Covina (1955–1957).

7. Robert Heisler interview; Lee Sandahl interview; "Macy's Sherwood Manor Ground Breaking Set," *Lodi News-Sentinel*, 16 May 1964, I:2; Katie Foxworth Lee, "Carved in Stone: Stone Bros. & Associates and M&M Stone, Inc. Have Put an Indelible Mark in Stockton, California's Growing Retail Scene," *Shopping Center Business* (May 2007).

8. Orrico quoted in Copeland, *The SAFECO Story, 1923–1980*, 172; Brad D. Inman interview (quoted); Richard M. Kunnath interview, 21 October 2008 (quoted); Michael R. Adamson, Memorandum of Conversation with Russell J. Osterman, Pasadena, CA, 29 July 2009.

9. Dean E. Stephan interview.

10. Conan "Doug" Craker interview; Brad D. Inman interview; Dean E. Stephan interview; "Prefabrication of Panels Is Key to Faster, Cheaper Walls," *ENR*, 10 September 1970, 35; Ray N. Atkinson, *Guy F. Atkinson Company of California: A Free Enterprise Success Story* (New York: Newcomen Society, 1985). Pankow Construction Company was, in fact, the construction entity for the projects discussed in this chapter. To minimize confusion, the narrative uses CPI, which became a holding company for several companies, as the next chapter shows, to refer to the company responsible for constructing projects on the Mainland.

11. Brad D. Inman, letter to the author, 23 February 2010.

12. "Perini Corp. Names George B. Roberts Vice President," *Business Wire*, 20 February 1991.

13. Richard M. Kunnath interview, 21 October 2008 (quoted).

14. Conan "Doug" Craker interview.

15. Conan "Doug" Craker interview; Robert Law interview; Thomas D. Verti interview (quoted); William "Red" Ward interview.

16. Brad D. Inman interview; Dean E. Stephan interview (quoted); Saxenian, *Regional Advantage*, 50–7.

17. Purdue University continues to contribute a large number of engineers and construction managers to the Pankow firm. At the end of 2008, the company employed 15 professionals who had graduated from Purdue in the previous fifteen years alone. During that time, many other Purdue interns, professional practice (co-op) students, and graduated engineers spent time with the company. In all, 21 Purdue alums worked for the company at the end of 2008 (J. J. Mollenkopf, letter to the author, 15 January 2009).

18. Dean Browning interview; Robert Law interview; "Tilt-Up and Site-Precast Buildings Highlighted at Concrete Conference," *Civil Engineering* 49 (May 1979): 16.

19. Joseph Sanders interview (quoted); "Employee in Focus: Joe Sanders," *CP News* 8 (Spring 1990) (quoted); "Executives in the News," *Daily Pacific Builder*, 25 August 1995.

20. Thomas D. Verti interview (quoted).

21. Kim Lum interview; Dick Walterhouse interview; Katherine Conrad, "Walterhouse Runs $100 Million Firm," *East Bay Business Times*, 26 January 2001; "Employee in Focus: Dick Walterhouse," *CP News* 7 (Spring 1989); "Executives in the News," *Daily Pacific Builder*, 25 August 1995.

22. Layton, *The Revolt of the Engineers*, 1 (quoted).

23. Dean E. Stephan interview (quoted); Ralph Van Cleave and George B. Roberts, "Estimating: Standard Procedures for Cost Control," a paper delivered at the annual meeting of CPI, San Francisco, 13–14 September 1974, *Annual Meeting 1974*, binder, CPBL, Pasadena, CA; Robert A. Hewitt, Minutes, Offsite Managers' Meeting, Vancouver, BC, 6–9 May 1978; "Overview of PBI Estimating History,"

presentation enclosed in binder, Offsite Managers' Meeting, Chicago, 4–6 May 1984. On the complexity of the Atkinson organization, see, "Dean of Contractors Dies at 93," *ENR*, 19 September 1968, 83.

24. Dean E. Stephan interview; "Overview of PBI Estimating History," presentation, enclosed in binder, Offsite Managers' Meeting, Chicago, 4–6 May 1984.

25. Robert Law, telephone interview with the author, 6 October 2010.

26. Longstreth, *City Center to Regional Mall*, 323 (quoted); Arthur Love interview (quoted).

27. Longstreth, *City Center to Regional Mall*, 323 (quoted); Welton Becket and Associates, *Vision . . . Through Supervision*, n.p. John Graham (1908–1991) is perhaps best known for the design of the Space Needle. In the 1980s, Pankow expanded two of Graham's regional mall projects, Ala Moana Shopping Center, Honolulu (opened in 1960) and Capitol Court, Milwaukee (opened 1956). In all, Graham's firm was responsible for the design of more than 70 regional shopping centers. Architect Albert C. Martin Jr., who collaborated with the May Company on the design of Lakewood Center, also took his cues from Graham. On the work of John Graham & Company, see, "A Report on the Work and Organization of John Graham & Co., Architects and Engineers, Seattle, Washington, 1950," copy in LC 7-2; A. Alexander Bül and Nicholas Ordway, "Shopping Center Innovations: The Past 50 Years," *Urban Land* 46 (June 1987): 22–5; "Center a Story above Grade," *ENR*, 25 February 1960, 41–8; Meredith L. Clausen, "John Graham Jr.," in *Shaping Seattle Architecture: A Historical Guide to the Architects*, ed. Jeffrey Karl Ochsner (Seattle: University of Washington Press, 1994), 258–63; idem, "Northgate Regional Shopping Center—Paradigm From the Provinces," *Journal of the Society of Architectural Historians* 43 (May 1984): 144–61; G. R. Cysewski, "Portland's Lloyd Center, *Traffic Engineering* 29 (December 1958): 11–13, 38; "Portland Gets Nation's Largest Shopping Center," *ENR*, 3 September 1959, 40–2; Arthur W. Priaulx, "Northgate—Suburban Shopping Center, Seattle, Washington," *Architect & Engineer* 182 (September 1950): 14–21; "The Architect's Place in the Suburban Retail District," *AF* 93 (August 1950): 116.

28. Arthur Love interview (quoted); Dean E. Stephan interview; "Imposing New Projects Demand Expanded Services," *BD&C* 13 (March 1972): 43.

29. Arthur Love interview (quoted).

30. Robert Heisler interview; "Project of the Quarter: Oxmoor Mall Expansion," *CPI News* 2 (Spring 1984): 3; Copeland, *The SAFECO Story, 1923–1980*, 175; John E. Kleber, ed., *The Encyclopedia of Louisville*, vol. 2000–2001 (Lexington: University of Kentucky Press, 2001), 682; Gregory A. Luhan, Dennis Domer, and David Mohney, *The Louisville Guide* (Princeton, NJ: Princeton Architectural Press, 2004), 127 (quoted); Charles J. Pankow, "Contractors Alternate Saves $2 Million on Louisville Building," *Journal of the American Concrete Institute* 70 (May 1973): 341–5. The Oxmoor name by which the estate was known as early as 1785 was taken from *The Life and Opinions of Tristram Shandy, Gentleman*, the classic novel by Laurence Sterne. Thomas Walker Bullitt died in 1991. He and his wife deeded

the house and 79 acres of land to the Filson Historical Society, leaving 439 acres of farmland for redevelopment.

31. Pankow, "Contractors Alternate Saves $2 Million on Louisville Building," 341–2, 345; Robert Law interview; Arthur Love interview; Luhan, Domer, and Mohney, *The Louisville Guide*, 128 (quoted). In keeping with critics' overall verdict on the architecture of Welton Becket's office buildings, Luhan et al. deemed the Citizens Fidelity tower to be "a serviceable example of mid-century modernism" (127).

32. Pankow, "Contractors Alternate Saves $2 Million on Louisville Building," 345 (quoted); Charles J. Pankow, "Site-Precasting," *Concrete Construction* 25 (April 1980): 334 (quoted).

33. Jon A. Benner, "Winmar Building," technical report, n.d., CPBL, Pasadena, CA; Don King, "First Security National Bank & Trust Building," technical report, n.d. [1974], CPBL, Pasadena, CA; Kris Reiswig and Ciaran Barry, "Pacific First Federal Center," technical report, n.d. [1981], CPBL, Pasadena, CA; "Project of the Quarter: 411 East Wisconsin Building," *CPI News* 3 (Winter 1985); Pankow, "Site-Precasting," 334 (quoted).

34. David Carlson, "What Next in Shopping Centers?" *AF* 112 (April 1960): 129.

35. Lynch quoted in, "Shopping Malls Are Where the Money Is," *Syracuse Post-Standard*, 17 August 1978; "2 Developers Cite Experience," *Syracuse Post-Standard*, 25 March 1976, special section; James J. Farrell, *One Nation Under Goods: Malls and the Seductions of American Shopping* (Washington, DC: Smithsonian Books, 2003), table 1 on xii; Homer Hoyt, "The Status of Shopping Centers in the United States," *Urban Land* 19 (October 1960): 3–6; Jerome J. Michael, "Renovation and Expansion of Shopping Centers," *Urban Land* 43 (March 1984): 8–11.

36. "2 Developers Cite Experience"; "New Center Under Way," *Syracuse Post-Standard*, 2 November 1973 (quoted); "Project of the Quarter: Oxmoor Mall Expansion," *CPI News* 2 (Spring 1984): 3; *Construction in Hawaii, 1981* (Honolulu: Bank of Hawaii, 1981), 18; *Construction in Hawaii, 1982* (Honolulu: Bank of Hawaii, 1982), 17; Copeland, *The SAFECO Story, 1923–1980*, 172. Though built in Hawaii, the Windward Mall project was run out of the Southern California region (Altadena office).

37. Dean E. Stephan interview (quoted); "Penn-Can Mall," technical report, n.d., CPBL, Pasadena, CA. CPI constructed three of the department stores associated with Washington Square (Charles Pankow, Inc., "Building In The Pacific Northwest," marketing brochure, CPBL, Pasadena, CA).

38. Jack Gould, "Management Re-Appraisal of Shopping Centers," *Urban Land* 23 (March 1964): 3–6; Michael, "Renovation and Expansion of Shopping Centers," 8.

39. Charles M. Kober, "Regrowth for Existing Shopping Centers," *Urban Land* 36 (February 1977): 3–9; Michael, "Renovation and Expansion of Shopping Centers."

40. John A. Casazza, "Shopping Center Expansion and Renovation," *Urban Land* 43 (June 1984): 2–6; James B. Douglas, "The Enclosed Mall and Other

Development Trends in the Shopping Center Business," *Urban Land* 21 (September 1962): 3–5; Kober, "Regrowth for Existing Shopping Centers."

41. "Capitol Court Mall," technical report, n.d. [1978], CPBL, Pasadena, CA; Gruen and Smith, *Shopping Towns USA*, table on 216–8, 236–7; Homer Hoyt, "Sales in Leading Shopping Centers and Shopping Districts in the United States," *Urban Land* 20 (September 1961): table 3 on 5; idem, "Sales Trends in Shopping Centers, 1958–1963," *Urban Land* 25 (April 1966): table 5 on 5, 9. Ed Schuster opened his downtown Milwaukee emporium in 1884 and later added a branch store along the Mitchell Street retail corridor on the city's south side.

42. Hoyt, "Sales in Leading Shopping Centers and Shopping Districts in the United States," table 3 on 5; idem, "Sales Trends in Shopping Centers, 1958–1963," table 5 on 5, 9; Robert Law, "South Shore Plaza," technical report, n.d., CPBL, Pasadena, CA.

43. "Capitol Court Mall," technical report, n.d. [1978]; Robert Law, "South Shore Plaza," technical report, n.d.; Robert Law and Roger Mutti, "South Shore Plaza, Sears Pad and Parking Structure," technical report, n.d.; Dick Walterhouse and Robert Law, "South Shore Plaza, Expansion," technical report, n.d. [1980] (all records of CPBL, Pasadena, CA).

44. Robert Law interview (quoted).

45. Robert Law, "South Shore Plaza," technical report, n.d., CPBL, Pasadena, CA; Copeland, *The SAFECO Story, 1923–1980*, 172. Construction of the Lord & Taylor's emporium began in May 1978.

46. "Capitol Court Mall," technical report, n.d. [1978], CPBL, Pasadena, CA.

47. Ibid. Pankow returned to Capitol Court in 1985. Located in an area in economic decline and faced with competition in outlying areas, the shopping center struggled for business, notwithstanding the redevelopment. T. A. Chapman's closed in 1983; Gimbels followed closely on its heels, shuttering its doors in 1984. Milwaukee-based Boston Store replaced Chapman's. Pankow redeveloped the Gimbels store for Target. The 125,000-square-foot project involved "a complete architectural and structural revitalization." It included raising the parking structure one level and constructing external stair towers and elevators to the mall at each end. The creation of the largest Target store in the state did little, however, to arrest the mall's decline. The anchor stores closed in turn: J. C. Penney in 1986, Boston Store in 1987, Sears in 1992, and, finally, Target in 1996. Efforts by Winmar to revive the property in the 1990s—none involving the Pankow firm—failed. The property was sold. The new owners demolished the mall in the spring of 2001 and replaced it with Midtown Center, an open-air "power center" (Kevin Smith interview; "Target Department Store—Capitol Court Mall," *CPI News* 3 [Spring/ Summer 1985]: 7; "Milwaukee's Capitol Court Center," Mall Hall of Fame website, 26 September 2007, mall-hall-of-fame.blogspot.com/2007_09_01_archive.html).

48. Richard M. Kunnath interview, 21 October 2008 (quoted).

49. James Pygman, "Tall Office Buildings," *Urban Land* 45 (January 1986): 32–3; Wheaton and Torto, "Office Construction Booms," 32.

50. Russell J. Osterman, "Making It 'Pencil,'" memorandum, enclosed in binder, Offsite Managers' Meeting, Monterey, CA, 2–5 May 1976.

51. Timothy P. Murphy interview; Thomas D. Verti interview.

52. Michael R. Adamson, Memorandum of Conversation with Russell J. Osterman, Pasadena, CA, 29 July 2009; Richard M. Kunnath interview, 9 July 2009 (quoted).

53. Therese Poletti and Tom Paiva, *Art Deco San Francisco: The Architecture of Timothy Pflueger* (Princeton, NJ: Princeton Architectural Press, 2008), 58–72.

54. Jon A. Benner, "San José Plaza I," technical report, n.d. [1972], CPBL, Pasadena, CA; Conan "Doug" Craker interview; Robert Heisler interview; Lee Sandahl interview; Russell L. Wahl interview.

55. "San José Plaza II," technical report, n.d. [1973], CPBL, Pasadena, CA; Ron Smith and Dan Mitchell, "PT&T Casting Yard," technical report, n.d., CPBL, Pasadena, CA; Conan "Doug" Craker interview; Robert Heisler interview.

56. J. K. Dineen, "Reinvention Under Way at Third and Folsom," *San Francisco Business Times*, 14 April 2008 (quoted).

57. Charles Pankow, Inc., "Headquarters Building of the Pacific Telephone & Telegraph Co.," brochure, n.d. [1976], CPBL, Pasadena, CA; Robert Law interview; Alan D. Murk interview, 15 December 2008; "Loan Committed for Building," *Oakland Tribune*, 20 July 1975, C4. CPI and Building Enterprise negotiated a lump-sum contract.

58. Charles Pankow, Inc., "Headquarters Building of the Pacific Telephone & Telegraph Co.," brochure, n.d. [1976], CPBL, Pasadena, CA; Robert Law interview.

59. Ibid.; Robert Heisler interview; "Three Casting Techniques Cut Building's Cost," *ENR*, 4 December 1975, 16.

60. Robert Law interview; Ron Smith and Dan Mitchell, "PT&T Casting Yard," technical report, n.d., CPBL; "Three Casting Techniques Cut Building's Cost" (quoted). The casting yard was operated as a separate job with its own superintendent and project engineer.

61. Charles Pankow, Inc., "Headquarters Building of the Pacific Telephone & Telegraph Co.," brochure, n.d. [1976], CPBL, Pasadena, CA. The Executive Centre project (see chapter 3) illustrates the risk that this latter tactic may entail. As Al Fink pointed out to the author on a tour of Pankow properties in September 2009, the City of Honolulu did not issue a building permit until crews had reached the 30th floor of Bishop Tower, after issuing a series of floor-by-floor extensions of the foundation permit to that point.

62. Ronald Smith, "Citizens Bank Building," technical report, July 1974, CPBL, Pasadena, CA; idem, "Jefferson Plaza," technical report, n.d., CPBL, Pasadena, CA; Lee Sandahl interview; Dean E. Stephan interview (quoted); Russell L. Wahl interview.

63. Jon T. Eicholtz interview; "Design Change Sets Slipforming Record," *ENR*, 26 November 1970; "Former THC Company Sold"; "Two Island Construction Firms Merge." In 1978 Pacific merged with DMA/Hawaii. The latter, founded

in 1972, had been a joint venture partner of Tecon Services, another Murchison subsidiary, in the construction of two local projects, Grosvenor Center and a Waikiki condominium.

Murchison's father was one of the so-called Big Four independent oil men who made millions of dollars in the East Texas and other fields during the interwar period. The others included Roy Cullen, H. L. Hunt, and Sid Richardson. Muchison Sr. became notorious for shipping so-called hot oil in excess of quotas set by the National Industrial Recovery Act and state law. The commercial empire of Muchison *fils* collapsed along with plummeting oil and real estate prices in the mid-1980s. He died in March 1987 (Bryan Burrough, *The Big Rich: The Rise and Fall of the Greatest Texas Oil Fortunes* [New York: Penguin, 2009], chap. 3, 83–6, 141–9, 293–325, 411–20).

64. Jon T. Eicholtz interview.

65. Robert A. Hewitt, Minutes, Offsite Managers' Meeting, Vancouver, BC, 6–9 May 1978; Minutes, Offsite Managers' Meeting, Albuquerque, NM, 31 March–1 April 1979 (quoted); Jon T. Eicholtz interview (quoted).

66. Helene Lesel, "Mini-Manhattan, Just West of Los Angeles," *LAT*, 14 November 2004, K2.

67. Jon T. Eicholtz interview.

68. "Project of the Quarter: Catalina Landing Office Complex," *CPI News* 3 (Spring/Summer 1985): 3 (quoted); "Project Start: Catalina Landing," *CPI News* 1 (Fall 1983): 4; Suzanne Dow Nakaki interview; Dean E. Stephan interview.

69. Jon T. Eicholtz interview.

70. Dean Browning interview; Jon T. Eicholtz interview; Mike Liddiard interview; "Project of the Quarter: 2101 Webster Street," *CPI News* 3 (Fall 1985): 3 (quoted); "Project Start: 2101 Webster," *CPI News* 2 (Winter 1984): 4; "Work In Progress: 2101 Webster," *CPI News* 2 (Summer/Fall 1984): 5; Jane Bowar Zastrow, "Bay Area Projects: Old Friends and New Faces," *Urban Land* 55 (October 1996): 114.

71. Robert E. Lang, *Office Sprawl: The Evolving Geography of Business*, Survey Series, Center on Urban and Metropolitan Policy, Brookings Institution, Washington, DC (October 2000); Wheaton and Torto, "Office Construction Booms."

72. Dean E. Stephan interview (quoted).

73. "Capitol Court Mall," technical report, n.d. [1978], CPBL, Pasadena, CA; Ronald Smith, "Jefferson Plaza," technical report, n.d., CPBL, Pasadena, CA (quoted); Dick Walterhouse and Robert Law, "South Shore Plaza, Expansion," technical report, n.d. [1980], CPBL, Pasadena, CA (quoted).

74. Conan "Doug" Craker interview; Luhan, Domer, and Mohney, *The Louisville Guide*, 165–6, 231; Pankow, "Contractors Alternate Saves $2 Million on Louisville Building," 343–4.

75. Conan "Doug" Craker interview.

76. Mike Liddiard interview (quoted); William "Red" Ward interview (quoted).

77. Conan "Doug" Craker interview; Mike Liddiard interview; William "Red" Ward interview.

78. "Capitol Court Mall," technical report, n.d. [1978], CPBL, Pasadena, CA (quoted); Kris Reiswig and Ciaran Barry, "Pacific First Federal Center," technical report, n.d. [1981], CPBL, Pasadena, CA (quoted); Ronald Smith, "Jefferson Plaza," technical report, n.d., CPBL, Pasadena, CA (quoted).

79. "Capitol Court Mall," technical report, n.d. [1978], CPBL, Pasadena, CA (quoted); Don King, "First Security National Bank and Trust Building," n.d. [1974], CPBL, Pasadena, CA; Kris Reiswig and Ciaran Barry, "Pacific First Federal Center," technical report, n.d. [1981], CPBL, Pasadena, CA.

80. Charles J. Pankow, "Automation in Reinforced Concrete," President's memo, *Concrete International* (October 1980): 5.

81. On the efforts of CUAIR, a forerunner of the Business Roundtable, established in 1972, see, Nelson Lichtenstein, *State of the Union: A Century of American Labor* (Princeton, NJ: Princeton University Press, 2002), 225–30; Kim Moody, *An Injury to All: The Decline of American Unionism* (New York: Verso, 1988), 127–35.

82. Minutes, Offsite Managers' Meeting, Monterey, CA, 2–5 May 1976 (quoted); Dean E. Stephan to Robert E. McCarthy, 3 March 1982, letter enclosed in binder, Offsite Managers' Meeting, Chicago, 4–6 May 1984; Conan "Doug" Craker interview; Albert W. Fink interview, 25 March 2009; Brad D. Inman interview.

83. Conan "Doug" Craker interview (quoted); Brad D. Inman, letter to the author, 22 June 2009; *Pankow Construction Co. v. Advance Mortgage Corporation v. Pacheco Village Properties and Leo S. Wou*, 618 F2d 611 (9th Cir 1980). Inman did not work on the project. He was transferred to the Winmar office building project in Lexington, Kentucky, soon after he was hired.

84. Michael R. Adamson, Memorandum of Telephone Conversation with Alan D. Murk, 8 May 2009; Brad D. Inman, letter to the author, 22 June 2009; *Pankow Construction Co. v. Advance Mortgage Corporation v. Pacheco Village Properties and Leo S. Wou*, 618 F2d 611 (9th Cir 1980).

85. Michael R. Adamson, Memorandum of Telephone Conversation with Alan D. Murk, 8 May 2009; Minutes, Offsite Managers' Meeting, Monterey, CA, 2–5 May 1976.

86. Michael R. Adamson, Memorandum of Telephone Conversation with Alan D. Murk, 8 May 2009; Darl Williams, Minutes, Offsite Managers' Meeting, Juarez, Mexico, 12–14 May 1975; Minutes, Offsite Managers' Meeting, Monterey, CA, 2–5 May 1976.

87. Dean E. Stephan interview (quoted); Darl Williams, Minutes, Offsite Managers' Meeting, Juarez, Mexico, 12–14 May 1975; Minutes, Offsite Managers' Meeting, Monterey, CA, 2–5 May 1976; Agenda, Offsite Managers' Meeting, Coronado, CA, 19–22 May 1977; Robert A. Hewitt, Minutes, Offsite Managers' Meeting, Vancouver, BC, 6–9 May 1978; Minutes, Offsite Managers' Meeting, Albuquerque, NM, 31 March–1 April 1979.

88. Contractor liability remained a top agenda item at the annual meetings of the company's offsite managers. For instance, the binder for the 1980 meeting contained a number of articles to provide background to a discussion of the topic, including, "Facing Up to Legal Problems in the Building Industry, *BD&C* 21 (April 1980): 70–1; Roy L. Wilson, "The Construction Claim," *Merit Shop Contractor* (April 1977); "A Claims Administration Program Can Save You Money," *Excavating Contractor* (August 1978); Peter Goetz, "Construction Contracts: Are They Etched in Stone?" *Merit Shop Contractor* (October 1978); Irving M. Fogel, "What You Should Know About Construction Claims," *Construction Contracting* 61 (August 1979): 28–9; H. Murray-Hohns, "Minimizing the Effects of Claims and Litigation," *Western Building Design* 16 (March 1980): 22. The 1986 offsite managers' meeting, held in Phoenix, discussed a recent white paper published by the National Society of Professional Engineers, "The Liability Crisis" (1985).

89. Michael R. Adamson, Memorandum of Conversation with Richard M. Kunnath, San Francisco, 9 July 2009. A graduate of the University of California, Berkeley, and the Boalt Hall School of Law, McCarthy became a senior partner in the firm in 1955. From 1981 to 1982, he served in the Office of Policy Development in the Reagan White House before the president named him to the board of directors of the Legal Services Corporation in a recess appointment.

90. Kim Lum interview (quoted).

Chapter 5

1. Richard M. Kunnath interview, 21 October 2008; Timothy P. Murphy interview (quoted); Kim Petersen interview; Dean E. Stephan interview (quoted).

2. Kim Petersen, letter to the author, 14 April 2010.

3. Dean E. Stephan interview.

4. "Dwindling Volumes Result from Slower Regional Growth," *ENR*, 19 January 1989, 54–7; "Many Markets Motionless or Worse," *ENR*, 25 January 1990, 42–4; Glenn R. Mueller, "What Will the Next Real Estate Cycle Look Like?" *Journal of Real Estate Portfolio Management* 8 (May 2002).

5. Julie Nakashima, "Third Largest Design-Build Firm," *Southern California Real Estate Journal*, 23 May 1988.

6. Richard M. Kunnath interview, 9 July 2009.

7. Alan D. Murk interview, 15 December 2008.

8. Thomas D. Verti interview (quoted).

9. Conan "Doug" Craker interview; Jon Eicholtz interview; Timothy P. Murphy interview (quoted); Kim Petersen interview.

10. Michael R. Adamson, Memorandum of Conversation with Doris Pankow, Steve Pankow, and Rick Pankow, Altadena, CA, 12 February 2009; Rick Pankow interview; Steve Pankow interview; Stremfel, "All for One," 40 (quoted).

11. Timothy P. Murphy interview.

12. Ibid.; Kim Petersen, letters to the author, 14 April 2010 and 15 April 2010.

13. Kim Petersen, letter to the author, 14 April 2010 (quoted).

14. Ibid.

15. W. Elliot Brownlee and C. Eugene Steuerle, "Taxation," in *The Reagan Presidency: Pragmatic Conservatism and Its Legacies*, ed. W. Elliot Brownlee and Hugh Davis Graham (Lawrence: University of Kansas Press, 2003), 168–73. On goals and achievements of tax reform during the Reagan years generally, see, also, Sheldon D. Pollack, "Tax Reform: The 1980s in Perspective," *Tax Law Review* 46 (Summer 1991): 489–536; C. Eugene Steuerle, *Contemporary U.S. Tax Policy*, 2d ed. (Washington, DC: Urban Institute Press, 2008), chaps. 5–7.

16. Ibid., 172–3; Myron S. Scholes and Mark A. Wolfson, "The Effects of Changes in Tax Laws on Corporate Reorganization Activity," *Journal of Business* 63 (January 1990): 154 (quoted), 155.

17. Arthur Andersen & Co., *Tax Reform 1986: Analysis and Planning* (Chicago: the company, 1986), 98–101, 110–11; Shannon P. Pratt, *Business Valuation: Discounts and Premiums*, 2d ed. (New York: Wiley, 2009), 277–8; Scholes and Wolfson, "The Effects of Changes in Tax Laws on Corporate Reorganization Activity," 142–3, 154 (quoted); *General Utilities & Operating Co. v. Helvering*, 296 U.S. 200 (1935).

18. Timothy P. Murphy interview; Kim Petersen interview (quoted); Kim Petersen, letter to the author, 14 April 2010.

19. Albert W. Fink interview, 13 February 2009; George F. Hutton interview (quoted); Kim Lum interview; Kim Petersen interview.

20. Richard M. Kunnath interview, 21 October 2008 (quoted); Thomas D. Verti interview (quoted).

21. George F. Hutton interview; Richard M. Kunnath interview, 21 October 2008 (quoted).

22. Richard M. Kunnath interview, 21 October 2008 (quoted).

23. Thomas D. Verti interview (quoted).

24. Laney quoted in Lucy Jokiel, "The Worst of Times," *Hawaii Investor* 13 (June 1993): 10.

25. "Construction Economics," *ENR*, 12 January 1984, 63–7; "Construction Economics," *ENR*, 24 January 1985, 45–50; Roger J. Hannan, "Construction Economics," *ENR*, 23 January 1986, 53–8; Roger J. Hannan and Rani Isaac, "Construction Economics," *ENR*, 22 January 1987, 48–56; Roger J. Hannan and Rani Isaac, "Construction Activity Gears Down, Heads toward Recession in 1989," *ENR*, 21 January 1988, 64–76.

26. *Construction in Hawaii, 1985*, 1–2; "New Projects," *CP News* 4 (Summer 1986); "Work in Progress," *CP News* 5 (Summer 1987); "Work in Progress," *CP News* 7 (Spring 1989). The Mandarin project was renamed Maile Court.

27. Conan "Doug" Craker interview; Kim Petersen interview.

28. Conan "Doug" Craker interview; George F. Hutton interview.

29. George F. Hutton interview; Timothy P. Murphy interview; Kim Petersen interview (quoted).

30. "Forecast '86," *California Builder and Engineer*, 13 January 1986, 30, 38; "Forecast '88," *California Builder and Engineer*, 11 January 1988, 32, 37, 45; "1989

California Heavy Construction Volume Will Revert to 1987 Levels," *California Builder and Engineer*, 11 January 1989, 35–42.

31. George F. Hutton interview; Kim Lum interview; Norman "Red" Metcalf interview; "Employee in Focus: Don Kimball," *CPI News* 3 (Winter 1985); "Employee in Focus: Norman 'Red' Metcalf," *CPI News* 3 (Fall 1985).

32. George F. Hutton interview; Kim Lum interview (quoted); Jean Collins, "25 Years and Still Building," *Pasadena Star-News*, 9 May 1988, 12.

33. Albert W. Fink interview, 13 February 2009 (quoted); Kim Lum interview.

34. Kim Lum interview.

35. George F. Hutton interview; Michelle R. Thompson, "Park Place Timed for Local Market," *Pacific Business News*, 13 May 1991, 31; ITOCHU Corporation website, www.itochu.co.jp/en/about/history.

36. *The Hawaii Journal: A Real Estate and Business Newsletter from Hawaii* (April 1980): 3 (quoted); Rainalter quoted in Thompson, "Park Place Timed for Local Market," 31; "Project of the Quarter: Honolulu Park Place," *CP News* 8 (Spring 1990). On Lacayo, see also note 100, page 416.

37. Albert W. Fink interview, 13 February 2009; "Project of the Quarter: Honolulu Park Place," *CP News* 8 (Spring 1990) (quoted); "Pulse: Contracts and Low Bids, Hawaii," *ENR*, 15 February 1990, 40; Thompson, "Park Place Timed for Local Market."

38. George F. Hutton interview; "Project of the Quarter: Marathon Plaza," *CP News* 6 (Spring 1988); "Project of the Quarter: Resort at Squaw Peak," *CP News* 9 (Fall 1990); "Project of the Quarter: YMCA," *CP News* 5 (Spring 1987).

39. George F. Hutton interview.

40. Ibid. (quoted).

41. Andrew Gomes, "Pier Review," *Honolulu Advertiser*, 9 March 2008 (quoted). On the development of Faneuil Hall Marketplace and Harborplace as spaces "animated by nostalgia," see Isenberg, *Downtown America*, chap. 7. On the redevelopment of ports as diversified, non-maritime places generally, see Ann Breen and Dick Rigby, *Waterfronts: Cities Reclaim Their Edge* (New York: McGraw-Hill, 1993); Peter Hendee Brown, *America's Waterfront Revival: Port Authorities and Urban Redevelopment* (Philadelphia: University of Pennsylvania Press, 2009); Bonnie Fisher et al., *Remaking the Urban Waterfront* (Washington, DC: Urban Land Institute, 2004).

42. Gomes, "Pier Review."

43. Ibid.

44. George F. Hutton interview; "Pulse: Contracts and Low Bids, Hawaii," *ENR*, 22 February 1993, 61; Michelle R. Thompson, "Aloha Tower Project Set for Spring Startup," *Pacific Business News*, 22 February 1993, A28.

45. Gomes, "Pier Review"; Jokiel, "The Worst of Times"; Thompson, "Aloha Tower Project Set for Spring Startup." On the bursting of the bubble and the subsequent "lost decade" in Japan, see Tim Callen and Jonathan D. Ostry, eds., *Japan's Lost Decade: Policies for Economic Revival* (Washington, DC: International

Monetary Fund, 2003); Gary R. Saxonhouse and Robert M. Stern, *Japan's Lost Decade: Origins, Consequences, and Prospects for Recovery* (Malden, MA: Blackwell Publishing, 2004); "To Lose One Decade May Be Misfortune," *Economist*, 2 January 2010, 52–3; "What Ails Japan?," survey, *Economist*, 30 April 2002; Christopher Wood, *The Bubble Economy: Japan's Extraordinary Speculative Boom of the 80s and the Dramatic Bust of the 90s* (New York: Atlantic Monthly Press, 1992).

46. Dean Browning interview; George F. Hutton interview; Hutton quoted in Thompson, "Aloha Tower Project Set for Spring Startup"; Jokiel, "The Worst of Times"; CPBL quoted in Jerry Tune, "Aloha Tower Project Has New Contractor," *Honolulu Star-Bulletin*, 16 April 1993, A1; Rainalter quoted in ibid., A8; Christine Rodrigo, "U.S. Pacific Builders: A Cut Above the Competition," *Pacific Business News*, 31 August 1992, A5. U.S. Pacific Builders entered the market at a fortuitous moment. It grossed $30 million in its first year and $37 million in 1991. As of August 1992, it had $65 million of work on its books. Deuchar took his cue from Pankow's reorganization to structure the business so that its key employees owned stock in the company.

47. Dean Browning interview (quoted); "New Faces/New Places," *CP News* 9 (Fall 1990). Browning's comments refer to the construction of the project, not its fate in the marketplace. The Marketplace performed weakly as a commercial property, owing primarily to inadequate parking. In 1995, ATA fell behind on the ground rents it owed the ATDC for the undeveloped land on the site. The following year, Mitsui Trust & Banking Company, the construction lender, foreclosed on the property. In 1997, the developer was forced into bankruptcy (Gomes, "Pier Review").

48. George F. Hutton interview (quoted).

49. Jon Eicholtz interview (quoted); Richard M. Kunnath interview, 9 July 2009; Mark J. Perniconi interview; Kim Petersen interview (quoted); Russell L. Wahl interview; Dick Walterhouse interview; Richard M. Kunnath, "Continuing to Bring Value to Our Clients," *CP News* 10 (Spring 1991) (quoted).

50. Richard M. Kunnath interview, 1 October 2008; "San Diego Branch Office Opens," *CP News* 9 (Spring 1990) (quoted); "Construction Begins on Four Seasons Resort Aviara," *CP News* 9 (Fall 1990).

51. Richard M. Kunnath interview, 1 October 2008 (quoted).

52. Ibid. (quoted); Richard M. Kunnath, telephone conversation with the author, 7 January 2011.

53. Conan "Doug" Craker interview (quoted); George F. Hutton interview; Brad D. Inman interview (quoted); Richard M. Kunnath interview, 21 October 2008 (quoted); Dean E. Stephan interview. Charlie Pankow did not fire Inman. Rather, he transferred him to Hawaii as vice president of construction, apparently without consulting George Hutton. Having little need for Inman's services, Hutton fired him 18 months later.

54. Pankow quoted in Collins, "25 Years and Still Building," 12.

55. "Project of the Quarter: Grand Financial Plaza," *CP News* 4 (Summer

1986); "Project of the Quarter: Shoreline Square," *CP News* 7 (Spring 1989); "Self-Climbing Forms Speed Long Beach Project," *California Builder and Engineer*, 6 July 1987; "Shoreline Square Topped Out," *LAT*, 8 November 1987, VIII:1; "Shoreline Square Update," *CP News* 5 (Fall 1987) (quoted). The Shoreline Square office building received the 1989 Outstanding Structural Design Award from the Los Angeles Tall Buildings Structural Design Council. For more on mixed steel and concrete framing systems for high-rise buildings, see Hal Iyengar, "Mixing Steel and Concrete," *Civil Engineering* 55 (March 1985): 46–9. Grand Financial Plaza was renamed Chase Plaza before its completion.

56. Kevin Smith interview (quoted); Joseph Korom, *Milwaukee Architecture: A Guide to Notable Buildings* (Madison, WI: Prairie Oak Press, 1995), 20 (quoted); "Project of the Quarter: 411 East Wisconsin Building," *CPI News* 3 (Winter 1985).

57. "Project of the Quarter: YMCA," *CP News* 5 (Spring 1987).

58. Ibid.

59. Ibid.

60. Ibid.

61. Ibid.; Mike Liddiard interview (quoted). Groundbreaking for the project took place on 20 March 1985. The YMCA opened in October 1986.

62. Albert W. Fink interview, 25 March 2009.

63. Ibid.; "Charles Pankow: A Landmark Project," *Hawaii Contractor* (Spring 1993): 44–5; Maria Torres, "Waikiki Landmark Achieves a Hawaii First," *Building Industry* 38 (August 1992); Weinstein quoted in Gordon Wright, "Twin-Tower Condo Provides a Gateway to Waikiki," *BD&C* 34 (December 1993): 46. The Landmark included an 11-story, 235,000-square-foot parking structure with 570 stalls.

64. Albert W. Fink interview, 25 March 2009.

65. Ibid.; Wright, "Twin-Tower Condo Provides a Gateway to Waikiki," 47–8.

66. Albert W. Fink interview, 25 March 2009; "Waikiki Landmark Condominium Completed," *Pankow News* 11 (Spring 1993): 1–2. Fink emphasizes that the architects worked closely with the structural engineers and contractor to incorporate the developer-approved revisions into the working drawings.

67. Chock quoted in Torres, "Waikiki Landmark Achieves a Hawaii First"; Feldman quoted in "Waikiki Landmark Condominium Completed," 2; Liddiard quoted in Torres, "Waikiki Landmark Achieves a Hawaii First"; "Project of the Quarter: Marathon Plaza."

68. Chock quoted in Torres, "Waikiki Landmark Achieves a Hawaii First"; Torres, "Waikiki Landmark Achieves a Hawaii First" (quoted); "Waikiki Landmark Condominium Completed," 2. Martin & Bravo, established in 1968, later became Martin & Chock.

69. Bravo quoted in Torres, "Waikiki Landmark Achieves a Hawaii First"; Doke quoted in Torres, "Waikiki Landmark Achieves a Hawaii First"; Feldman quoted in "Waikiki Landmark Condominium Completed," 2; Liddiard quoted in Torres, "Waikiki Landmark Achieves a Hawaii First."

70. Brent F. Howell, "Under New Ownership: Regional Shopping Centers Face New Challenges," *Urban Land* 45 (June 1986): 2–5, 5 (quoted). Howell was first vice president and national marketing director for Commercial Properties with Coldwell Banker Commercial Real Estate Services in Los Angeles.

71. As a developer and contractor whose clients included grocery store chains Alpha Beta, Lucky, Ralphs, Safeway, and Vons, Ernest W. Hahn was an early practitioner of design-build. He negotiated contracts that refunded project savings to clients and involved architect, contractor, owner, and leasing agent in a team effort "to insure low costs, maximum efficiency, and adherence to tight schedules," an approach the *Los Angeles Times* deemed "unique" (Bob Boich, "His Forte: Building Shopping Centers," *LAT*, 12 August 1962, II:1).

72. Suzanne Dow Nakaki interview; "Ernest W. Hahn, Pioneer of the Modern Mall, Dies," *LAT*, 29 December 1992, A3; Norm Husk, "Tyler Mall Expansion: Keeping Tenants Happy during Construction," *Urban Land* 51 (March 1992): 26–9; Hans Ibold, "Mall Is Giving Way to New Vision after Losing Appeal," *Los Angeles Business Journal*, 30 October 2000, 50; Kober, "Regrowth for Existing Shopping Centers," 4–5.

73. Robert Law interview; Suzanne Dow Nakaki interview (quoted); Husk, "Tyler Mall Expansion"; "New Technology Allows 'Suspended' Second Level Tyler Mall Addition," 9 (Fall 1990).

74. Husk, "Tyler Mall Expansion" (quoted).

75. Robert Law interview (quoted); Suzanne Dow Nakaki interview (quoted); Husk, "Tyler Mall Expansion."

76. Husk, "Tyler Mall Expansion," 29.

77. Dwyer quoted in "Nation's Fifth Largest Shopping Center Expansion," *CP News* 10 (Winter 1992); "Expansion/Renovation Completed On Major U.S. Mall," *Pankow News* 11 (Spring 1993); "Work in Progress," *CP News* 6 (Spring 1988); "Work in Progress," *CP News* 7 (Summer 1989); "Work in Progress," *CP News* 8 (Summer 1990); "Brea Mall Expansion/Renovation," *CP News* 6 (Spring 1988); "Project of the Quarter," *CP News* 7 (Summer 1989); Gruen and Smith, *Shopping Towns USA*, table on 216–8, 244–5.

78. Lipsy quoted in Leslie Eimas, "Tough Times at the Malls," *Syracuse Herald-Journal*, 28 April 1992, A6; "Birth, Death, and Shopping: The Rise and Fall of the Shopping Mall," *Economist*, 22 December 2007, 103 (quoted); Steve Kerch, "Tough Times Visiting Shopping Centers, Too," *Chicago Tribune*, 2 June 1991; William Neikirk, "The Party's Over for Shops and Shoppers," *Chicago Tribune*, 30 December 1991, 11.

79. Beyard quoted in Leslie Eimas, "Tough Times at the Malls," A6; "Pankow Wraps Up Roosevelt Field Mall—Phase III," *Single Source* 15 (Spring 1997); "Pankow Starts Mall Renovation," *Single Source* 15 (Summer 1997); "Pankow Nears Completion of Walt Whitman Mall Renovation," *Single Source* 16 (Fall 1998).

80. Charles J. Pankow, "Looking Forward," *CP News* 6 (Spring 1988).

Chapter 6

1. "Charles J. Pankow, ASCE Member," obituary, *Transactions of the American Society of Civil Engineers* 169 (2004): 816–7; Dennis McLellan, "Charles Pankow, 83; Founded Firm that Built MTA Complex," *LAT*, 19 January 2004; Katherine Seligman, "Charles Pankow—Art Collector, Civil Engineer," *San Francisco Chronicle*, 16 January 2004. The "President's memo" was a monthly feature of the ACI publication, *Concrete International*.

2. Charles J. Pankow, "On-Site Precasting and Tilt-Up" and "Slipform Construction of Buildings," chaps. 33 and 34 in *Concrete Construction Handbook*, ed. Joseph J. Waddell, 2d ed. (New York: McGraw-Hill, 1974); Dean E. Stephan, "On-Site Precasting" and "Slipform Construction of Buildings," chaps. 33 and 34 in *Concrete Construction Handbook*, ed. Joseph J. Waddell and Joseph A. Dobrowolski, 3d ed. (New York: McGraw-Hill, 1993); idem, "On-Site Precasting" and "Slipform Construction of Buildings," chaps. 31 and 32 in *Concrete Construction Handbook*, ed. Joseph A. Dobrowolski, 4th ed. (New York: McGraw-Hill, 1998).

3. Vincent P. Drnevich, letter to the author, 13 October 2011; Seely, "Research, Engineering, and Science in American Engineering Colleges," 367–79; Wisnioski, "Engineers and the Intellectual Crisis of Technology," 101–5.

4. *Bulletin of Purdue University: Catalogue Number* (1941), 190 (quoted), 191–6; Schools of Engineering, Purdue University, *Announcement for the Year 1970–71* (West Layfayette, IN: Purdue University, 1970), 48–50; Seely, "Research, Engineering, and Science in American Engineering Colleges," 379.

5. Schools of Engineering, Purdue University, *Announcement for the Year 1970–71*, 43–50.

6. Vincent P. Drnevich interview; Robert Law, telephone interview with the author, 6 October 2010; College of Engineering, Purdue University, engineering. purdue.edu/Engr/AboutUs/History.

7. Vincent P. Drnevich, letter to the author, 13 October 2011; College of Engineering, Purdue University, engineering.purdue.edu/Engr/AboutUs/History. The School of Civil Engineering also established an advisory council, in 1991. Robert Law, for one, has served in this capacity.

8. Vincent P. Drnevich, letter to the author, 13 October 2011 (quoted); idem, "The Senior Design Process at Purdue University," *Proceedings of the 2005 American Society for Engineering Education Annual Conference and Exposition*; Vincent P. Drnevich and John B. Norris, "Assigning Civil Engineering Students to Capstone Course Teams," *Proceedings of the 2007 American Society for Engineering Education Annual Conference and Exposition*.

9. A second concrete research facility named for Charlie Pankow was dedicated on 2 October 2009. The Charles Pankow Concrete Materials Laboratory was built and equipped with $2 million from a $4.7 million donation from the Pankow family. Research conducted at the laboratory focuses on developing more durable and environmentally friendly concrete (Emil Venere, "New Purdue Lab Aims to Aid Nation's Aging Infrastructure," University News Service, Purdue University,

27 October 2009, www.purdue.edu/newsroom/general/2009/story-print-deploy-layout_1_1478_1478.html).

10. Richard M. Kunnath interview, 9 July 2009; "Design-Build Continues to Grow in U.S.," *Civil Engineering* 66 (December 1996): 18–19.

11. Dean E. Stephan interview (quoted); "Amid Controversy, Construction Management Blossoms," *BD&C* 13 (February 1972): 35; "Change: The Building Team Is Getting Together for a Change," 34–6; "The Changing Role of the General Contractor," *BD&C* 12 (April 1971): 43; Jane Edmunds, "The Pendulum Swings Toward Design-Construct: A Committee of 100 Report," *Consulting Engineer* (October 1984): 73.

12. "Amid Controversy, Construction Management Blossoms," 35–7; "Delivery Options: A Wide Range of Choices," *BD&C* 21 (February 1980): 66–9; Edmunds, "The Pendulum Swings Toward Design-Construct," 73–4; "How Owners Decide: Whom to Hire, Why, and How to Build," *BD&C* 21 (April 1980): 86–9.

13. "Amid Controversy, Construction Management Blossoms," 35, 36 (Reed quoted).

14. Robert Law interview (quoted); Dean E. Stephan interview (quoted); Thomas D. Verti interview (quoted); Charles Pankow, Inc., marketing brochure, n.d. (c. 1977), CPBL, Pasadena, CA (quoted); Horowitz quoted in "The Changing Role of the General Contractor," 45; Peters quoted in "The Changing Role of the General Contractor," 47.

15. Horowitz quoted in "The Changing Role of the General Contractor," 45; O'Neil quoted in "The Changing Role of the General Contractor," 45; Perini quoted in "The Changing Role of the General Contractor," 44.

16. Edmunds, "The Pendulum Swings Toward Design-Construct," 74.

17. Kendall quoted in Christopher Olson, "Design/Build Effectively Serves a Growing Market Segment," *BD&C* 30 (October 1989): 101.

18. Chell quoted in "The Changing Role of the General Contractor," 46.

19. Thomsen quoted in "Project Delivery and Owner Requirements," *BD&C* 26 (February 1985): 52.

20. Gordon Wright, "Design/Build: Single-Source Option Gains Wider Acceptance," *BD&C* 28 (April 1987): 64–71; "One Contract Replaces Two or More in Design/Build Work," *Concrete Construction* (November 1985): 909–10.

21. Dean E. Stephan interview.

22. Wilberg quoted in Wright, "Design/Build: Single-Source Option Gains Wider Acceptance," 71.

23. Charles J. Pankow, "Pankow Celebrates 35 Years," *Single Source* 16 (Spring 1998); Wright, "Design/Build: Single-Source Option Gains Wider Acceptance," 64 (quoted), 71 (Stephan cited, not quoted in article); Daniel C. Brown, "Speed and Quality with Design/Build," *BD&C* 26 (January 1985): 58 (quoted).

24. Christopher Olson, "Design/Builders Rapidly Expand Their Market," *BD&C* 25 (June 1984): 72; idem, "Design/Build Effectively Serves a Growing Market Segment," 101.

25. Olson, "Design/Build Effectively Serves a Growing Market Segment," 101–2.

26. Brown, "Speed and Quality with Design/Build," 58–60; Michael J. Miller, "Design/Build Process Helps First-Time Owner," *BD&C* 22 (July 1981): 156–9; Gordon Wright, "How a Design/Build Contest Saved a Year," *BD&C* 21 (April 1980): 46–53; idem, "Design/Build: Single-Source Option Gains Wider Acceptance"; idem, "Design/Build's Impact Continues to Grow," *BD&C* 29 (February 1988): 59–67.

27. William Quatman, *Design-Build for the Design Professional* (New York: Aspen Law and Business, 2001), 31–3.

28. Milton F. Lunch, "ASCE Report Addresses Design/Build Benefits and Concerns," *BD&C* 33 (July 1992): 25; Quatman, *Design-Build for the Design Professional*, 28–31; William G. Quatman, "Company Standard Form Design-Build Contracts," in *Design-Build Contracting Handbook*, ed. Robert Franck Cushman and Michael C. Loulakis (New York: Aspen Law and Business, 2001), 201–11; Wright, "Design/Build's Impact Continues to Grow."

29. James Denning, "Design-Build Goes Public," *Civil Engineering* 62 (July 1992): 76.

30. Richard M. Kunnath interviews, 1 and 21 October 2008; profile of Preston Haskell, DBIA website, www.dbia.org/about/awards/brunelleschi/haskell2002.htm; "Design/Build Group Organizes, Opens Washington Office," *BD&C* 34 (November 1993): 8.

31. Richard M. Kunnath interview, 1 October 2008.

32. Richard M. Kunnath interview, 21 October 2008 (quoted); "Design/Build Delivery Accelerates," *Civil Engineering* 68 (January 1998): 14; DBIA website, www.dbia.org.

33. Richard M. Kunnath interview, 21 October 2008; "Design/Build Group Organizes, Opens Washington Office"; Quatman, *Design-Build for the Design Professional*, 33–4.

34. Kim Lum interview.

35. Ibid.

36. Denning, "Design-Build Goes Public," 76.

37. Stewart quoted in Denning, "Design-Build Goes Public," 77; Wiernicki quoted in Wright, "Design/Build's Impact Continues to Grow," 66; Wiernicki quoted in Denning, "Design-Build Goes Public," 77; "Change: The Building Team Is Getting Together for a Change," 34–6; Olson, "Design/Builders Rapidly Expand Their Market," 75 (quoted), 76; Wright, "Design/Build: Single-Source Option Gains Wider Acceptance," 71.

38. Winchell quoted in Wright, "Design/Build: Single-Source Option Gains Wider Acceptance," 66–7; Olson, "Design/Builders Rapidly Expand Their Market," 75; Wright, "Design/Build's Impact Continues to Grow," 59.

39. Denning, "Design-Build Goes Public," 77–8; Milton F. Lunch, "Design/Build Gets Boost from the Corps of Engineers," *BD&C* 35 (May 1994): 29; Charles

Robinson, "Design/Build a Wise Choice for Center Construction," *Honolulu Star-Bulletin*, 12 August 1997; Wright, "Design/Build: Single-Source Option Gains Wider Acceptance," 66–71; Wright, "Design/Build's Impact Continues to Grow," 64–7. Robinson was director of marketing for CPBL in Hawaii.

40. Denning, "Design-Build Goes Public," 78–9; Lunch, "Design/Build Gets Boost from the Corps of Engineers."

41. Buehler quoted in Wright, "Design/Build: Single-Source Option Gains Wider Acceptance," 71.

42. Richard M. Kunnath interview, 9 July 2009; American Institute of Architects, *Design-Build-Bid Task Force Report* (1975); Jeffrey L. Beard, Edward C. Wundram, and Michael C. Loulakis, *Design-Build: Planning through Development* (New York: McGraw-Hill Professional, 2001), chap. 9; Denning, "Design-Build Goes Public," 77; Frank W. Chitwood, "The RFP and Selection Process," in *The Architect's Guide to Design-Build Services*, ed. G. William Quatman and Ranjit Dhar (New York: Wiley, 2003), 89–91; Quatman, *Design-Build for the Design Professional*, chap. 14; David B. Rosenbaum, "Can't We All Just Get Along?" *ENR*, 16 October 1995, 13.

43. "UC Berkeley Project Now Under Design/Build May Become Model," *CM Magazine* (December 1995): 16.

44. Lum quoted in "Charles Pankow Builders: Design-Build Revolutionaries?" *Building Industry* 43 (July 1997): 67.

45. Dean Browning interview (quoted); Albert W. Fink interview, 25 March 2009 (quoted); "University of Hawaii Special Events Arena," information sheet provided to *Design-Build Dateline* magazine, n.d., CPBL, Pasadena, CA; Cindy Luis, "First Special Event at Hawaii's New Arena," *Honolulu Star-Bulletin*, 21 September 1994, D1; David Yount, *Who Runs the University? The Politics of Higher Education in Hawaii, 1985–1992* (Honolulu: University of Hawaii Press, 1996), 141–4, 141 (quoted).

46. Bill Bramschreiber interview; Albert W. Fink interview, 25 March 2009 (quoted).

47. Dean Browning interview (quoted); Albert W. Fink interview, 25 March 2009 (quoted); Browning quoted in Christine Rodrigo, "Precast Concrete Cuts Costs," *Pacific Business News*, 23 August 1993, A6.

48. Bill Bramschreiber interview (quoted); Waihee quoted in Luis, "First Special Event at Hawaii's New Arena"; Yoshida quoted in Bill Ewon, "Night to Remember for Rainbow Sports," *Honolulu Star-Bulletin*, 22 October 1994, C4; "University of Hawaii Special Events Arena," information sheet provided to *Design-Build Dateline* magazine, n.d., CPBL, Pasadena, CA; Cindy Luis, "Home Sweet Dome," *Honolulu Star-Bulletin*, 22 October 1994, C1; Rodrigo, "Precast Concrete Cuts Costs."

49. Michael R. Adamson, Memorandum of Conversation with Crodd Chin, Emeryville, CA, 15 September 2008; "Pankow Builds Boalt Hall Expansion," *Daily Pacific Builder*, 18 January 1996.

50. Kunnath quoted in "UC Berkeley Project Now Under Design/Build May Become Model," 17.

51. Michael R. Adamson, Memorandum of Conversation with Crodd Chin, Emeryville, CA, 15 September 2008; Woodruff Minor, *The Architecture of Ratcliff* (Berkeley, CA: Heyday Books, 2006), 170 (quoted). The Life Sciences Building, designed by George W. Kelham, the university's supervising architect, and completed in 1930, was a "brooding relic of institutional growth," as Minor describes it. The largest building on campus, it measured 500 feet by 250 feet. Upon completion of the project, the building was renamed the Valley Life Sciences Building on behalf of its principal donor. See, also, Harvey Helfand, *The Campus Guides: University of California, Berkeley* (Princeton, NJ: Princeton Architectural Press, 2001), 147–52.

52. Michael R. Adamson, Memorandum of Conversation with Crodd Chin, Emeryville, CA, 15 September 2008; "UC Berkeley Project Now Under Design/ Build May Become Model," 17 (quoted).

53. Michael R. Adamson, Memorandum of Conversation with Crodd Chin, Emeryville, CA, 15 September 2008; Kunnath quoted in "UC Berkeley Project Now Under Design/Build May Become Model," 17; Kunnath quoted in "Pankow Builds Boalt Hall Expansion," 4.

54. Dean E. Stephan interview (quoted); Richard Cook quoted in Elyse Umlauf, "Cool Analysis Betters the Odds in the Competition Gamble," *BD&C* 33 (June 1992): 64. The GSA, with more experience than almost all other public agencies, at any level, was an exception to charges that public sector design-build competitions lacked transparency. It received praise for its explicit RFPs and equitable review processes.

Cook was a principal in Stowell Cook Frolichstein, a Chicago-based architectural firm.

55. Denning, "Design-Build Goes Public," 79; Michael C. Loulakis, *Design-Build for the Public Sector* (New York: Aspen Law and Business, 2003), 137–49; Rosenbaum, "Can't We All Just Get Along?"; Umlauf, "Cool Analysis Betters the Odds in the Competition Gamble," 65 (quoted). The reports cited in these sources also include *Design-Build in the Federal Sector* (Washington, DC: ASCE, 1992); *AIA/ AGC Recommended Guidelines for Procurement of Design-Build in the Public Sector* (Washington, DC: AIA and AGC, 1995). Both *Design-Build RFQ/RFP Guide for Major Public Sector Projects* and *Design-Build RFQ/RFP Guide for Small-to-Medium Projects* were authored by Edward C. Wundram, an architect with more than four decades of experience in the design and procurement of major public facilities projects. In 1980, he founded The Design Build Group Consulting, which manages this type of project and administers design-build competitions related to them.

56. Richard M. Kunnath interview, 9 July 2009.

57. "Project Delivery Method Market Share for Non-Residential Construction," figure, RSMeans Business Solutions, a division of Reed Construction Data, May 2011, www.dbia.org/pubs/research/rsmeans110606.htm.

58. "Hard Lessons Come Home: Shoddy Design and Construction Blamed for Much Damage," *ENR*, 16 January 1995, 28–33; Ayman S. Mossallam, P. R. Chackrabarti, and Ernie K. Lau, "Concrete Connections," *Civil Engineering* 69 (January 1999): 43–5.

59. Ibid.; Dean E. Stephan, "Lessons of Northridge only Now Being Assessed, Says Master Builder," *Daily Pacific Builder*, 21 September 1995 (quoted).

60. Dean E. Stephan interview (quoted).

61. Stephan quoted in Stuart F. Brown, "Building Business Buildings Better," *Fortune*, 8 September 1997.

62. Michael Lewis writes: "The new new thing . . . is not necessarily a new invention. It is not necessarily a new idea. [It] is a notion that is poised to be taken seriously in the marketplace" (*The New New Thing* [New York: Penguin, 2000], xvii).

63. Nakaki quoted in Craig A. Shutt, "Hybrid Precast Frame Meets Seismic Challenges," *Ascent* (Spring 1997): 16; Seagren quoted in ibid., 15; Stanton quoted in ibid., 16; Richard Spaulding, "New Structural Design May Be Ultimate in Quake Resistance," *San Diego Daily Transcript*, 17 October 1994; Gordon Wright, "Targeting Improved Seismic Performance," *BD&C* 38 (January 1997): 56–8.

64. Brown, "Building Business Buildings Better"; Lew quoted in Shutt, "Hybrid Precast Frame Meets Seismic Challenges," 14; Seagren quoted in Todd L. Whitlock, "All Shook Up," *Urban Land* 59 (June 2000): 31.

65. Dean E. Stephan interview (quoted); Martha Blastow, "Momentary Connection," *Concrete Products* 103 (September 2000), 19–23; John Stanton, William C. Stone, and Geraldine S. Cheok, "A Hybrid Reinforced Precast Frame for Seismic Regions," *PCI Journal* 42 (March/April 1997): 24–5. In 1991, Hawkins left the University of Washington to head the Department of Civil and Environmental Engineering at the University of Illinois. His biographical profile may be found online, www.cee.illinois.edu/node/250.

66. Suzanne Dow Nakaki interview; Blastow, "Momentary Connection."

67. Ibid.; Shutt, "Hybrid Precast Frame Meets Seismic Challenges," 16.

68. Stephan quoted in Spaulding, "New Structural Design May Be Ultimate in Quake Resistance"; Stanton, Stone, and Cheok, "A Hybrid Reinforced Precast Frame for Seismic Regions," 20–32, 24 (quoted); Dean E. Stephan, "Technology Transfer—The Innovative Task Group," President's memo, *Concrete International* (January 1995): 7; Wright, "Targeting Improved Seismic Performance," 56–7. Stone, Cheok, and Stanton first reported the results of the tests in, "Performance of Hybrid Moment-Resisting Precast Beam-Column Concrete Connections Subjected to Cyclic Loading," *ACI Structural Journal* 92 (March 1995): 229–49.

69. Suzanne Dow Nakaki interview (quoted); Hamburger quoted in Khanh T. L. Tran, "Designer Sets Tower Project in Concrete," *Wall Street Journal*, 7 July 1999, B10; Stanton, Stone, and Cheok, "A Hybrid Reinforced Precast Frame for Seismic Regions," 31 (quoted); Stephan, "Lessons of Northridge only Now Being Assessed, Says Master Builder" (quoted); ICBO, *Uniform Building Code* (Whittier, CA: ICBO, 1994).

70. Suzanne Dow Nakaki interview (quoted); Whitlock quoted in David B. Rosenbaum, "Record-Height Concrete Building Uses Quake-Resistant Precast," *ENR*, 14 June 1999, 14; Wright, "Targeting Improved Seismic Performance," 57.

71. Blastow, "Momentary Connection," 22–3; Shutt, "Hybrid Precast Frame Meets Seismic Challenges," 14–15.

72. Thornton quoted in Shutt, "Hybrid Precast Frame Meets Seismic Challenges," 18; Al Fink, "Desiring to Maintain Quality, Pankow Opens Mid-State Precast," *Single Source* 18 (Summer 2000); "Pankow Wraps Up Roosevelt Field Mall—Phase III," *Single Source* 15 (Spring 1997); Richard G. Weingardt, "Charles H. Thorton: Towering Builder of Towering Buildings," *Structure Magazine* (June 2008): 71 (quoted).

73. "Eugene Parking Garages Completed," *Single Source* 15 (Spring 1997); "Stanford Parking Garage—On Schedule," *Single Source* 17 (Summer 1999).

74. Blastow, "Momentary Connection," 19; "Record-Setting 3rd & Mission Apartments to Break Ground!" *Single Source* 17 (Summer 1999); Rosenbaum, "Record-Height Concrete Building Uses Quake-Resistant Precast"; Elizabeth Seifel, "Bay Area Models of Urban Infill Housing," *Urban Land* 62 (September 2003): 144. Third and Mission was the original name of the project. The 60-story Millennium Tower, completed in 2009, is currently San Francisco's tallest residential structure.

75. Douglas R. Porter, "Downtown San Francisco's New Plan: Something for Everyone—Almost," *Urban Land* 45 (February 1986): 34–5; Jane Bowar Zastrow, "Bay Area Projects: Old Friends and New Faces," *Urban Land* 55 (October 1996): 47–50, 114–5.

76. Gerald D. Adams, "The Rebirth of Civic San Francisco," *Urban Land* 55 (October 1996): 52–6; Jacobs, *The Good City*, 152–5; Witte quoted in Larry Flynn, "New Heart for San Francisco," *BD&C* 43 (April 2002): 34.

77. Seifel, "Bay Area Models of Urban Infill Housing," 143 (quoted), 144.

78. Larry Flynn, "Framing the Moment," *BD&C* 42 (August 2001): 32–4; Rosenbaum, "Record-Height Concrete Building Uses Quake-Resistant Precast"; Laurie A. Shuster, "Keeping It Together," *Civil Engineering* 70 (March 2000): 45–6.

79. Flynn, "Framing the Moment," 34; Rosenbaum, "Record-Height Concrete Building Uses Quake-Resistant Precast"; Shuster, "Keeping It Together," 45–6.

80. Joseph Sanders, interview with the author, 25 August 2010; Hanson quoted in Tran, "Designer Sets Tower Project in Concrete"; Robert E. Englekirk, "Design-Construction of The Paramount—A 39-Story Precast Prestressed Concrete Apartment Building," *PCI Journal* 47 (July/August 2002): 61–6; Rosenbaum, "Record-Height Concrete Building Uses Quake-Resistant Precast" (quoted); Shuster, "Keeping It Together," 47.

81. Englekirk, "Design-Construction of The Paramount," 66–9; Flynn, "Framing the Moment," 33–4, 34 (Fong quoted). Using a precast concrete frame rather than one made of structural steel also permitted designers to prescribe (lower) floor heights more appropriate for a residential building, enabling them to reduce the height of The Paramount without sacrificing the size of the units.

82. Richard M. Kunnath, telephone conversation with the author, 7 January 2011.

83. Ibid.; "39-Story Paramount Tower Tops Out," *Single Source* 19 (Summer 2001).

84. Shuster, "Keeping It Together," 47.

85. Michael R. Adamson, Memorandum of Telephone Conversation with Robert Law, 21 July 2009; Richard M. Kunnath interview, 2 July 2010; Charles J. Pankow, "Comments from the Chairman," *Single Source* 19 (Summer 2001).

86. Richard M. Kunnath, telephone conversation with the author, 7 January 2011; Charles J. Pankow, "Comments from the Chairman," *Single Source* 19 (Summer 2001).

Chapter 7

1. Pankow quoted in Collins, "25 Years and Still Building," 12.

2. Pankow quoted in Nakashima, "Third Largest Design-Build Firm"; Richard M. Kunnath interview, 9 July 2009 (quoted).

3. Richard M. Kunnath interview, 9 July 2009 (quoted).

4. Ibid.

5. Ibid.; Dean Browning interview; Dick Walterhouse interview.

6. Richard M. Kunnath interview, 9 July 2009.

7. Ibid.

8. Albert Josselson interview; Richard M. Kunnath, interviews of 9 July 2009 and 2 July 2010 (quoted). Josselson was Charlie Pankow's primary care physician.

9. Dick Walterhouse interview (quoted); "Executives in the News," *Daily Pacific Builder*, 25 August 1995.

10. Richard M. Kunnath interview, 1 October 2008; Katherine Conrad, "Walterhouse Runs $100 Million Firm," *East Bay Business Times*, 26 January 2001, 28.

11. Richard M. Kunnath interview, 1 October 2008; Wally Naylor interview.

12. Richard M. Kunnath interview, 1 October 2008.

13. Ibid.; Wally Naylor interview (quoted).

14. Wally Naylor interview.

15. Richard M. Kunnath interview, 1 October 2008 (quoted); Wally Naylor interview (quoted).

16. Wally Naylor interview; Dean E. Stephan interview.

17. Richard M. Kunnath interview, 1 October 2008; Wally Naylor interview.

18. Wally Naylor interview; Dick Walterhouse interview.

19. Kim Lum interview; Wally Naylor interview; Dean E. Stephan interview (quoted); "Charles Pankow Builders: Design-Build Revolutionaries?"

20. Dean E. Stephan interview (quoted); Dick Walterhouse interview.

21. Richard M. Kunnath interview, 1 October 2008.

22. Ibid. (quoted); Dean E. Stephan interview (quoted).

23. Richard M. Kunnath interview, 1 October 2008 (quoted); Dean E. Stephan interview (quoted); Dick Walterhouse interview (quoted).

24. Dean E. Stephan interview (quoted).

25. Dick Walterhouse interview.

26. "Pankow Special Projects Continues Growth from $10 million in 1991 to $100 million in 2001," *Single Source* 19 (Summer 2001).

27. Wally Naylor interview; Dick Walterhouse interview; "Charles Pankow Builders: Design-Build Revolutionaries?"; "Pankow Hires Two and Promotes Seven," *Single Source* 18 (Summer 2000); Conrad, "Walterhouse Runs $100 Million Firm."

28. Dick Walterhouse interview.

29. Conrad, "Walterhouse Runs $100 Million Firm"; "Deco Diva Reborn," *LAT*, 17 May 2007, Home:1, 10–11; "Eastern Columbia Lofts: Redevelopment Winner," *California Construction* (December 2007); David Silva, "Old Way of Doing Business in L.A. No Longer Applies," *California Construction* (January 2006); Kevin Smith, "Pasadena Firm Turning Historic Building into Luxury Lofts," *San Gabriel Valley Tribune*, 3 August 2005. Chase Plaza, which CPBL completed in 1986, was another example of the conversion of commercial space to residential use. In 2005 its top 11 floors were converted into 132 one- and two-bedroom upmarket apartments (Danny King, "Downtown Bubble Seen As Ready to Burst," *Los Angeles Business Journal*, 5 April 2004).

30. "30-Story Paramount Tower Tops Out"; "Pankow Special Projects Continues Growth from $10 million in 1991 to $100 million in 2001"; "Pankow Responds to Residential Demand" (all found in *Single Source* 19 [Summer 2001]).

31. Kim Lum interview.

32. Richard M. Kunnath interview, 2 July 2010.

33. Dean E. Stephan interview.

34. LePatner, *Broken Buildings, Busted Budgets*, 14.

35. On the "sheer irrelevance" of downtown Los Angeles and the lack of conviction in the architecture of buildings constructed as part of efforts to redevelop it, see Banham, *Los Angeles*, 183–93.

36. Fulton, *The Reluctant Metropolis*, 275 (quoted). See, also, Louis Sahagun, "Civic Center II Developer of Union Station Wants Government, Private Mix," *LAT*, 29 May 1991, B1.

37. Steven M. Nakada, "Gateway Center: Design That Sells Transit," *Urban Land* 57 (May 1998): 78–9; Richard Simon, "Urban Jewel or Height of Folly?" *LAT*, 24 September 1995, B1.

38. Brad Berton, "LACTC Ignores Critics and Sticks with Decision on Headquarters Site," *Los Angeles Business Journal*, 1 March 1993; Fulton, *The Reluctant Metropolis*, 126 (quoted); "MWD Signs Contracts for Headquarters Building at LA's Union Station," *Business Wire*, 11 April 1995; "Practicality Is Focus of Metropolitan Water District's New Headquarters Building," *Business Wire*, 21 August 1998; Ann Rackham, "Catellus Files Suit to Block Deal on New Headquarters," *Los Angeles Business Journal*, 17 May 1993; Simon, "Urban Jewel or Height of Folly?" B4. Even as the RTC agreed to build its headquarters at Union Station as part of what would become Gateway Center, the LACTC voted

to negotiate with developer Ray Watt for office space in his proposed Watt City Center project on West Seventh Street.

39. Fulton, *The Reluctant Metropolis*, 135–42.

40. Jeffrey L. Covell, "Catellus Development Corporation," in *International Directory of Company Histories*, ed. Jay P. Pederson, vol. 24 (Detroit: St. James Press, 1999), 98–9; Nakada, "Gateway Center." Three-fourths of Catellus's industrial, two-thirds of its office, and more than 80 percent of its retail properties were located in California, a legacy of the holdings of the land-grant Southern Pacific Railroad. On the accumulation of these holdings, see Richard J. Orsi, *Sunset Limited: The Southern Pacific Railroad and the Development of the American West, 1850–1930* (Berkeley: University of California Press, 2005), chap. 3.

41. Nakada, "Gateway Center," 78 (quoted), 79 (quoted); "Gateway Center Reaches Completion," *Pankow News* (Spring 1996) (quoted); Simon, "Urban Jewel or Height of Folly?" B4.

42. Richard M. Kunnath interview, 2 July 2010; Joseph Sanders, interview with the author, 25 August 2010; Dean E. Stephan interview.

43. McLarand, Vasquez, Emsiek and Partners, "MTA Headquarters Tower," project sheet, http://mvpi-architects.com/files/pr/79_MTA.pdf (quoted); Bungale S. Taranath, *Wind and Earthquake Resistant Buildings: Structural Analysis and Design* (New York: Marcel Dekker, 2005), 737–8.

44. Simon, "Urban Jewel or Height of Folly?" B1 (quoted), B4 (Kopp quoted), B4 (Hayden quoted).

45. Fulton quoted in Simon, "Urban Jewel or Height of Folly?" B1; Fulton, *The Reluctant Metropolis*, 146–50.

46. Joseph Sanders interview; Thomas D. Verti interview (quoted); McLarand, Vasquez, Emsiek and Partners, "MTA Headquarters Tower," project sheet; Simon, "Urban Jewel or Height of Folly?" B4. Catellus and the RTD each appointed three members to the review board.

47. Wodraska quoted in "Practicality Is Focus of Metropolitan Water District's New Headquarters Building"; Terry McDermott, "Knee-Deep Disputes for 'Water Buffaloes,'" *LAT*, 1 November 1998, A1 (quoted).

48. Fulton, *The Reluctant Metropolis*, chap. 4, 123 (quoted); Valle, *City of Industry*, 151–8; David Zetland, "The End of Abundance: How Water Bureaucrats Created and Destroyed the Southern California Oasis," *Water Alternatives* 2 (2009): 350–69.

49. Friedrichs and Schindler quoted in Southern California Ready Mixed Concrete Association and California Cement Promotion Council, "Functional Beauty in Southern California 182 Days Early and $2 Million under Budget," promotional brochure, n.d., CPBL, Pasadena, CA; Verti quoted in Sharon Leiter-Weintraub, "Built to Suit," *Concrete Products* 102 (April 1999): 32N; "MWD New Los Angeles Headquarters Dedicated," *California Construction* (January 1999): 26; "Practicality Is Focus of Metropolitan Water District's New Headquarters Building." CPBL completed construction of the office tower and wing in 10

months; the parking garage in 6 months; and the tenant improvements in another 6 months.

50. "MWD Signs Contracts for Headquarters Building at LA's Union Station"; "Practicality Is Focus of Metropolitan Water District's New Headquarters Building" (quoted).

51. Thomas D. Verti interview; Leiter-Weintraub, "Built to Suit," 32N; Southern California Ready Mixed Concrete Association and California Cement Promotion Council, "Functional Beauty in Southern California 182 Days Early and $2 Million under Budget."

52. Wodraska quoted in "Practicality Is Focus of Metropolitan Water District's New Headquarters Building"; Patricia Bayer, *Art Deco Architecture: Design, Decoration, and Detail from the Twenties and Thirties* (New York, H. N. Abrams, 1992), 87–115; Carla Breeze, *American Art Deco: Modernistic Architecture and Regionalism* (New York: Norton, 2003), 13–34, 223–53; Leiter-Weintraub, "Built to Suit," 32N–32P; Suzanne Tarbell Cooper, Amy Ronnebeck Hall, and Frank E. Cooper Jr., *Los Angeles Art Deco* (Chicago: Arcadia Publishing, 2005): 9–24.

53. Ivey quoted in Leiter-Weintraub, "Built to Suit," 32P; Liske quoted in ibid., 32N; Joseph Sanders interview; Kevin Smith interview; William "Red" Ward interview.

54. Michael R. Adamson, Memorandum of Conversation with Steven M. Nakada, Los Angeles, 30 July 2009.

55. Fulton, *The Reluctant Metropolis*, 275 (quoted). On the "de-malling" of America, cf. Hugh Cook, "What's in Store for Retail Facilities," *BD&C* 37 (December 1996): 42–4; "Malls Mutate, Breed Outdoor, Lifestyle Concepts," *BD&C* 42 (March 2001): 12; David Salvesen, "The De-Malling of America," *Urban Land* 60 (February 2001): 72–7; Kirsten Young, "Mall Boom Is Ending as Saturation, Slump Arrest Development," *Women's Wear Daily*, 4 June 2001.

56. Bogaard quoted in Ibold, "Mall Is Giving Way to New Vision after Losing Appeal," 50; Michael R. Adamson, Memorandum of Telephone Conversation with Bill Trimble, Department of Planning, City of Pasadena, 15 June 2009; Susan Salter Reynolds, "De-Malling the Mall," *LAT*, 18 December 2000, E3; John Woolard, "Generation-Old Indoor Mall Reshaped as 'Urban Village,'" *Los Angeles Business Journal*, 23 July 2001, 30; Kirsten Young, "Fix in the Mix: Pasadena Plans a New Retail-Residential Complex Downtown to Give Development a Boost," *Women's Wear Daily*, 25 July 2000, S10.

57. Ibold, "Mall Is Giving Way to New Vision after Losing Appeal"; Gerald M. Trimble, "Regrowth in Pasadena," *Urban Land* 36 (January 1977): 4–14, 5 (quoted), 13 (quoted).

58. Ibold, "Mall Is Giving Way to New Vision after Losing Appeal"; Morris Newman, "The Mall Is Dead! Long Live the Mall!" *Grid* 3 (May 2001): 85; Woolard, "Generation-Old Indoor Mall Reshaped as 'Urban Village.'"

59. Michael R. Adamson, Memorandum of Telephone Conversation with Bill Trimble, Department of Planning, City of Pasadena, 15 June 2009; Woolard,

"Generation-Old Indoor Mall Reshaped as 'Urban Village,'" 30 (quoted); Newman, "The Mall Is Dead! Long Live the Mall!" 85 (quoted).

60. William D'Elia, "Spurring Revitalization in Old Pasadena," *Urban Land* 55 (October 1996): 28–30; Newman, "The Mall Is Dead! Long Live the Mall!"

61. Michael R. Adamson, Memorandum of Telephone Conversation with Bill Trimble, Department of Planning, City of Pasadena, 15 June 2009.

62. Michael R. Adamson, Memorandum of Conversation with Steven M. Nakada, Los Angeles, 30 July 2009; Reynolds, "De-Malling the Mall," E1, E3 (Froese quoted); Froese quoted in Ibold, "Mall Is Giving Way to New Vision after Losing Appeal." In 1998 TrizecHahn sold its portfolio of regional shopping centers to Rouse Company and Westfield America for a total of $2.54 billion.

63. Michael R. Adamson, Memorandum of Conversation with Steven M. Nakada, Los Angeles, 30 July 2009; Bogaard quoted in Ibold, "Mall Is Giving Way to New Vision after Losing Appeal"; Reynolds, "De-Malling the Mall"; "Getting Started: TrizecHahn, EE&K, Pankow Work Together to Remake Plaza Pasadena," *California Construction* (September 2000): 46.

64. Nesbit quoted in "Getting Started," 46; Newman, "The Mall Is Dead! Long Live the Mall!"; Young, "Fix in the Mix." For discussion of how Southern California Anglos promoted an imagined Spanish past through the built environment, see Phoebe S. Knopp, *California Vieja: Culture and Memory in a Modern American Place* (Berkeley: University of California Press, 2006), esp. chap. 4; Kevin Starr, *Material Dreams: Southern California Through the 1920s* (New York: Oxford University Press, 1990), esp. chaps. 9–10.

65. Michael R. Adamson, Memorandum of Conversation with Steven M. Nakada, Los Angeles, 30 July 2009; Froese quoted in Ibold, "Mall Is Giving Way to New Vision after Losing Appeal"; Froese quoted in Eric Lassiter, "Housing Built Over Retail Stores Gaining Popularity," *Los Angeles Business Journal*, 29 January 2001, 34; "Profile on Post Properties," *Shopping Center World* 30 (September 2001): T9.

66. Lassiter, "Housing Built Over Retail Stores Gaining Popularity"; Morris Newman, "Coming Soon to Hollywood: A Mixed-Use Building," *New York Times*, 10 December 2000.

67. Todd D. Gish, "Building Los Angeles: Urban Housing in the Suburban Metropolis," PhD diss., University of Southern California, 2007.

68. Greg Hise and Todd D. Gish, "City Planning" in *The Development of Los Angeles City Government: An Institutional History*, ed. Hynda L. Rudd et al. (Los Angeles: Los Angeles Historical Society, 2007), 329–69.

69. Todd Gish, "We've Always Been Dense," *LAT*, 16 September 2007, M9. On subdividing and suburbanizing Los Angeles, see Robert Fogelson, *The Fragmented Metropolis: Los Angeles, 1850-1930* (Cambridge, MA: Harvard University Press, 1967); Greg Hise, *Magnetic Los Angeles: Planning the Twentieth-Century Metropolis* (Baltimore, MD: Johns Hopkins University Press, 1997).

70. Michael R. Adamson, Memorandum of Telephone Conversation with Bill

Trimble, Department of Planning, City of Pasadena, 15 June 2009; Reynolds, "De-Malling the Mall," E3.

71. Michael R. Adamson, Memorandum of Conversation with Steven M. Nakada, Los Angeles, 30 July 2009.

72. Ibid.; Paul Napolitano, "Paseo Colorado Pasadena Project Incorporates a New Steel Design into Existing Structure," *California Construction* (November 2000): 44; Newman, "The Mall Is Dead! Long Live the Mall!"; "Profile on Pankow Builders," *Shopping Center World* 30 (September 2001): T8.

73. Kevin Smith interview.

74. Michael R. Adamson, Memorandum of Conversation with Steven M. Nakada, Los Angeles, 30 July 2009; Dave Barista, "Hollywood's Epic Remake," *BD&C* 43 (April 2002): 24–30; David Bodamer, "Trizec Sells Hollywood and Highland, Cements Position as Office REIT," *Commercial Property News*, 16 March 2004, 3; Philip S. Hart and Maureen McAvey, "Hollywood's Time to Shine," *Urban Land* 54 (September 2005): 116–21; Stephanie Keyser, "Projects Reflect Culture, Heritage," *Chain Store Age* 76 (May 2000): 130; Chris Jones, "Chicago-Based Trizec Properties to Keep Las Vegas Shopping Center," *Las Vegas Review-Journal*, 20 December 2002; Danny King, "Hollywood and Highland Sale Price Cuts CRA Income," *Los Angeles Business Journal*, 1 March 2004. The Hollywood and Highland project was intended to evoke Hollywood's golden era. Babylon Court, its center plaza, was partially a recreation of the set of D. W. Griffith's epic film "Intolerance." The project also incorporated the historic Mann Chinese Theater and featured the Kodak Theatre as the new and permanent home of the Academy Awards.

75. Verti quoted in "Profile on Pankow Builders."

76. Michael R. Adamson, Memorandum of Conversation with Steven M. Nakada, Los Angeles, 30 July 2009; Newman, "Coming Soon to Hollywood: A Mixed-Use Building"; Lizbeth Scordio, "Statistics Persuaded Investor to Develop Housing," *Los Angeles Business Journal*, 21 February 2005.

77. Robert Law interview; Suzanne Dow Nakaki interview; Michael R. Adamson, Memorandum of Telephone Conversation with Robert Law, 21 July 2009; Michael R. Adamson, Memorandum of Conversation with Steven M. Nakada, Los Angeles, 30 July 2009 (Nakada quoted); Greg Aragon, "The Next Act in the New Hollywood: Pankow Builders Completes Major Mixed-Use Project," *California Construction* (August 2004); Christine Rombouts, "Historic Reuse," *Urban Land* 62 (October 2003): 65–72. The surviving structure was built in 1938; it had been abandoned since 1993. In 1996 it suffered a major fire at the hands of "night people." No effort had been made to secure the structure after it had been abandoned, or to repair and rehabilitate it after it had burned (Williams, *The Story of Hollywood*, 358, 372–3).

78. Michael R. Adamson, Memorandum of Conversation with Steven M. Nakada, Los Angeles, 30 July 2009; Joseph Sanders, interview with the author, 25 August 2010; Joseph Sanders, letter to the author, 26 October 2010; Aragon, "The Next Act in the New Hollywood."

79. Rick Pankow interview (quoted).

80. Andrea Colli, *The History of Family Business, 1850–2000* (Cambridge and New York: Cambridge University Press, 2003), 65–9, citing Andrea Colli and Mary B. Rose, "Family Firms in Comparative Perspective," in *Business History Around the World*, ed. Franco Amatori and Geoffrey Jones (Cambridge and New York: Cambridge University Press, 2002), 339–52, and Guido Corbetta, "Family Business," *International Encyclopaedia of the Social and Behavioral Sciences*, ed. Neil J. Smelser and Paul B. Baltes, vol. 8 (Oxford: Elsevier Science, 2001), 5319–24; Harold James, *Family Capitalism: Wendels, Haniels, Falcks, and the Continental European Model* (Cambridge, MA: Belknap Press of Harvard University Press, 2006), 377–84.

81. Richard M. Kunnath interview, 2 July 2010; Kim Petersen, letter to the author, 23 July 2010. In 1991 Kunnath had taken George Hutton's place on the Pankow Management, Inc., board after the latter's resignation. Pankow and Stephan were the other directors.

82. Richard M. Kunnath interview, 2 July 2010; "Kunnath and Verti Take the Reins at Pankow Firms," *Design-Build* 2 (June 1999): 75.

83. Richard M. Kunnath interview, 2 July 2010; "Dwyer and Turner Promoted and Named Vice Presidents," *Single Source* 16 (Fall 1998); "Names in the News," *California Construction* (January 1999): 33.

84. Michael R. Adamson, Memorandum of Telephone Conversation with Robert Law, 21 July 2009; Kim Lum interview; Richard M. Kunnath interview, 2 July 2010 (quoted); "Current Pankow Projects," *Single Source* 19 (Summer 2001); "Profile on Pankow Builders." Because "[we] made more than we should have" on other projects, notes Kunnath, Pankow avoided booking a loss for the 2001 and 2002 fiscal years.

85. Richard M. Kunnath interview, 2 July 2010; William "Red" Ward interview; Bill Hughes, vita, WEST Builders website, www.buildwest.net/west/about/management.php.

86. Richard M. Kunnath interview, 2 July 2010; WEST Builders website, www.buildwest.net/west/about/history.php (quoted).

87. Richard M. Kunnath interviews, 2 July 2010 and 17 October 2011.

88. Ibid.

89. Richard M. Kunnath interview, 17 October 2011; Kim Petersen interview.

90. Ibid.; Verti quoted in Kevin Felt, "Altadena, Calif., Construction Company Keeps Building Despite Death of Founder," *San Gabriel Valley Tribune*, 5 May 2004.

91. Richard M. Kunnath interview, 2 July 2010.

92. Ibid.; Michael R. Adamson, Memorandum of Conversation with Steven M. Nakada, Los Angeles, 30 July 2009.

93. Richard M. Kunnath interview, 2 July 2010 (quoted); Michael R. Adamson, Memorandum of Conversation with Steven M. Nakada, Los Angeles, 30 July 2009.

94. Richard M. Kunnath interviews of 2 July 2010 and 17 October 2011; Kim Petersen interview; Verti quoted in Kevin Felt, "Altadena, Calif., Construction Company Keeps Building Despite Death of Founder."

Epilogue

1. Richard M. Kunnath interview, 17 October 2011 (quoted); US Census Bureau, "Value of Construction Put in Place: Seasonally Adjusted Annual Rate," spreadsheet, various years and months, www.census.gov/construction/c30/c30index.html.

2. Greiner, "Evolution and Revolution as Organizations Grow."

3. Richard M. Kunnath interview, 17 October 2011.

4. Ibid.

5. Ibid; William "Red" Ward interview.

6. Richard M. Kunnath interview, 17 October 2011 (quoted); Dick Walterhouse interview; "Pankow Hires Two and Promotes Seven," *Single Source* 18 (Summer 2000).

7. Richard M. Kunnath, telephone conversation with the author, 7 January 2011; Charles J. Pankow, "Comments from the Chairman," *Single Source* 19 (Summer 2001).

8. Sanders quoted in Flynn, "Framing the Moment," 34; Whitlock quoted in Shuster, "Keeping It Together," 47; "Construction Begins on Westside Media Center," *Single Source* 17 (Summer 1999); "History of the Development of the PHMRF," timeline, n.d., CPBL, Pasadena, CA; "Pankow Starts Westside Media Center," *California Construction* (December 1999): 47.

9. Richard M. Kunnath, telephone conversation with the author, 7 January 2011.

10. Amy Eagle, "A Seismic Shift: State Regulation Prompts Major New Project at Los Angeles Hospital," *Health Facilities Management* 20 (February 2007): 14–20; "Construction Leaders," *California Construction* (June 2005); "History of the Development of the PHMRF," timeline, n.d., CPBL, Pasadena, CA; "Ready to Rumble," *Concrete Products* 110 (November 2007): 88. Simi Valley Hospital, one of White Memorial's sister facilities, had been one of 23 hospitals that had sustained damage during the Northridge earthquake sufficient to force the suspension of some or all services.

CPBL did not build the Citizens Business Bank Arena, perhaps an indication of an increased willingness of other contractors to use the PHMRF.

11. Mid-State Precast website, www.midstateprecast.com.

12. Richard M. Kunnath interview, 17 October 2011 (quoted).

13. Ibid. (quoted).

14. Timothy P. Murphy interview; Judy L. Vawter interview; Christie Brown, "The War between the Collectors," *Forbes*, 26 August 1991, 126–7; Jerry Carroll, "An Opulent Mansion Becomes House Divided," *San Francisco Chronicle*, 20 February 1989, B3. Catalogs of exhibits include the following: Heide Van Doren Betz and Rex A. Wade, *Greek and Russian Icons from the Charles Pankow Collection* (Honolulu: University of Hawaii Art Gallery and Department of Art, 1984); Charles Pankow, *The Charles Pankow Collection of Egyptian Antiquities* (San Francisco: Van Doren Gallery, 1981); Van Doren Gallery, *Egyptian Antiquities from the Charles Pankow Collection* (San Francisco: Van Doren Gallery, 1981).

15. Brown, "The War between the Collectors," 126; Carroll, "An Opulent Mansion Becomes House Divided," B4 (quoted); Marsha Ginsburg, "Le Petit Trianon: The Little Pleasure Palace," *San Francisco Chronicle*, 12 December 2004, K1. The mansion was added to the National Register of Historic Places in 1984.

16. Steven C. Beering interview; Steven C. Beering, *The Indomitable Spirit of Purdue*, Newcomen Publication No. 1367 (New York: Newcomen Society, 1992); "Steven Beering Interview, Part 1," conducted by Katherine Markee, 15 November 2006, Archives and Special Collections, Purdue University Libraries, West Lafayette, IN. Dr. Yang had been named dean in 1984. He served until 1994, when he became chancellor of the University of California, Santa Barbara.

17. Steven C. Beering interview (quoted); Judy L. Vawter, letter to the author, 21 March 2009.

18. Steven C. Beering interview (quoted); Rick Pankow interview.

19. Michael E. Porter and Mark R. Kramer, "The Competitive Advantage of Strategic Philanthropy," *Harvard Business Review* 80 (December 2002), reprint R0212D.

20. Richard M. Kunnath interview, 21 October 2008.

21. Ibid. (quoted); Timothy P. Murphy interview; Charles Pankow Foundation, "Fact Sheet," May 2010, www.pankowfoundation.org/ps.mediacenter.cfm?ID=15. A list of grants may be found on the Foundation's website, www.pankowfoundation.org/grants.cfm.

Six icons from the collection were returned to the Orthodox Church in Cyprus after the Cypriot government interceded to prevent their sale in the auction. The Church proved ownership of the icons, at least three of which had been stolen after the Turkish invasion of the island in 1974 ("Priceless Stolen Icons Returned to Cyprus," *Agence France Presse*, 26 January 2007). A group of businessmen approached the Foundation through Sotheby's to acquire the Russian icons and repatriate them (Judy L. Vawter interview). Le Petit Trianon was also sold. It was initially listed at $29 million (Ginsburg, "Le Petit Trianon").

22. Richard M. Kunnath interview, 21 October 2008 (quoted); Timothy P. Murphy interview; Charles Pankow Foundation, "Fact Sheet" (quoted).

23. Richard M. Kunnath interview, 21 October 2008 (quoted).

Conclusion

1. Charles Pankow, Inc./Builders, marketing brochure, n.d. (c. 1985), CPBL, Pasadena, CA.

2. Robert Law interview.

3. Charles Pankow, Inc., marketing brochure, n.d. (c. 1977), CPBL, Pasadena, CA.

4. As Richard M. Kunnath notes, the Pankow firm has worked with architects who did not embrace design-build, but not on design-build assignments.

5. Timothy P. Murphy interview.

6. Charles Pankow, Inc., profile placed in *The 1974–75 ENR Directory*.

7. Ibid.

8. Strikes in the steel industry provide strong evidence, as Nelson Lichtenstein argues, that the so-called labor-management accord said to have governed labor relations from the end of World War II until the 1980s "is a suspect reinterpretation of the postwar industrial era" (*State of the Union*, 98 [quoted], 132–40).

9. Thomas D. Verti interview.

INDEX

James Campbell Estate Building, 23, 115, 120; architectural plans for, 116–17; preparation for the construction of, 116–20; problem of water intrusion in, 121; redesign of, 121
Janss, Peter, 408–9n12
Janss Corporation, 119, 408–9n12
Japanese Americans, 92
Jefferson Plaza, 195–96
Jerde, Jon, 315
Jischke, Martin, 341–42
Joe West Hall, 104–5, 190; use of slipforming in the construction of, 104
John A. Martin & Associates, 236, 279–80, 281, 305
John F. Kennedy Towers, 99
John Hancock Mutual Life Insurance, 109
John S. Bolles & Associates, 99, 194
Johnson, John Butler, 31
Johnson, Maurice D. S., 179
Johnson, Clarence L. ("Kelly"), 46, 164, 339, 392n35
Jones, A. Quincy, 61–62
Jones, Colby, 242
Jordan Marsh, 186
Joselyn Art Museum, 44
Journal of the American Concrete Institute, 101, 254
jump (climbing) forms, use of, 236–37

Kahn, Louis I., 117
Kaimana Lanais condominiums, 142
Kaiser, Henry J., 126–27, 402n20; and the Bishop Estate, 127; death of, 128; and the development of Hawaii Kai, 127–28
Kaiser Community Homes, 127
Kaiser Hawaii Kai Development Company, 127–28

Kaiser Hawaii Kai housing development, 127–28; construction of a shopping center in, 128; initial sales in, 128; number of developers involved in, 128
Kaiser Permanente, 236
Kalamazoo, Michigan, revitalization of, 117, 118
Kapalua Bay Hotel, 223
Katz, Richard, 306, 307
Kauahikaua & Chun Architects, 271
Kauluwela Co-op, 210
Kawaiahao Plaza, 146
Kaweah Delta District Hospital, 169, 170
Kelham, George W., 435n51
Kelker, R. F., 303
Kendall, Charles, 261, 262
Kerner, Alex, 79
Key Engineering, 206
Key Mechanical Industries (KMI), 57, 58, 167, 194
Kiewit, George, 44, 45
Kiewit, Peter, Jr., 44, 45, 49, 60–61; conflict of with Charlie Pankow, 67–71; delegation of authority by, 46; leadership abilities of, 46; pride of in his job estimating abilities, 45
Kiewit, Peter, Sr., 23, 44
Kiewit, Ralph, Jr., 44, 45, 61
Kiewit, Ralph, Sr., 61
Kimball, Don, 131, 147, 225
King, Bart, 12
Knoll, H. B., 38
Kohl, George, 168
Koko Kai Shopping Center, 128
Kopp, Quentin, 306, 307
Korner, Herbert, 85
Kramer, Mark R., 342, 343
Kulka, Felix, 85, 120
Kunnath, Richard M. ("Rik"), 26, 82, 125, 166, 170–71, 175, 188, 190, 212, 216, 222–23, 234, 235, 273–74,